A Beautiful Ending

A Beautiful Ending

THE APOCALYPTIC IMAGINATION AND THE MAKING OF THE MODERN WORLD

John Jeffries Martin

Yale UNIVERSITY PRESS

New Haven & London

Published with assistance from the foundation established in memory of
Calvin Chapin of the Class of 1788, Yale College.

Copyright © 2022 by John Jeffries Martin.
All rights reserved.
This book may not be reproduced, in whole or in part, including illustrations,
in any form (beyond that copying permitted by Sections 107 and 108 of the
U.S. Copyright Law and except by reviewers for the public press), without
written permission from the publishers.

Yale University Press books may be purchased in quantity for educational,
business, or promotional use. For information, please email sales.press@yale.edu
(U.S. office) or sales@yaleup.co.uk (U.K. office).

Set in Adobe Garamond type by IDS Infotech Ltd.
Printed in the United States of America.

Library of Congress Control Number: 2021943630
ISBN 978-0-300-24732-9 (hardcover : alk. paper)
A catalogue record for this book is available from the British Library.

This paper meets the requirements of ANSI/NISO Z39.48-1992
(Permanence of Paper).

10 9 8 7 6 5 4 3 2 1

for Olivia

If you are holding a sapling in your hand and someone tells you, "Come quickly, the Messiah is here!," first finish planting the tree and then go to greet the Messiah.

—RABBAN YOHANAN BEN ZAKKAI, first century

If the Last Hour ensues while one of you has a seedling in his hand, and he is able to plant it before rising, then he should do so.

— saying of the PROPHET MUHAMMAD, 570–632

If I knew the world was to end tomorrow,
I would still plant an apple tree today.

—attributed to MARTIN LUTHER, mid-twentieth century

Contents

Introduction, 1
1 The Apocalyptic Braid, 13
2 Publishing the Apocalypse, 35
3 Christopher Columbus, 56
4 Conquest and Utopia, 75
5 The Last World Emperor, 92
6 Antichrist and Reformation, 112
7 "No One Knows the Hour," 133
8 Battles for God, 150
9 The Spiritual Globe, 169
10 Cannibals, 187
11 The Restitution of All Things, 206
12 Crossing the Pillars of Hercules, 225
Epilogue, 242

CONTENTS

Acknowledgments, 251
Notes, 255
Illustration Credits, 307
Index, 311

Introduction

A philosopher brought me to the Pillars of Hercules.

Getting there went smoothly. I took an overnight flight from New York to Tangier and, after two days in that exquisite Moroccan city, boarded a ferry to Spain. The trip across the Strait of Gibraltar lasts just over an hour. This channel, one of the world's most important frontiers and crossways, is a narrow passage that lies between two continents and two seas. Here Africa and Europe reach out toward each other, almost touching, and the waters of the Atlantic and the Mediterranean flow into each other. Above all, this is a spot where boundaries blur. Crossing over to Spain in the morning, the ferry braved heavy winds and strong rains. I could see only a few hundred yards. When I came back to Morocco in the early evening, at sunset, the sky was clear and the sea calm. I felt I could see forever.

The ancient Greeks called the promontories that border this channel—the Rock of Gibraltar to the north and Jebel Musa (Mount Moses) to the south—the Pillars of Hercules. So conceived, these outcroppings have played a central role in the imagination of the peoples of the Mediterranean and of Europe for thousands of years. Traditionally poets and philosophers viewed these columns as a moral boundary. Five centuries before Christ, the Greek poet Pindar had alluded to the Pillars as a limit to human striving; and a century later Plato, in his dialogue the *Timaeus,* evoked them as a boundary between the ordered civilization of the Greeks and the mysterious, unknown, and dangerous world that lay beyond. Other ancient and medieval writers viewed them in much the same way. In the fourteenth century, the Italian poet Dante related the story of Ulysses and his shipmates who, in their quest for knowledge, crossed the Pillars out onto the open sea only to meet their doom. Legend had it, moreover, that the warning *Non plus ultra* ("Go no further") was inscribed on the Pillars, reinforcing a sense of limits on human exploration, whether geographic or intellectual.

But in the early modern period, following Columbus's landfall in the Indies, Europeans' conception of the Pillars shifted from barrier to portal. Merchants, sailors, explorers, naturalists, and missionaries started passing through them on their way to distant continents. Throughout the sixteenth century many writers and artists celebrated these new connections, but it was the early seventeenth-century English statesman and philosopher Francis Bacon who captured this new view most vividly in the frontispiece to *The Great Instauration*, his manifesto for the new "science," published in 1620. Here, a galleon looming toward us appears to be returning from a long voyage, while another ship, off on the horizon, navigates choppy seas. Most conspicuously, the approaching ship crosses through the Pillars of Hercules. Beneath the image is the Latin epigram *Multi pertransibunt & augebitur scientia:* "Many will pass through and knowledge will be increased."

In his frontispiece Bacon—the philosopher who lured me to the narrow channel that divides Europe from Africa—offers a compelling sense of the ways in which Europe's growing overseas trade and exploration contributed to an intellectual revolution. Increasingly aware that their own knowledge now exceeded that of such thinkers as Plato and Aristotle, who had not even known of the "New World," early modern scholars began to rely less on ancient authorities and more on their own experience and their expanding knowledge of the world, its peoples, its flora, and its fauna. For Bacon, crossing the Pillars offered the promise of new and expanding knowledge of the world and of nature. Bacon, that is, signaled a shift to modernity itself.

Yet such an interpretation, while not incorrect, nonetheless misses Bacon's central point. While we are likely today to read Bacon's epigram *Multi pertransibunt & augebitur scientia*—with its evocation of the role of commerce in the making of science—in a modern and secular key, Bacon's contemporaries would have immediately recognized this motto as scriptural, derived from the Book of Daniel, where the phrase "many will pass through and knowledge will be increased" pointed not to the development of secular knowledge but rather to the End of History and the final sorting of humanity into heaven and hell.

Crossing the Pillars of Hercules. Frontispiece to Francis Bacon's *Instauratio magna* (1620); engraving by Simon van de Passe.

Indeed, these crossings, as the archangel Michael proclaims in Daniel, would serve as evidence that "many of those who sleep in the dust of the earth shall awake, some to everlasting life, and some to shame and everlasting contempt. Those who are wise shall shine like the brightness of the sky, and those who lead many to righteousness, like the stars for ever and ever" (Daniel 12:2–3).[1] Nor was this the only moment in which Bacon portrayed the growth of commerce and knowledge as signs of the Last Days. To the contrary, throughout his writings he turned again and again to the notion that his own age inaugurated the "autumn of this world" and that a great renewal—a return to Eden—was at hand. Indeed, for Bacon, the overseas discoveries and the explosion of knowledge in his day constituted nothing less than the culmination of God's divine plan for humanity. Occurring in the "autumn of this world" or in the "last ages of the world," they announced not modernity but the End of History—an ending that Bacon and most of his contemporaries did not fear but welcomed. Bacon was, therefore, an apocalyptic thinker, one deeply hopeful that the growth of commerce and knowledge were signs that the End was near.

Such a perspective may seem odd at first. After all, in our lifetimes, and at least since the Second World War, the Apocalypse is rarely viewed as a hopeful event—much less as something for which we might long. To the contrary, the Apocalypse is generally seen as a pending catastrophe—indeed so catastrophic that it could end human life altogether. On this front the accelerated pace of global warming has convinced many researchers, scientists, and public policy experts of an apocalyptic future—one that will require huge investments in mitigation and adaptation if we are to stave off extinction altogether. But, in the modern era, the most persistent horseman of the Apocalypse has been the threat of nuclear war and its potential, in the midst of a fiery holocaust and its immediate aftermath, to wipe out human life in a matter of months. Such an Apocalypse would be an absolute event, a total erasure of human history and the ultimate collapse of meaning.

But men and women have not always viewed history in this light. To the contrary, from antiquity through the early modern period—and certainly in Bacon's lifetime—a radically different view of the Apocalypse held sway. Certainly, this earlier view, like our own, often included a fear of the End. In the ancient, medieval, and early modern world, that is, men and women viewed events such as pestilence and war, for example, or natural disasters such as earthquakes and floods, as signs of the Last Days. Yet, crucially, the apocalyptic imagination—as it developed first in Judaism and then in Christianity and Islam—invariably combined such fears with a sense of profound hope. As a result, visions of the End were rarely simple. Rather, they foresaw, following the catastrophes or the tribulations, a period of abundance, peace, and unity. Thus, at least for the faithful, the Apocalypse gave hope that, on the horizon of time, they would experience a Beautiful Ending. In short, in traditional societies—as counterintuitive as this might seem to us who live in the shadow of the fear of a human-made apocalypse—men and women yearned for a Beautiful Ending, convinced that, after the tribulations, the End of History would bring forth a golden age.

To a large extent this belief was based on the conviction that it was God and not human beings who would bring the world to an end; and, in this divinely ordained ending, a world of justice and a world without suffering would come into being. From their earliest history, Jews—often suffering under a repressive ruler—had awaited the Messiah. Similarly, early Christians had believed that Jesus's Second Coming was imminent. And Muslims too—from the beginnings of their faith—had longed for the coming of the Mahdi (the Islamic tradition's messianic figure) and "the Hour." For each of these traditions, the final rolling up of history promised—just beyond a horizon of suffering—a Beautiful Ending. Collectively these three faiths, often reinforcing and influencing one another, formed an "apocalyptic braid," a tradition of overlapping and interwoven ideas about the End of History that reached first from the ancient and late antique worlds down to the early modern period, and then, from the early modern down to our own time.

Despite the antiquity of these traditions, it was only in the twelfth and thirteenth centuries, in the wake of the Crusades, that apocalyptic ideas began to make themselves broadly felt, as believers—often encouraged by their religious leaders—attended more and more to portents and prophecies that suggested that the End Times were at hand. But it was in the late fifteenth century that apocalyptic beliefs became a defining feature of the broader culture. At this time events themselves prompted expectations that the End of History was near: the Ottoman conquest of Constantinople in 1453; the appearance of a bright starlike object (we now know this was Halley's Comet) in the skies over Europe in 1456; and perhaps especially the conjunction of events in 1492: the defeat of the last of the Muslim emirates in Iberia, the expulsion of the Jews from Spain, and Columbus's first landfall in the Caribbean.

But history is rarely shaped by events alone. And, indeed, deep-seated changes in European and Mediterranean societies from the fifteenth through the seventeenth century also played a decisive role in the intensification of apocalyptic fears and hopes in the early modern period. The invention and rapid spread of the printing press, especially, did much to intensify the sense that the Last Days were imminent. In the late fifteenth century, tens of thousands of books, almanacs, broadsides, pamphlets, calendars, prognostications, and prophecies—many of them produced in inexpensive and often illustrated editions aimed at a popular audience—poured off the presses and reached an increasingly literate public, creating the first mass culture shaped by visions of the End. The expanding role of commerce to which Bacon had pointed and that historians have traditionally labeled—from a Eurocentric perspective—the "age of discovery" was equally decisive. From the very beginning European merchants and missionaries often interpreted their encounters with new lands and peoples in the Americas, Asia, and Africa in an apocalyptic key, nourishing dreams of utopia, of a more just world, perhaps even of a world in which suffering would be no more. Indeed, the expansion of early modern empires was almost always inspired and legitimated by messianic hopes. Emperors and sultans alike portrayed themselves as "Last World Emperors," assigned by God the role of ushering in the End. And these political theologies proved contagious, leading not only to imperial but to anti-imperial forms of apocalyptic utopianism, as popular groups, attentive as ever to the signs of nature and the great events of their day that they believed disclosed to them the workings of God, grew convinced that their ideals too would lead to the Millennium or the Hour, the triumph of justice and the end of suffering.

That this was the case was due above all to the way in which hopes for the End of History—whether one imagined the end as the coming of the Messiah, or the Second Coming of Jesus, or the return of the Mahdi—were influenced by a radically different understanding of history from our own. Above all, early modern men and women understood history not only as the study of the past—not only, that is, as chronicle or as *res gestae*—but rather as a living sacred force that enabled them to make sense of their present with attention not only to the past but also to the future.[2] For them, in short, history reached from Creation to the End of Time. The world in which men and women lived—the *saeculum*—could perhaps be understood or even analyzed on the basis of archives, documents, and memories of a verifiable past. But eternity—which to many was true historical time—could not be subdivided into past, present, and future. And, within this frame, early modern men and women were extraordinarily receptive not only to learned priests, imams, and rabbis, who could interpret recent and current events in light of scripture, but also to the scores if not hundreds of peasant and street-corner preachers and prognosticators, alchemists and astrologers, wandering friars and Sufis, whose prophecies electrified those trying to make sense of their anxieties and fears. The stars would fall from the heavens and the Elect would slip into eternity.

Within each tradition, that is, believers could find in their scriptures and traditions not only signs of present tribulations but also revelations of what remained to be fulfilled before the End would come. And, in this context, many men and women came to view themselves as actors in a sacred drama. We see this clearly in the case of Columbus, whose explorations were animated largely by his conviction that, in bringing Christ to the "heathen," he would bring all nations to Christ, a necessary prelude, as the Gospel of Matthew makes clear, to the Second Coming: "And this good news of the kingdom will be proclaimed throughout the world, as a testimony to all the nations; and then the end will come" (Matthew 24:14). Yet Columbus was only the first of tens of thousands of missionaries, merchants, and settlers who embarked on crossings—in conditions and with risks that are virtually impossible to imagine today—confident that they were engaged in something much larger than themselves and that their voyages were part of a providential plan in which they would play a role in bringing about the End of History.

Apocalypticism also fueled the growth of empire. In the courts both of the Sultan Süleyman and Charles V, for example, astrologers and advisors assured their respective princes that God had chosen them for a special role. Such a vision, undoubtedly, not only motivated many of the efforts from the imperial center to conquer more and more territory, but also justified and legitimated imperial and royal authority. Yet, on the political front, the apocalyptic imagination not only encouraged the growth of empire; as we shall see, hopes for a Beautiful Ending also shaped various utopian dreams of community. What is essential is the recognition that the history of early modern politics—which scholars have tended to interpret in an almost exclusively secular key—were shaped and shaped profoundly by the political theologies of the age. Indeed, it is these theologies and the popular forms they often assumed that best explain the growth of popular political passions in the early modern world. Each religion, after all, viewed itself as the locus of the Truth. Many Christians used this view to proselytize against Jews and Muslims and, increasingly, against the peoples of the New World. And Muslims, also driven by apocalyptic fantasies, waged war against Christians, while Jews also expressed hostility to Christianity and Islam. These religious conflicts did much to shape the geopolitical structures of the early modern state system.

And, finally, in the development of the new natural philosophies of the sixteenth and early seventeenth centuries, apocalypticism and millenarian hopes inspired the research and writings of such figures as Paracelsus, Tommaso Campanella, the Rosicrucians, and Bacon. For these philosophers, the conquest of nature was never, that is, a purely secular quest. To the contrary, it too was both a path to the recovery of an Edenic state and even, at times, a prelude to millenarian dreams. At the same time the apocalyptic had negative consequences. Men and women often believed that they enjoyed a divine mandate for their actions and many proved willing to kill and to die for their cause, whether it was a matter of destroying a political or religious enemy or uprooting paganism in the worlds across the sea. The apocalyptic, in short, was ambiguous, fostering innovations both peaceful and violent. Both sides would contribute significantly to the making of the modern world.

To be sure, what we mean by "the modern world" depends very much on who and where we are. The notion that it should be possible to provide one single standard for modernity—and that standard has almost always celebrated such "accomplishments" of western Europe and the United States as the triumph of capitalism and secularism and, above all, the creation of democratic institutions as the yardsticks against which the very category should be measured—can no longer stand. For the overthrow of European colonial power throughout much of Africa, Asia, and Latin America in the twentieth century has been accompanied by a new, global scholarship that has made it clear that many of the features of the modern that the West has traditionally celebrated—from technology to "rationality"—were often the very instruments that brought many peoples around the globe into subjection. Accordingly, the idea of the modern is inevitably contested, since what appeared to be liberating to the architects of empire and modern capitalism was often based on the oppression and exploitation of the colonies that Europeans ruled. Thus, in using the phrase *the modern world*, I have sought not to lose sight of the fact that individuals and societies experienced the changes around them very much according to whether they were primarily beneficiaries or victims of the transformations taking place in their world. Certainly the incorporation of the Americas into a world system along with the development of empires and new forms of commercial capitalism on a global scale, the shattering of traditional religious institutions and beliefs and the emergence of confessional states, and the search for new canons of truth and the acceleration of the circulation of ideas not only within but across continents were aspects of the modernity that was emerging in the early modern period. At the same time, these developments must be viewed from both within and outside Europe. What I attempt to show in this book is that the energies unleashed by the intensification of apocalypticism in this period did much to enable the shift to this new regime. The rapid changes and innovations that characterized early modernity, that is, not only led to an intensification of apocalyptic hopes but were also fueled by them. For modernity did not grow out of secularization. God had not been pushed aside. Indeed, God was both present and revealed in the scriptures and in the dreams and visions of living prophets. To be sure, some continued to look to God to rescue them from their fears and bring about the fulfilment of their hopes. But for most—and this is critical—the divine revelations made it clear that both individuals and communities had roles to play in bringing about a Beautiful Ending.

An approach that locates the fundamental drive to modernization in religious beliefs and expectations breaks radically from more familiar historical narratives of secularization. Beginning with writers as diverse as Jacob Burckhardt and Karl Marx in the nineteenth century and Max Weber in the twentieth, many have viewed the decline of religion and the loss of faith in divine providence as pivotal factors in the making of the modern world. And in so doing they have laid great stress on the more familiar cultural features of the Renaissance and the early modern age, from the emergence of humanism and secular thought to the first fumblings toward what would come to be called the New Science. I do not deny that these features—the focal points of most traditional narratives—contributed to the making of modernity. At the same time, I believe it essential that we recognize that religious faith, far from being marginalized, in fact played a major role in the making of the modern world. And, within the religious beliefs of Jews, Christians, and Muslims, the apocalyptic imagination was especially decisive.[3] After all, with its visions of the future, the apocalyptic constituted a powerful set of ideas that enabled men and women—often under difficult social and economic circumstances—to confront an uncertain, even frightening future with the confidence that their own actions could help ensure a Beautiful Ending.

From the late fifteenth through the early seventeenth century, the apocalyptic imagination played a central, animating role in many of the more significant transformations of the period. In dreaming of a Beautiful Ending, men and women found the energies that would undergird many facets of an emerging modernity—a modernity I have called "providential." Traditionally, when scholars have looked back on such phenomena as the global expansion of empire and the beginnings of European colonialism, the forging of new forms of political community that drew on utopian and republican ideals to counter the intensification of hierarchies in this age, and the explosion of new forms of knowledge—especially the growth of science—they have tended to view such developments as the result of new economic and political forces, on the one hand, and, on the other, of new intellectual currents that emphasized reason and rationality over faith and religious ideals.

While such a view of these changes captures what we might call their "modernity," the features of this world that foreshadow or seem to foreshadow our own, we must not discount the role that religious belief—and in particular, the apocalyptic imagination—played in these transformations. And here a framework of "providential modernity," as I seek to convey in this book, offers a more capacious perspective, one that excavates the way in which hopes for a Beautiful Ending, always subtending the terrors of present tribulations, underlay the varied intellectual, political, and even economic enterprises of the period. Faith in divine providence did not, as perhaps we too often assume, lead to passivity. Rather, in focusing the attention of Christians, Jews, and Muslims on a sacred narrative, it inspired new forms of agency as men and women sought to realize the prophecies of their scriptures and their traditions. Modernity was, in the end, a providential project.

In the early twentieth century the German philosopher and literary critic Walter Benjamin captured the sense of the great gap between the modern and the earlier messianic concept of time in his interpretation of Paul Klee's painting *Angelus Novus* of 1920. Strikingly, Benjamin's text is only loosely connected to the work of art. Klee, whom Benjamin had known in Berlin, had been parsimonious. In his canvas we see only an awkward and startled creature: part bird, part man. By contrast, Benjamin sees far more: this creature, he tells us, is "the angel of history." And this angel is not gazing upon nothingness, as we might assume from looking at the painting; rather his face "is turned towards the past. . . . Where we perceive a chain of events," Benjamin continues—here gesturing to a sequence of historical developments that are not represented in Klee's work—"he sees one single catastrophe which keeps piling wreckage upon wreckage and hurls it in front of his feet." Then, Benjamin invests the angel with a messianic mission: "the angel would like to stay, awaken the dead, and make whole what has been smashed." Yet the mission, Benjamin concludes—again introducing elements that are not in the canvas—is frustrated: "A storm is blowing from Paradise; it has got caught in his wings with such violence that the angel can no longer close them. The storm irresistibly propels him into the future to which his back is turned, while the pile of debris before him grows skyward. This storm is what we call progress."

The angel of history. Paul Klee's *Angelus Novus* (1920). Walter Benjamin purchased this painting in 1921. After Benjamin's suicide, his friend Gershom Scholem displayed it in his study in Jerusalem.

Personally I have always found Benjamin's insight, forged in a Europe ravaged by war—that progress was not the outgrowth of an enlightened linear transition from barbarity to civilization but rather a "storm"—no less than brilliant. There had, of course, been "storms" in earlier periods of history: the Crusades, the Black Death, the Wars of Religion, and so on. But prior to

World War I none of these either individually or together had dislodged the angel of history from a providential and comforting role. In the Renaissance and early modern period—the era I explore in this book—artists had depicted angels not as awkward creatures but as beautiful and graceful divine messengers. Religious paintings in the age of the Reformation asserted over and over again that a repentant humanity would experience the Heavenly Jerusalem and a Beautiful Ending. Equally crucially, individuals in this period had understood time and history not as open-ended but as sacred. Events—whether wars or floods—were not meaningless but symbols and evidence of a divine purpose, even if that purpose was not yet revealed to everyone. Accordingly, the early twentieth-century recovery of a deepened sense of the messianic and the apocalyptic in thinkers such as Benjamin has played a significant role in enabling this book, in which I have sought above all to enter empathically into a period in which people carried out their lives under the aegis not of an idea of Progress but rather of Providence. More important, it was in studying Benjamin and several of his contemporaries—above all the philosopher Ernst Bloch and the historian Gershom Scholem—that I finally understood, as paradoxical as it may seem, that it is only by grasping this perspective that it becomes possible to retrieve the wellsprings and passions that motivated individuals in the fifteenth through seventeenth centuries to attempt to change the world. Whether they were crossing new seas and inventing new technologies, or creating new political institutions and maritime colonies, or offering new theologies and theories of nature, men and women in this age—not only Christians but also Jews and Muslims—acted with the hope that their actions would accelerate the coming of a Beautiful Ending. Providence was a source of agency. As a consequence—and this is the central claim of this book—the apocalyptic imagination and the hope and energies it inspired played a fundamental role in the shaping of the modern world.

CHAPTER I

The Apocalyptic Braid

> For I am about to create new heavens and a new earth.
>
> —ISAIAH 65:17

> And God himself will be with them; he will wipe every tear from their eyes. Death will be no more; mourning and crying and pain will be no more; for the first things have passed away.
>
> —BOOK OF REVELATION 21:3–4

> Faces that Day shall be blessed, contented by their endeavoring, in a lofty garden, wherein they hear no idle talk. Therein lies a flowing spring. Therein are raised couches, goblets placed, cushions arrayed, and carpets spread.
>
> —QUR'AN, SURA 88:8–16

In the early modern world people heard the Apocalypse.

In Florence in 1490 they came to the Church of San Marco, a Renaissance gem, to listen. There, standing high in the pulpit, was the preacher everyone was talking about: the Dominican Girolamo Savonarola. And what the people said was true: Savonarola, his dark eyes peering out from beneath his hood, could unpack the most obscure mysteries of a divine book, the Book of Revelation. A fellow Dominican praised him for his uncommon gifts that enabled him to make sense of the text and of its descriptions of "ruins, voices, candelabras, trumpets, precious stones, crowns, terrible thrones, and marvelous ladies, dragons, angelic battles, and many other things, all pregnant with spiritual and sacred meaning." "He also described so elegantly and worthily gates and walls and lightning bolts over the earth," the Dominican continued, "that he enticed all those to whom this seemed incredible and especially those who were already devout and curious to hear such things." But there was more. Savonarola did not speak in the ordered way of other preachers. To the contrary, for many he was divinely inspired, a voice of God, since "it seemed that the spirit put into his mouth that which he was saying."[1] The following year—Savonarola had been so successful—the Church of San Marco could no longer accommodate the crowds. In February he moved to Santa Maria del Fiore, the spacious cathedral whose Renaissance dome, then as now, defines the Florentine skyline.

Prophets walk a fine line between the strange and the familiar. On the one hand, they claim divine inspiration. In their visions and dreams God himself or other divine messengers reveal to them—often in symbolic and figurative language—what has previously lain hidden. On the other hand, prophets draw on images that are already familiar to their audiences. In the late Middle Ages and the Renaissance, Florentines, like many others who lived in towns and cities across Europe and the Mediterranean, absorbed images of the Apocalypse everywhere: not only from scripture and not only from the sermons of friars and street-corner preachers but also from vivid church paintings and sacred narratives. Prophets, therefore, do not invent their visions. To the contrary, prophets offer the faithful a way of organizing or making sense of powerful images that speak to both their hopes and fears. Moreover, when prophets are able to breathe a sense of emotion and intensity into their words—Savonarola did just this, his piercing eyes often tearing up—their words could be especially powerful. But above all, prophets, capturing the anxieties of an age, offer hope.

Certainly men and women not only in Italy but throughout Europe and the Mediterranean had reason to be anxious. To many, after all, it felt as though the world they and their ancestors had known—everything that had been familiar to them—was slipping away. They weren't wrong. Across the second half of the fifteenth century a series of invasions and conquests had radically altered the political landscape of the Mediterranean, reducing the power of city-states such as Florence and laying the foundations for new empires. For Christians, the most calamitous event had been the Turkish conquest of Constantinople in 1453—a deathblow to what remained of the Byzantine Empire. The Turks, it was reported, had been ruthless, destroying and desecrating Christian sanctuaries and torturing, raping, and massacring many of the inhabitants, while taking still others into captivity. The fall of this great city—the city of Constantine, the Roman emperor who had first given political support to Christianity—sent shockwaves through Europe.

Aeneas Silvius Piccolomini, the future Pope Pius II, captured the horror as news of the city's capture reached Italy. "So much blood was shed," he wrote, "that it flowed in streams through the streets."[2] Over the late fifteenth century the terror only crescendoed, as the Turks carried out successful incursions in the Mediterranean, the Balkans, Hungary, and even Italy. The Sultan Mehmed I took the Venetian colony of Morea in 1458, Serbia in 1459, Bosnia in 1463, Albania in 1479, and Herzegovina in 1483. And in 1480 the Ottomans struck into the heart of western Europe, seizing Otranto, a town on the Adriatic coast. It was a brutal assault, and when reports of a massacre—perhaps exaggerated—reached Rome, there was fear that all of Italy "would go up in flames."[3]

But the Ottoman Empire was not the only large state expanding into the Mediterranean in this period. In 1472, the crown of Aragon, after a protracted civil war, asserted its authority over Barcelona, until then a largely independent city. In 1492, the combined monarchies of Aragon and Castile defeated the Muslims in Granada, the last remaining outpost of Islam in western Europe. And in 1494, the French King Charles VIII invaded Italy and brought the Italian city-states, which until then had been the most vital polities in Europe, to their knees. Power, it was increasingly clear, no longer lay in the cities throughout the Mediterranean but rather in the newly organized territorial states—monarchies and empires—that were capable of marshalling resources and projecting military power on a scale never before seen.[4] As a result, it was the ballooning empire of the Ottomans that now posed an existential a threat to Christendom, while the newly organized and expansive kingdoms of Spain and France, both buoyed by imperial ambitions, could combine their forces with the papacy and Venice in counteroffensives against the Turks whose expansion they feared.

Like many of his contemporaries, Savonarola struggled to make sense of these changes, above all the success of the Turks, interpreting the rapid expansion of their power as a divine scourge or chastisement for a Christendom that had lost its way in sin and corruption. In his sermons he thundered against an overly worldly papacy, urged its reform, and called upon those who flocked to hear him to repent of their sins: their greed, their vanity, their sexual license.[5] But he also found hope in the emergence of another empire: that of King Charles VIII of France, who, Savonarola maintained on the basis of a rich medieval prophetic tradition, was the Second Charlemagne whose role it would be to install a new pope who would in turn crown him emperor. Then King Charles and the new pope—the angelic pastor—would reform the Church and lead a crusade against the infidel. Savonarola cast Florence in the role of a New Jerusalem but his vision was global. The Christians, he prophesied, would win a great victory against the Turks. The Jews would convert to Christianity. Charles would "unite the world in one flock under a single shepherd."[6]

Anxieties also swept through the Ottoman world as the century came to a close. Certainly the fall of Granada must have put the court of Sultan Bayezid II on edge.[7] And Ottoman anxieties only intensified with the report two years later of Charles VIII's descent into Italy. Charles, after all, had made no secret of his plan not only to conquer Naples but to use it as a base for a crusade that would end both Mamluk control of Jerusalem and Ottoman control of Istanbul. What made the prospect of a crusade all the more worrisome was that the papacy held prisoner Cem Sultan, Bayezid's brother and archenemy. When their father, Mehmed II, had died some fifteen years earlier, the two brothers battled each other for the succession before Cem was spirited off to France and then Italy as a high-value hostage. King Charles VIII certainly knew Cem's value and managed to take possession of this Turkish prisoner in Rome in January 1495. And, even though Cem died—probably of natural causes, though many believed he had been poisoned—the following month on the outskirts of Naples, these maneuvers generated concern of an impending French invasion. There were not only anxieties in the Topkapi Palace in Istanbul but also in the provinces. Venetian spies and ambassadors shared reports that the Turks along the Dalmatian and Albanian coasts were so afraid of an impending attack that they fled the shore and sought refuge in the mountains.[8]

Yet, like the Christians, the Ottomans too were able to draw on a spiritual vision that offered hope. The Topkapi Palace was an intricate complex that served above all to elevate the sultan to a nearly divine status. And Bayezid, always attentive to ceremonial, was rarely seen in public.[9] Moreover, among Bayezid's courtiers were a number of astrologers and scholars whose major goal was to assure not only the sultan and other elites but also the people that Bayezid and the Ottomans would ultimately prevail. They had taken Constantinople and they would soon take Rome. Bayezid's court historians, poets, astronomers, and religious advisors invested him with a pivotal role in history. The calendar itself demanded it, especially since the Christian year 1494–95 coincided with the beginning of the tenth century in Islam.[10] And, because it was widely believed that the tenth century would be the last before the End, surely Bayezid would play a providential role.[11] The court astrologers not only prepared *taqwan*s or annual prognostications for each year, they also veered into eschatology. The courtier Uzun Firdevsi, drawing on both the science of the stars and the Islamic prophetic tradition, described the sultan as the *ṣāḥib-qirān* ("the master of the auspicious conjunction")—a title that not only invested the sultan with mystical authority but also portrayed him as a world conqueror.[12] And, early in the next century, largely in response to the messianic claims of Shah Isma'il in Iran, courtiers embellished Bayezid's role even more. The court scholar Mirim Çelebi who, like Firdevsi combined astrology and the prophetic writings of Islam, came to describe Bayezid not only as the *ṣāḥib-qirān* but also as the Mahdi.[13] These titles were not mere honorifics. To the contrary, in the pious atmosphere of Bayezid's court, they pointed to a deeper reality. According to widespread beliefs, it would be the messianic figure of the Mahdi who would lead his people in victory against the Christians. The Mahdi would restore justice, while the Prophet 'Isa—Jesus, who had not died on the cross, as the Christians taught, but had gone to live in heaven—would return to earth to restore peace and harmony among the devout. Like their Christian contemporaries, Muslims at the end of the fifteenth century were riven by fear and hope; but they too, again like the Christians, dreamed of a Beautiful Ending: a just world on earth before the final consummation of history.

Despite their tribulations, Jews in this period were also full of hope for a Beautiful Ending. Experience convinced them of this. When in 1492 the Spanish monarchs had issued an edict that required Jews either to submit to baptism or to leave their homelands, many Jews—at least eighty thousand, perhaps many more—went into exile. The largest group of these refugees went to neighboring Portugal, only to face mass forced conversions in this kingdom in 1497. In the wake of this repression, Iberian Jews settled in North Africa and Italy as well as in the Ottoman Empire. There they made their homes at first along the Mediterranean coast, from Split and Valona in Dalmatia to Salonika in Greece. Eventually, important Jewish communities developed inland in Sarajevo and Belgrade in the Balkans, and in Safed in Palestine. These were not the only expulsions that Jews experienced. In the late fifteenth and early sixteenth centuries, the cities of Geneva, Nuremberg, Ulm, Brandenburg, Regensburg, and Reutlingen expelled the Jews from their midst, with the majority of the German or Ashkenazic Jews seeking refuge in eastern Europe, especially in the Kingdom of Poland-Lithuania.[14]

It was in the shadow of these forced conversions that Isaac Abravanel, who had prospered as a courtier, financier, and scholar in the royal courts of Portugal and then of Spain, published a series of messianic works that would resonate powerfully in the Jewish communities throughout the early modern period. In Monopoli, a quiet town on the Adriatic coast, Abravanel composed works in which he interpreted the tribulations of his own age—not only the expulsion of the Jews from Spain and their forced conversion in Portugal but also the wars and tensions between the Ottomans and the Christians—as evidence that the End of History was at hand. His *Wells of Salvation,* published in 1497, prophesied a great conflagration in the Mediterranean. The Muslims would conquer Italy, obliterate Rome, and then defeat the Christian armies in Palestine. But then the Jews, gathering their forces from throughout the world, would defeat the Muslims and the Messiah would appear. At the dawn of the messianic era, Jews—who had suffered for so long—would prosper in the land of Israel and, in a great inversion, would now rule other nations. Above all, it would be an era of peace, holiness, and justice—a restoration of all that the diaspora had taken away from the Jews. Breaking with Maimonides, the celebrated medieval Jewish philosopher who had counseled his fellow Jews "not to calculate the end," Abravanel maintained that such a calculation would strengthen his fellow Jews. "And if, as I believe, the hour of redemption is arriving, if we are at the end of our exile and if our salvation is near," Abravanel wrote, "then—in order that he [the Messiah] will appear—the time has come to act in response to God's promise and to draw up this work as a messenger of good news, and to announce our deliverance."[15] Then, based on his careful exegesis of The Book of Daniel, he predicted that the Messiah was soon to come, possibly—though Abravanel was not consistent in the dates he predicted—as early as 1503.[16]

Meanwhile in the small Istrian port of Isola, a town under Venetian rule, the Ashkenazic rabbi and kabbalist Asher Lemlein, whose family had been expelled from Reutlingen in 1495, did much to stir up hopes for the Messiah among the Jews of northern Italy and the Holy Roman Empire. Lemlein led an ascetic life and was often subject to visions. He too called upon his fellow Jews to repent. The Messiah, he assured them, would soon come and would gather up his people and bring them back to Jerusalem where their tribulations—for too long they had suffered under the Christians in Europe—would end. The sources are fragmentary, but, in them, it is possible to glimpse the excitement that his teachings created among the Jews of northern Italy and Germany. Synagogues throughout Italy sent representatives to Isola to meet him. In Westphalia, an old man was so convinced of Lemlein's prophecies that he smashed the oven he used to bake matzo, since he was certain that on the next Passover all Jews would be gathered in Israel, and history would come to an end. Looking back on this movement years later the Christian Hebraist and cartographer Sebastian Münster, recalled that, in 1502, "the Jews did penance in all their dwelling places and in all the lands of exile in order that the Messiah might come," adding that for "almost a whole year, young and old, children and women did penance in those days, the like of which had never been seen before."[17]

While apocalyptic ideas had never been as intense or as widely shared as they were in the early modern period, hope for a Beautiful Ending, at the heart of divine revelation, had always been a defining belief in Judaism, Christianity, and Islam. The hope for a divine rescue of humanity—the belief that God or the gods might bring about a radical shift in the order of things and end suffering and poverty—is likely universal, even if it often assumes different forms. In the ancient world, such beliefs first emerged with clarity in the teachings of Zoroaster. This figure, also known as Zarathustra, is elusive. We don't even know in which century he lived—only that he was active in Iran at some point between 1500 and 1000 B.C.E. But his emphasis on what he saw as a great cosmic struggle between Good and Evil that would end, ultimately, in the triumph of the Good and a world without suffering or death resonated. These teachings circulated in the oral cultures of the ancient world long before the emergence of Judaism, Christianity, and Islam.[18] Eventually each of these faiths would absorb earlier hopes about the eventual triumph of the Good; and, collectively, these ideas would be woven into a prophetic tradition—the apocalyptic braid that would stretch from antiquity through the Middle Ages and into the early modern world, and indeed down to our own time.

The strands of the apocalyptic braid—at times distinct, at times overlapping—stood out like bright threads in the sacred texts of each tradition: in the Torah, both oral and written, of the Jews; in the holy scripture of the Christians; and, among Muslims, in both the Qur'an and the hadiths or early sayings of the Prophet Muhammad and his companions. Within the Tanakh or Hebrew Bible apocalyptic hopes were most often expressed by the prophets who placed their faith in a *mashiyah* or Messiah. In the earlier books of the Hebrew Bible, the term Messiah—literally "the anointed one"—referred to either a king or a priest who was anointed by God to lead the people of Israel. Over time, however, Jews began to hope for a Messiah—whom they typically believed would be a descendant of the House of King David—who would deliver them from their sufferings. It was the Book of Daniel, a product of the second century B.C.E., that transformed the Messiah into a divine messenger, one who would break miraculously into history, bringing it to a Beautiful Ending. "As I watched in the night visions, I saw one like a human being coming with the clouds of heaven. And he came to the Ancient One and was presented before him. To him was given dominion and glory and kingship that all peoples, nations, and languages should serve him" (Daniel 7:13–14). And the vision continues, making it clear that this new ruler was bringing about a new order outside of history: "His dominion is an everlasting dominion that shall not pass away, and his kingdom is one that shall never be destroyed" (Daniel 7:14). Daniel does not use the term Messiah, but more than any other book in the Hebrew Bible it is Daniel that laid the foundation for the apocalyptic hope that a Messiah would come and, with him, a new dispensation that would enable Israel to escape the horrors of history.[19]

In the two or three centuries after the Book of Daniel was first written down, Galilee teemed with prophets, many of them itinerant preachers, who drew large numbers of followers, even as others dismissed them as pretenders. But some gained credibility. A certain John—who would become known as "the Baptist" to early Christians and who ate "locusts and wild honey"—preached that the End of Time was coming. Jesus of Nazareth, another Jew, made a similar promise, calling on those who gathered around him to follow him in order to enter the Kingdom of God. To do so, his followers had to renounce their families and traditions and the sectarian divisions within Judaism—as each was called upon to "love your neighbor as yourself." The Gospels—written down some two or three generations after Jesus's execution—portrayed him as a radical apocalyptic prophet, intent on overturning the existing social order and convinced that the "Kingdom of God" was at hand.[20] First, there would be great suffering. The Gospel of Mark reported that Jesus warned his followers that "nation will rise against nation, and kingdom against kingdom; there will be earthquakes in various places; there will be famines . . . For in those days there will be suffering, such as has not been from the beginning of the creation that God created until now, no, and never will be" (Mark 13:8–19). "But in those days," Jesus continued, now drawing on the Book of Daniel, "after that suffering the sun will be darkened, and the moon will not give its light, and the stars will be falling from heaven, and the powers in the heavens will be shaken. Then they will see 'the Son of Man coming in the clouds' with great power and glory. Then he will send out the angels, and gather his elect from the four winds, from the ends of the earth to the ends of heaven" (Mark 13:24–27). Similarly, in his letters St. Paul, a devout Jew who became an apostle only after Jesus's crucifixion, expressed the widespread hope that that those who accept Jesus as their savior will experience the coming of the Kingdom of God. As Paul wrote to the small Christian community in Thessalonica, "the Lord himself . . . will descend from heaven, and the dead in Christ will rise first. Then we who are alive, who are left, will be caught up in the clouds together with them to meet the Lord in the air; and so we will be with the Lord forever" (1 Thessalonians 4:16–17).

While both the Gospels and the Letters of St. Paul frequently promised a Beautiful Ending to those who followed Jesus, no single early Christian work was as powerful on this front as the Book of Revelation. Indeed, it was the original Greek title of this work—the Apocalypse, which means "revelation"—that did the most to shape Christian visions of the End. The text presents a series of visions: of heavenly candelabra, of charging horsemen, of trumpets, angels, monsters, demons, great battles, and of the Lamb of God and the Heavenly Jerusalem that ultimately promised Christians, who remained faithful to God, a Beautiful Ending. Following tribulations and the harrowing of the final battle against Satan, the text offered its readers an enticing vision of the End. "Then I saw a new heaven and a new earth; for the first heaven and the first earth had passed away, and the sea was no more. And I saw the holy city, the New Jerusalem, coming down out of heaven from God, prepared as a bride adorned for her husband. And I heard a loud voice from the throne saying,

> See, the home of God is
>   among mortals.
> He will dwell with them;
> they will be his peoples,
> and God himself will be
>   with them;
> he will wipe every tear from
>   their eyes.
> Death will be no more;
> mourning and crying and pain
>   will be no more,
> for the first things have
>   passed away. (REVELATION 21:3–4)

But others found in the text the promise also of a heavenly kingdom on earth—one that would last for a thousand years, a Millennium, before the End of Time—and this second reading would play a major role in shaping the apocalyptic and utopian longings of Christians in early modern Europe.

Islam, a faith forged largely in dialogue both with Jews and Christians of Arabia in the sixth and early seventh centuries, also promised a Beautiful Ending. The Qur'an, its foundation and most holy book, had come together rapidly, as Muhammad shared his revelations and his early followers made an effort to record them, with scribes working to collect the words of the Prophet from "parchments, scrapula, leafstalks of date palms, and from the memories of men."[21] "Truly those who believe," the Archangel Gabriel had promised Muhammad, "and those who are Jews, and the Christians, and the Sabeans—whosoever believes in God and the Last Day and works righteousness shall have their reward with their Lord. No fear shall come upon them, nor shall they grieve" (Qur'an, sura 2:62).[22] Another sura adds: "Excellent indeed is the abode of the reverent. They shall enter the Gardens of Eden with rivers running below. Therein they shall have whatsoever they will. Thus does God recompense the reverent, those whom the angels take while they are in a state of goodness. They will say, 'Peace be upon you! Enter the Garden for that which you used to do'" (Qur'an, sura 16:30–32).

While the Qur'an promised salvation, it did not offer a redeemer. But in the hadiths—collections of sayings of the Prophet and his companions that were collected and compiled in the centuries after Muhammad—a redeemer figure and a more explicitly apocalyptic vision of the End of History emerged. To the Shi'a, it was to be the figure of the Mahdi, the "Rightly Guided One," who would eventually return, defeat the Antichrist, and establish a millenarian kingdom of justice that would precede the final consummation of history.[23] But other Muslim groups, most notably the Sunni, while not rejecting the figure of the Mahdi, tended to view him primarily as a military conqueror who would restore political order and whose rule would be relatively brief. While Sunni traditions about the Last Days varied, many viewed the Mahdi as the one who would restore the true faith but placed their apocalyptic hopes on the figure of 'Isa or Jesus of Nazareth. Already in the Qur'an 'Isa had been portrayed as the most important prophet before Muhammad. Then, gradually in the first few centuries after Muhammad's death, 'Isa would come to be seen as playing a redemptive role. In Islam he was not, as the Christian believed, the Messiah but rather a prophet whom God had taken up into hiding and whom God would send again to restore the true faith. Both the Twelfth Imam of the Shi'a and the 'Isa of the Sunni were fully present in Creation, even if they had not yet returned. Notwithstanding these differences in emphasis both Muslim communities, the Shi'a as well as the Sunni, developed their apocalyptic visions, as the Jews and Christians had, in the face of suffering and oppression and in the conviction that God is just and that, ultimately, the devout would enjoy a Beautiful Ending.

None of these traditions was contained within the scriptures alone. To the contrary, the scriptural texts proved powerful precisely because their poetic language reverberated within the context of a predominantly oral culture. We are likely today to think of scriptures as written texts; and, while scholars in the ancient world and the early Middle Ages did copy out and study their sacred texts with great care, most men and women experienced scripture in an oral form: in readings of the Torah in their synagogues, in the readings of selections from the Gospels in their parish churches, and in recitations of the Qur'an in mosques. The poetic form these "texts" had assumed, along with their use of powerful stories and images, invested them with particular power. In particular, the divine visions that they conveyed—whether Daniel's dreams, the visions of John of Patmos, or the revelations of Gabriel to Muhammad— were, after all, far from quotidian. Rather, they presented those who listened to the cantors, monks, and imams with a vast imaginary that was both terrifying and reassuring. The apocalyptic provided a poetry that grappled with the primal dimensions of the human experience. Stars would fall from the sky, mountains would split open, armies would invade; but then a savior would come or return and establish justice and peace. The dead, so the Book of Daniel promises, will "awake" from their "sleep" in the dust of the earth. And such images spilled out into paintings, sacred plays, and popular stories.

Inevitably these longings would resonate, cascading through history, though their intensity could fluctuate. As Christianity began a rapid ascent as the dominant religion in the late Roman Empire, and as its adherents came to occupy more and more positions of political authority, the messianic intensity of Paul and the early Church ebbed. Especially following the Church's embrace by Emperor Constantine, Christians came to place less and less emphasis on their hopes for the End Times. On this front, St. Augustine of Hippo's massive *City of God,* completed in the early fifth century, played a pivotal role. It shifted the drama of the Christian's life from the community to the individual believer, though there were occasional expressions of apocalyptic dreams and even millenarian movements in the early and central Middle Ages. Among Jews, by contrast, rabbis and other community leaders continued to foster expectations for a Messiah. This was the message of the Talmud, the immense collection of Jewish law and teachings that the rabbis of late antiquity had compiled to help guide their communities. According to one of its sixty-three tractates, the *Sanhedrin,* history was a sacred unfolding. First there had been two thousand years of desolation; then two thousand years of Torah; and finally, there would be two thousand years of the Messiah. On the question of when the Messiah would arrive, the rabbis differed, but they did not disagree on the importance of awaiting him. It was to be a hopeful waiting: after wars and destruction, the Messiah would bring peace and holiness to the Jews, gathering them again in Jerusalem.[24] Similarly, within Islam there were intense apocalyptic movements in the seventh to the tenth century. But, in general, it would be in the late Middle Ages, in the wake of a growing emphasis on mysticism within all three faiths, that apocalyptic expectations would intensify.

This intensification, which began in the twelfth century and would continue down through the fifteenth, created a new, influential corpus of writings that, ultimately, added a decidedly optimistic view of the End of History to the apocalyptic braid. In Latin Christendom, for example, many monks and scholars sought to offer a deeper understanding of the shifts taking place, often in brilliant exegetical works in which they gave particular attention to the Apocalypse. The major figure in expressing these new hopes was Joachim of Fiore, abbot of a Cistercian monastery located in the gentle mountains of Calabria, in southern Italy. Joachim articulated what would prove to be the single most influential explication of the Apocalypse. In the introduction to his *Exposition of the Apocalypse,* he portrayed himself as having been at first entirely blocked in his effort to understand the book, but then, on Easter morning 1184, he wrote, "something happened to me as I was meditating on this book . . . suddenly something of the fulness of this book and of the entire agreement of the Old and New Testaments was perceived by a clarity of the understanding of my mind's eye."[25]

For the next year and a half he wrote intensely, developing his revolutionary and inspirational ideas about the shape of history. Many monastic writers before Joachim had viewed the birth of Christ as the institution of a new age. Before Christ, in the Age of the Father, God had been wrathful, quick to punish the sins of humanity. But after Christ, in the Age of the Son, humanity had entered a new era of grace and forgiveness. Joachim's central contribution was to offer a trinitarian reading of history. In particular Joachim discerned a third era: the Age of the Holy Spirit.[26] This final stage constituted a Beautiful Ending, a period of peace that would follow the defeat of the Antichrist and precede the Last Judgment in which the Saved would live in peace and harmony on earth. It would be, Joachim argued, a period "without war, without

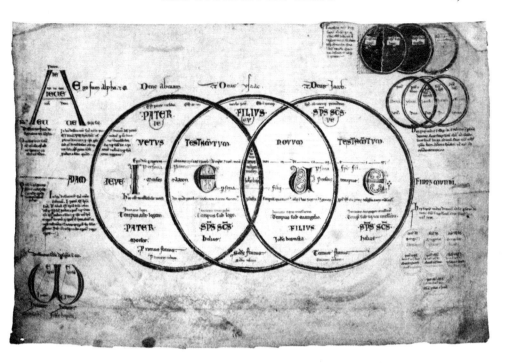

A dynamic vision of history. The Age of the Father (*Pater*) has given way to the Age of the Son (*Filius*), and the Age of the Holy Spirit (*Spiritus Sanctus*) is about to dawn. From Joachim of Fiore's *Liber figurarum,* an early thirteenth-century manuscript.

scandal, without worry or terror, since God shall bless it and he shall sanctify it, because in it, he shall cease from all of his labor that he has accomplished."[27] Moreover, in this final stage, God would convert the Jews and the Greeks—there would be one flock and one shepherd. These ideas gained the favorable attention of several popes, who saw in his teachings a view of history that reinforced their own ambitions for Rome. Over the course of the late Middle Ages, Joachim's influence would continue to grow, often attracting the poor and the dispossessed to his vision of an earthly kingdom in which their sufferings would cease. In the *Divine Comedy,* Dante placed Joachim in paradise and described the abbot as *di spirito profetico dotato*—"endowed with a prophetic spirit."

But it was the Frenchman John of Rupescissa, a fourteenth-century Franciscan, who would offer the most radical vision of the Beautiful Ending. Like many in his order John was deeply critical of its leadership for its materialism and moral laxity. But it was John's visions and prophecies, which increasingly linked what he saw as the corruption of his order to the coming of the Antichrist, that alarmed his superiors. In December 1344 Guillaume Farinier, the provincial of the order, had John arrested, confining him in a dungeon in the Franciscan house at Figeac, a town in the south of France. There, locked in an underground pit, John somehow managed to survive the damp, muddy conditions, and even the Black Death. Perhaps his increasingly hopeful visions sustained him. In 1349 he related the most powerful of his revelations in a short treatise entitled *The Book of Hidden Events*, which he composed in his defense during his trial at Avignon. "On the Feast of St. Jacob the Apostle," he wrote, "while I was praying and standing during vigil, and holding a rod in my hand, all of a sudden, in the flash of an eye, my mind was opened and I came to understand the future through an intellectual vision."[28] It was a vision, moreover, that astonished Rupescissa himself, especially in the revelation that "the age must last at least a thousand years before the end of the world, computing these thousand years from the death of the Antichrist. In an instant, it was revealed to my intellect that this conclusion was most certain, infallible, and necessary, as it had been written down in Holy Scriptures. I was greatly astounded by this revelation, since it seems incompatible with the commentaries and sayings of the saints."[29]

John was correct that the vision of the thousand-year reign of the holy overturned prior teachings about the End of History. St. Augustine had interpreted the *mille annis* of the Book of Revelation as symbolic and the age of the Church. Even Joachim had never embraced the idea of a literal Millennium. By contrast, in his *Book of Hidden Events* John wrote that "it is evident what is said there concerning the martyrs of the time of the Antichrist, namely that 'those who did not adore the Beast nor his image' will be resurrected in their bodies and live and reign in their bodies for a thousand years, that is from the day of the death of the Antichrist down to the coming of Gog and the end of the world."[30] These ideas would then find a broader readership through his *Guide for the Tribulation*, a text he completed in 1356.

In the late Middle Ages hopes for the End also intensified among Jews, especially those who were caught up in the great mystical flourishing of the Kabbalah. First emerging in the late twelfth and thirteenth centuries, Kabbalah—a Hebrew term for "tradition" or for "what is received"—contained a system of symbols that enabled the believer to grasp—or to try to grasp—the very nature of reality, of divinity, of creation, and of the ultimate End of History. The thirteenth-century Castilian mystic Moses of León, working with others, created the vast treasury of symbols that would fascinate generations, producing several books, most importantly the *Zohar* or *Book of Splendor*. They asserted that this was the written form of ancient oral traditions that could unlock the secrets of the faith that both the Talmud and the growing rationalistic traditions of the rabbis failed to disclose.

Kabbalah is immensely complex and multifaceted, but, for many, its promise of deep mystical experiences and union with God prompted messianic expectations. In 1263 the kabbalist Moses ben Nahman, a rabbi in Gerona in northeast Spain, had offered a description of the Messiah and how to recognize when he has truly come. "When the time of the end will have come," Moses wrote, "the Messiah will, at God's command, come to the Pope and ask of him the liberation of his people, and only then will the Messiah be considered really to have come, but not before that."[31] In 1270 in Barcelona, Abraham Abulafia became a student of the Kabbalah and subsequently experienced visions and began to prophesy. His spiritual explorations became deeply mystical, and gradually he came to believe that he himself was the Messiah. In 1280, he undertook a mission to Rome. It was his intent to meet with Pope Nicholas II on New Year's Eve in the Basilica of St. Peter, to confer with him about his prophecies, and perhaps to have himself crowned as the Messiah. But Abraham never met the pope. Nicholas, who had threatened to have Abraham arrested and executed, died just before Abraham entered the city, where he was briefly held as a prisoner by the Franciscans. But he continued to harbor hopes for the End of History. He believed that his teachings would play a role in accelerating redemption.[32]

Finally, there were similar shifts in the Muslim world. In the late twelfth and early thirteenth centuries the Sufi mystic and philosopher Muhyi ad-Din ibn al-'Arabi breathed new life into Islam. Born into a wealthy family in Andalusia, Ibn 'Arabi—rather like St. Francis—rejected his wealth for a spiritual life. Through his nearly incessant travels—first to the Maghreb and later to the eastern Mediterranean—he continued his studies and ascetic practices while writing almost continuously. His ideas quickly found a receptive audience among many Muslims who had grown disenchanted with the rigidities of the imams. And, like many mystics, Ibn 'Arabi saw the center of the religious life in holiness and inner intention or sincerity rather than in external forms or creeds. What mattered most was the believer's inner journey toward God; this emphasis on the inner life opened him up to a recognition—a recognition that he saw as rooted in the Qur'an—that no one faith tradition could claim to be the only way to salvation. A true understanding of God, Ibn 'Arabi wrote in the concluding lines of his *Bezels of Wisdom,* would entail a recognition that just as "the color of the water is the same as that of its container," so God—when he is not domesticated or possessed—can be found "in every form and in every belief."[33]

This universalizing spirituality also underlay Ibn 'Arabi's vision of the End of History. To be sure, he accepted the hadith teachings on the Mahdi. Midway through *The Meccan Revelations*—an immense work of 560 chapters that served as an extended and spiritualized commentary on the Qur'an—he looked forward to a Beautiful Ending. The Mahdi would return, and with the support of seventy thousand Muslim troops, conquer Constantinople without spilling so much as a drop of blood. Then, shortly afterward, now joined with seventy thousand Jewish troops, they would defeat the *Dajjal* or the Christians who, in the wake of the Fourth Crusade, had ruled Acre. Then Jesus would come and the Mahdi would rule—though Ibn 'Arabi did not know for how long—over a just and abundant world before the final consummation of history.[34] Furthermore, in drawing on these prophecies, Ibn 'Arabi refused to view them in either triumphalist or purely political terms. To the contrary, the Mahdi he envisioned could not be a political opportunist or a purely secular lord. Rather he would be the one who would "breathe the spirit back into Islam," and his helpers would be deeply spiritual men, saints in their knowledge of the divine will, who would ensure that the Mahdi will not be one who is followed but rather be himself a follower of the Prophet. Ibn 'Arabi's Beautiful Ending was, therefore, a capacious one, in which the sincerity of the faithful brought forth a spiritual unity—a unity enabled by the emphasis placed on hidden and inner realities of faith rather than on its external expression.[35]

Nor were hopes for the End of History limited to scholarly and spiritual elites. From the time of the Crusades on, a number of men and women in western Europe found inspiration for social change in the millenarian teachings. While for the most part these movements were sporadic and regional, in the late fourteenth and fifteenth centuries, a movement that its enemies called the heresy of the Free Spirit gained a popular following not only in the Low Countries and the Rhine valley but also in central Germany, and even Silesia.[36] But such popular messianic movements were even more pronounced among Muslims in North Africa, where in this same period a number of political upheavals broke out in which a religious or political reformer claimed he was the Mahdi. At times these movements were transformative. In the early twelfth century, after years of theological studies in Cordova and Baghdad, the Berber tribesman Muhammad ibn Tumart returned to his homeland in the Atlas Mountains. There, claiming that he was the Mahdi, he was able to mobilize thousands and managed to overthrow the Almoravids. The Almohads—the regime he established—stretched from Andalusia through Morocco and then eastward into Algeria, its monarchs ruling over a complex and sophisticated civilization.

It was movements such as these that intrigued the fourteenth-century scholar 'Abd al-Rahman Ibn Khaldūn, who offered a penetrating analysis of the political and religious upheavals of his age in his *Muqaddimah* (Prolegomenon), a sweeping work in which he sought to offer an explanation of the rise and fall of political regimes in the Muslim Mediterranean. In the course of his career as a judge, Ibn Khaldūn had traveled widely in Iberia and North Africa; and he had been struck by the stark contrasts between such wealthy cities as Granada, Marrakesh, and Tunis, which were often homes to ruling dynasts and supported a rich court culture, and the predominantly nomadic tribal societies that lay further inland in the Maghrib and Ifriqiyah—the two great regions of North Africa that reached from Morocco in the West to Libya in the East. In the great cultural clash between these two worlds—each largely shaped by its environment—Ibn Khaldūn was able to discern, beneath the whirl and chaos of historical events, the major social, economic, and cultural forces that had shaped the history of the rise and fall of the dynasties that had dominated the Muslim world. He insisted that there were patterns in history and sought to explain the events of the past and the present in more philosophical terms. "On the surface," he wrote, "history is no more than information about political events, dynasties, and occurrences of the remote past."

In place of this traditional approach, Ibn Khaldūn argued that "the inner meaning of history ... involves speculation and an attempt to get at the truth, subtle explanation of the causes and origins of existing things, and deep knowledge of the how and why of events."[37] Developing this approach, Ibn Khaldūn essentially articulated a cyclical view of history in the Muslim Mediterranean where, in his view, wave after wave of Bedouin—fierce and pious warriors whose tribes enjoyed remarkable social cohesion—would overwhelm a reigning dynasty and come to rule in its stead, though this new dynasty would quickly lose its piety and social cohesion and thus would easily be replaced by another tribal movement from below. Moreover, one of the key forces that he identified in enhancing the social cohesion of the rising group was Mahdism, the recurrent expectation of Muslims that they were living at the End of History, a set of beliefs that, as Ibn Khaldūn made explicit, pulsated with revolutionary potential.

As a devout Muslim, Ibn Khaldūn shared the expectations of the End. In writing of the Hour, he notes that none of what is written on this matter in the Qur'an "has anything, as far as I know, ambiguous about it. In effect, one doesn't find there any equivocation or anything like that. It is simply a question of the time of these events, of which Allah has reserved knowledge for himself, as this is expressly indicated in His book and expressed through the mouth of His prophet."[38] Yet, while accepting the authority of the Qur'an, Ibn Khaldūn cast doubts on the hadiths, many of which—as he argued in his *Muqaddimah*—were spurious. He was especially critical of Shi'a teachings, but found much in the Sufi teachings about the Mahdi to be worthy of respect. What cannot be doubted is that he shared the hope for the End of History—a time in which the Mahdi along with Jesus would return, defeat the Antichrist, and establish a period of peace and abundance at the End of Time.[39] He appears to have accepted that God is the agent of all change. Yet he rejected the notion that it was possible to know when the End would come.

At the end of the fifteenth century a skilled illustrator, working in his shop in the German city of Nuremberg, spilled his heart out in his depiction of the End of History. In bold images, Albrecht Dürer—at this point a young artist, still unknown to most of his contemporaries—captured the terror and hope of Christianity's most astonishing prophecy: the Revelation of the Lord to St. John of Patmos. Dürer's illustrations depict the opening of the seals and the sounding of the trumpets. The Four Horsemen of the Apocalypse—who dash menacingly across the sky—bring conquest, war, famine, and disease down upon an already prostrate humanity. No other image captured the fears and insecurities of the early modern world more powerfully than this woodcut. The top figure, bearing a bow and wearing a Turkish headdress, offered a powerful portrayal of the widespread fear of the Turks who would continue to pose a threat to western and central Europe throughout the sixteenth and seventeenth centuries. But equally fearsome were the other three horsemen whose presence haunted Europe and the Mediterranean in these centuries. War—the figure of the second rider, who wields a sword—was a nearly constant presence in Europe in these centuries, as was the threat of famine, caused not merely by food shortages but also by the great lords exploiting the poor, a threat vividly depicted by the third horseman, the well-fed nobleman who bears a balance, an instrument of economic power, as though it were a weapon, in his right hand. And, finally, there was disease or pestilence, represented by the emaciated, frightening figure of the fourth horseman.[40]

The Four Horsemen of the Apocalypse. Engraving by Albrecht Dürer (1498).

As Dürer's woodcut makes clear, the Apocalypse served as a powerful framework through which men and women could visualize the constant and overlapping threats of conquest, war, famine, exploitation, and pestilence that—even if they are constant and unwelcome companions of humanity—were particularly present in that epoch. From the late fifteenth to the mid-seventeenth century hardly a year passed without war, and famine and plague were also endemic. Crucially, while there was nothing new about war itself, warfare—with increasingly deadly weapons and now organized on a much larger scale than ever before—became more and more threatening to ever larger parts of the population. Moreover, in this predominantly agrarian world, a period of drought often spelled disaster for the food supply, and famines occurred at least on a regional scale as often as once in every four years. But an even more constant threat was plague. Before the second half of the seventeenth century not a year passed without plague, at least somewhere in certain regions or localities of Europe, killing thousands; and, on seventeen occasions, this disease became pandemic, exploding across the entire continent, resulting in hundreds of thousands, if not millions of deaths. Learned opinion judged the recurrent visitations of famine and plague as divine punishments for a sinful humanity. "It is confirmed, constant, and received opinion in all Ages amongst Christians," the French surgeon Ambroise Paré observed in the sixteenth century, "that the Plague and other Diseases which violently assail the life of Man, are often sent by the just anger of God punishing our offences."[41]

But Dürer's images, while often terrifying, also offered hope. The Lamb of God brings harmony. St. Michael slays the dragon. An angel locks away Satan. And a Heavenly Jerusalem, the promised home of peace and salvation, descends upon the Earth.[42] History did not, therefore, end purely in chaos, catastrophe, and fear—even if these terrors of history seemed inescapable. To the contrary, the Apocalypse had revealed a far more hopeful ending in which Satan would at last be locked away and the Heavenly Jerusalem would descend.[43] Of course, this was the Christian story, but Jewish and Islamic visions of the End also offered a framework in which hope would triumph over disaster.

The powerful apocalyptic narratives in the early modern world were never purely about fear. To be sure, preachers at times stoked the fears of those who gathered in churches and in public squares. But they did so, almost always, to encourage repentance. And they also, almost invariably, offered a message of hope. In short, the power of the apocalyptic imagination lay above all in its

The Heavenly Jerusalem. Jerusalem descends and an angel binds Satan for a thousand years. Engraving by Albrecht Dürer (1498).

capacity to help men and women, who were genuinely anxious and afraid, both to confront their fears and to believe that it was still possible to be hopeful about the future. If the scriptures and traditions had proven so right about the current disasters—for in the early modern world their predictions of deadly wars, plagues, and famines had so often come true—then there was little reason to doubt that they were also right about the future. History in this period was rarely only the story purely of the past. To the contrary, history was a narrative which invested the present and the future with sacred meaning.

For many, of course, the promise of a release from fear and suffering was transcendent. They did not look forward to a better life on earth but rather to a blissful afterlife. But for many others, prophecies, ancient and modern, pointed to a better life on earth—whether it was through a return to Eden or the golden age or through the shaping of a new world, a utopia, and a period of peace and harmony on earth—before the final consummation of history. Finally, in the face of such hope, it was difficult to be passive. The scriptures teemed with prophecies of things that must happen before the divine promise of a harmonious and peaceful world would be fulfilled. As a result, men and women not only repented in preparation for the End, they also frequently took actions to help accelerate the promised messianic age. Fear was a powerful emotion, but hope was equally powerful, if not more so, for it was hope that moved and energized men and women to work hard to fulfill the remaining prophecies that would enable them to realize the Beautiful Ending promised in their scriptures. It would be hope, as we shall see, that made it possible for Columbus and thousands of others like him to cross the oceans; that inspired the conquest of the Sultan Süleyman and the forging of a global Islamic empire; and that encouraged such Jewish mystics as Isaac Luria to engage in ritual acts of *tiqqun* to repair not only the world but indeed the entire cosmos.

CHAPTER 2

Publishing the Apocalypse

> Printing is the ultimate gift of God and the greatest one. Indeed, by means of it, God wants to spread word of the cause of the true religion to all the Earth, to the extremities of the world.
>
> —MARTIN LUTHER, *Table Talk*

In 1452 or 1453 Johannes Gensfleisch—only later did he adopt his mother's last name, Gutenberg, as his own—was still tinkering, perfecting his new system of "movable type," when he produced what was likely the first printed book in Europe: a text in rhyme containing the Sibylline prophecies that the End was near. Only a single leaf of the work, which would have originally been some thirty pages long, survives.

We know about Gutenberg's edition of this prophetic poem by pure chance. Even though it was produced with a revolutionary method, at the time of its production no one thought to preserve it, to place it in a library, or to lock it in a vault. Someone—we don't know who or when or under what circumstances—tore out its pages and used them for other ends. Then one of these pages, through one of history's unexpected wormholes, somehow survived and was eventually used as a folder or file for documents in the University Archives of Mainz. There, in the late nineteenth century, an archivist noticed its curious Gothic lettering and thought it might be of some value. A local doctor then purchased it, perhaps for a private collection, but shortly afterward gave it to a newly established museum in the city. Scholars at once recognized the typeface as consistent with Gutenberg's earliest experiments in printing. And almost immediately a historian, realizing that this single leaf involved a discussion of the End of Time, baptized the text the *Fragment vom Weltgericht* (Fragment of the Last Judgment). A few years later, in 1908, a

The Last Judgment (ca. 1552–53). The oldest surviving remnant of a book printed with movable type in Europe.

scholar identified the surviving page as a portion of a fourteenth-century poem: *The Book of Sibyls*.[1] Gutenberg's first book was an apocalyptic poem.

In the ancient world the sibyls had been oracles, mad and inspired women who prophesied the future. Plato praised them in his *Phaedrus*. The Roman poet Virgil, in his *Eclogues,* has the Cumaean Sibyl prophesy the return of the golden age. Early Jews and Christians incorporated the pagan sibyls into their dreams of a Beautiful Ending.[2] As a result the sibyls took their place alongside Christian prophets in predicting the Last Days; by the fifteenth century they were so popular that it is impossible to know precisely which version of the *Sibyllenbuch* Gutenberg printed. Nonetheless what is certain is that one of the very first books—if not the first book—Gutenberg printed was a German text focused on the End of History. Of course, it is possible that Gutenberg chose this theme in the hopes of making a profit. Or, possibly, as one scholar has recently argued, Gutenberg was genuinely taken with the sibyl poem, since in it the prophetess emerges as an impresario in decoding letters and their hidden meanings and offered Gutenberg a symbol of the potential power of print.³ But, above all, it seems likely that his decision was based on the anxieties and hopes of the moment. The Turkish threat was growing, and a young emperor, Frederick III, seemed Germany's greatest hope. Shortly after publishing the *Sibyllenbuch,* moreover, Gutenberg printed the *Turkish Calendar,* a work that portrayed the Turk as the Antichrist and called upon Christian rulers to fight against them.⁴ These two texts, both printed in the vernacular, connected prophecy with astrology—a connection that would continue to shape early modern efforts to make sense of the present in light of the future. Then in his Bible—published in 1454–55 and the work for which Gutenberg is best known—he shared the full sweep of sacred history, from the Creation to the End of Time. Like the sibyl, Gutenberg would act as a kind of prophet, or rather a purveyor of prophecies. From the very beginning Gutenberg's project made it clear that the printing press and the prophetic would forevermore be joined.⁵

From the time of the fall of Rome down through the Middle Ages, the production of a book had required the labor of scribes—most often monks who copied out the texts by hand in *scriptoria,* candle-lit rooms monasteries dedicated to this purpose. While a late medieval shift from parchment and vellum to paper (a Chinese invention, introduced into Europe by the Arabs in the twelfth or the thirteenth century) had brought the costs of book production down substantially, it still remained an immensely labor-intensive and costly undertaking. To be sure, some businessmen in the fourteenth and fifteenth centuries had set up secular copy-houses that could produce multiple copies of books relatively quickly, but in a traditional monastic setting the production of a single Bible would generally have taken two years. Gutenberg's machine had accelerated the rate of production by a factor of nearly two hundred.

Gutenberg was not the first to print with moveable type.⁶ The Chinese had used movable type many centuries earlier, but the complexity of the Chinese writing system—especially the enormous number of *hanzi,* or characters—made the process both cumbersome and expensive.⁷ Even in Europe itself, it seems, Gutenberg was not alone in pushing to develop this new system. Laurens Janszoon Coster, a burgher in the Dutch city of Haarlem, appears to have printed with movable type several decades before Gutenberg; and other skilled craftsmen were likely experimenting with similar techniques in this same period.⁸ Ultimately, however, it was Gutenberg's invention that would serve as the basis for a revolution in communications.

As a goldsmith Gutenberg had learned how to form molds for coins and medallions. And it was on these skills that he drew in order to cast the type he needed for printing. There is no scholarly consensus on his method, though it is certain he was able to produce multiple copies of the necessary characters relatively quickly. In all likelihood Gutenberg first etched individual letters into a hard metal, and next pressed this form into a copper mold into which he could repeatedly pour a lead-based alloy. When this alloy hardened, it formed a metallic letter that could be used, along with other similarly produced individual fonts, to typeset a page. But this was only one step in a far more complex process.[9] What is clear is that Gutenberg attended with care to every detail of his invention. Most of its elements were already well known. For instance, the hand press, a technology used since antiquity for pressing olives, had already been used for printing, but it had been limited to block printing, generally highly illustrated books with little text. Moreover, to create such a text, craftsmen had to carve the letters into the wooden block used for printing, an extremely time-consuming process. To make it suitable for printing with metallic type, Gutenberg modified the press by adding a *forme* or a frame to hold the type set to produce the desired page. To make the system effective he also had to modify both the nature of the ink used and the quality of the paper—though Gutenberg did print some of his Bibles on vellum, in all likelihood to satisfy the refined taste of certain buyers.

Despite the innovations, the work required immense patience. As the type wore out, new letters had to be cast. Typesetters had to take exceeding care in preparing the printed page, arranging the metallic fonts in just the right order, carefully adjusting the margins, and checking with a mirror (for the fonts were inverted letters) that they had accurately copied the manuscript they were using as their model. Another worker then inked the fonts, while yet another pinned a clean sheet of paper in a frame that was carefully lowered over the typeset fonts below. The workers then slid this frame and the fonts below the actual press. It was the pressman himself, with a pull of a lever, who then lowered a platen or fixed plate that brought the paper into firm contact with the inked letters, thus printing the page. Gutenberg worked with only black ink. To complete the page, illuminators used red and other colored inks to decorate the work, making it appear almost identical to the hand-made manuscripts upon which it was largely modeled.

Gutenberg's first printed works—among them the *Sibyllenbuch* and the *Turkish Calendar*—were evidence that the new method would work. But it was the Bible that would make his reputation and demonstrate the power of

The bustle of the print shop (ca. 1591). Type-setters on the left work with fonts, while a compositor sets them with a frame. In the background a porter carries wares, a worker inks the letters, and a pressman prints the page. A patrician publisher oversees the shop while a diminutive editor works by the light of a lantern in the attic.

his invention. In 1454 an imperial diet was held in Frankfurt to determine how to coordinate the western response to the advance of the Turks into eastern Europe—they had conquered Constantinople the previous year. Gutenberg himself (or perhaps one of his business-partners) traveled to Frankfurt to show samples of the masterpiece in progress—samples in the form of quires (gatherings of folios)—to the prelates and princes who had converged on this city only some forty kilometers from Mainz. This marketing was successful, as a letter from the Italian legate Aeneas Silvius Piccolomini—later Pope Pius II—to the Spanish Cardinal Juan Carvajal makes clear: "The lettering is extremely clear and accurate," Piccolomini wrote, and then quipped, "even without spectacles, your Grace would be able to read it without difficulty." Had he known of the cardinal's interest earlier, he would have bought him a copy. But now the first print run of either 158 or 180 copies (he has heard both figures) has likely sold out. As he notes, copies "were being committed to buyers even before the book was finished."[10]

On all accounts, Gutenberg's Bible was a great success. Yet, even at this stage, it was not immediately obvious that print would take hold. On the one hand, many readers continued to proclaim their preference for manuscripts over their printed counterparts. This was the argument of Johannes Trithemius, abbot of Sponheim, a Benedictine abbey not far from Mainz. In his *In Praise of Scribes* of 1492, Trithemius argued that the printed book was less desirable because it was ephemeral. "The printed book," he wrote, "is made of paper and, like paper, will quickly disappear. But the scribe working with parchment ensures lasting remembrances for him and for his text." And, as he adds later in the treatise, the scribe's labor "will render mediocre books better, worthless ones more valuable, and perishable ones more lasting."[11] But the widespread unease about books stemmed primarily from anxieties raised by the prospect of circulating ideas to larger and larger readerships. In 1473 Filippo di Strato, a Venetian scribe from the island of Murano, warned that printing presses "corrupt susceptible hearts."[12] And, of course one obvious corruption was heresy. As Pope Alexander VI warned in 1501, "it will be necessary to maintain full control over the printers so that they may be prevented from bringing into print writings which are antagonistic to the Catholic faith."[13]

Yet, despite various expressions of resistance, printing spread relatively quickly. Many welcomed the new technology. As we have seen, Gutenberg's Bible had sold out, and clerics like the Cardinal Carvajal were excited by the prospect of owning a printed book. In 1470, moreover, the Florentine Leon Battista Alberti recalled a conversation with his fellow humanist Gregorio Dati at the Vatican in which they had "approved very warmly the German inventor who has recently made it possible, by making the imprints of certain letters, for three men to make more than two hundred copies of a given original text in one hundred days."[14] The new technology was adopted especially rapidly in Germany, France, Italy, and the Low Countries. By the late fifteenth century, there were over two hundred presses on the continent. Most of these were in urban centers located along "Europe's spine," a belt of towns and cities—an urban archipelago—reaching from the Baltic and the North Sea down through Flanders, Germany, Switzerland, and on into northern and central Italy.[15] It was this urban network above all that enabled the transfers in technology and skills that hastened the rapid spread of the printing press. It was a German merchant, for example, who set up the first publishing house in Venice, which rapidly became the single most important center of printing. Moreover, knowledge transfers among these cities fostered specialization. Merchants found it in their interest to print books for relatively well-defined markets. Some specialized in law; others in theology; others, such as the brilliant bookman Aldus Manutius of Venice, in the reprinting of classical texts. Printers, moreover, developed particular distribution networks. Finally, also as a consequence of specialization within this network, by the beginning of the sixteenth century, a dozen or so cities dominated the book trade: Augsburg, Nuremberg, Cologne, Strasbourg, Basel, and Leipzig in Germany; Rome, Venice, Florence, and Milan in Italy; and Paris and Lyon in France—again all cities located up and down Europe's spine.[16] Was it purely a coincidence that Mainz was at the center of this network?

By the late fifteenth century, therefore, the growth of print had begun to transform the ways in which textual materials were created and distributed, putting far more texts into circulation than ever before.[17] And the impact of this new medium—in a period defined to a large degree by urbanization, growing literacy, and religious reform—was transformative on almost all levels. In the world of scholarship it accelerated the speed with which both learned humanists and natural philosophers could exchange ideas and share new discoveries.[18] But the press also affected the broader culture, especially in the religious sphere.[19] Indeed, although it was only one factor among many, the printing press played a significant role in deepening the hold of the apocalyptic imagination on men and women in the early modern world.

In the late Middle Ages, the Christian Bible—copies of which were extremely expensive to produce—had circulated primarily among monks, clerics, and aristocrats. But, with print bringing the cost down, the early modern period made it possible for growing numbers of merchants and artisans to own a Bible. In the sixteenth century alone more than two thousand editions of Christian Bibles were published in western and central Europe. Many of these editions, especially in Catholic lands, were in Latin. But this period also witnessed the growing influence of vernacular bibles, with such works as Martin Luther's German Bible, Tyndale's Englished New Testament, Lefèvre d'Étaple's French Bible, Antonio Brucioli's Italian translation, and the Geneva Bible—the most influential English translation before the King James of 1611—bringing the Word of God directly to an increasingly literate public. While the Bible itself made such apocalyptic texts as the Book of Daniel and the Book of Revelation immediately available to readers, the Book of Revelation was often printed separately and it came to be a bestseller on its own. Albrecht Dürer's 1498 edition—which, as we have seen, captured the terrors and the hopes of the Apocalypse in dramatic images—was especially successful.[20] Dürer produced two editions: *Die heimlich Offenbarung Ioannis* (in German) and *Apocalipsis cum figuris* (in Latin).[21] For this work as for others, he had obtained the services of an agent to ensure its distribution "from one country to another, and from one city to another."[22] As was his practice, he claimed personal authorship of the illustrations in this work with his initials, A.D.: "Albrecht Dürer" but also "Anno Domini" ("in the year of Our Lord") inscribed at the bottom of each plate—one of history's most famous signatures.[23]

Many evangelicals not only approved of but even encouraged these translations. In his preface to his edition of the New Testament of 1526 the Dutch humanist of Rotterdam, Erasmus, dreamed of a world in which ordinary women would read the Gospels and the Letters of St. Paul in their own tongue, and in which ploughmen, weavers, and travelers would pass their time reciting verses from the Bible.[24] But it was the Protestants, beginning with Luther's doctrine of *sola scriptura*—by Scripture alone—who placed the Bible at the center of the Christian life. "A simple layman armed with scripture," he thundered in a debate with a Catholic theologian in 1519, "is to be believed above a pope or a council without it."[25] And, only a few years later, Ulrich Zwingli, the leader of the Reform in Zurich, concurred, emphasizing the transformative power of the Bible. "When the Word of God shines on the human understanding," he wrote, "it enlightens it in such a way that it understands and confesses the Word and knows the certainty of it."[26] By contrast, many in the Catholic Church were anxious about the laity reading the scriptures without the mediation of the clergy. During the Fourth Session of the Council of Trent, in 1546, the holy fathers not only held up the Vulgate— the Latin translation of the Bible that the Church had used throughout the Middle Ages—as the most authoritative version of Scripture, but also insisted on the Church's exclusive authority to interpret the text, noting that the faithful were to be guided not only by Scripture but also by "the unwritten tradition which, received by the Apostles from the mouth of Christ Himself, the Holy Ghost dictating, has come down to us, transmitted as it were from hand to hand."[27]

Yet, in Christian communities, the holy scriptures constituted only one layer or one strand of the apocalyptic imaginary. In learned circles, in addition to the Bible, a wide variety of commentaries on the Apocalypse circulated. Many of these were medieval texts such as Joachim of Fiore's *Exposition of the Apocalypse* and John of Rupescissa's *A Guide for the Tribulations* that now reached a broad public for the first time. Others were modern works, reaching from late fifteenth- and early sixteenth-century commentaries by Savonarola, Luther, and Bullinger, down to seventeenth-century works by such English reformers as Joseph Mede and Gerrard Winstanley.[28] In this new spiritual environment, readers were eager for the assistance of learned guides in making sense of the difficult symbols and imagery of the Bible and the ways in which they pointed to the End of History.

Within Judaism, it was primarily the activity of a number of learned Jewish or formerly Jewish scholars, working in Venice with the wealthy young Flemish merchant Daniel Bomberg, who did the most to make the sacred writings of the Jews accessible to a larger public. Following their chance encounter in Venice in late 1514 or early 1515, the Augustinian friar Felice da Prato, a recent convert from Judaism, persuaded Bomberg (a Christian) to put his resources into the publishing of Hebrew books. It was a bold proposal. Not only was there growing hostility toward the Jews in Venice—a hostility that would lead to the establishment of the Ghetto in 1516—but there was an explicit antipathy toward Jews in many other parts of Europe as well. In 1509 the theologian Johannes Pfefferkorn—a former follower of Asher Lemlein who had converted to Catholicism after the Messiah had failed to come—demanded that the emperor confiscate and burn all Jewish books.[29] Yet many contemporary humanists, who valued the Hebrew texts as important for a deeper understanding of Christian traditions and scriptures, opposed this proposed destruction of Hebrew books. Bomberg clearly shared this humanist perspective. Moreover, even with the establishment of the Ghetto, the restrictions on Jewish mobility and employment in the city were not absolute, and Bomberg was able to persuade the Venetian senate to allow Jews, who knew Hebrew and Aramaic well, to work at his press as compositors, typesetters, and proofreaders. The senate even exempted them from the obligation of having to wear yellow berets outside the Ghetto walls.[30] Indeed, over the next few decades, Bomberg's shop would gain renown as a meeting place for leading Hebrew scholars—not only Felice da Prato, but also the Talmudist and judge Hiyya Me'ir ben David and Rabbi Jacob ben Hayyim ibn Adoniyahu, who had immigrated to Venice from Tunisia.

The first major publication produced by Bomberg's press was a new, four-volume edition of the Hebrew Bible in 1517, edited by Felice da Prato, with commentaries. This successful edition was reprinted in 1521.[31] During the same period, Bomberg, in collaboration with Hiyya Me'ir ben David, produced an even more massive work: the entirety of the Babylonian Talmud, the first edition of which, filling fifteen folio volumes, was completed in 1519–23.[32] Then, prompted largely by Jacob ben Hayyim ibn Adoniyahu, whose knowledge of the Bible was unparalleled, Bomberg brought out a new edition of the Jewish holy book. This work, published in 1524–25, would rapidly establish itself as the standard text of the Hebrew Bible for the early modern period.[33]

While Bomberg had not been the first to publish the Torah or tractates from the Talmud—the Soncino family of Brescia had taken the lead on both these initiatives in the late fifteenth century and had even continued their work later in Salonika and Istanbul—his undertaking proved especially transformative.[34] Speaking only of the Torah, the sixteenth-century Polish rabbi Shlomo Luria, known as the Maharshal, captured its inexhaustible complexity with a famous image: "Were all the heavens above scrolls and all the oceans ink," he wrote, "they would not suffice to expound even one passage and all the doubts arising from it and all the innovations emerging from it."[35] To address the complexity of Jewish writings, Bomberg not only printed the texts of the Hebrew scriptures and the Talmud but also offered readers the interpretative tools they needed to make sense or try to make sense of these writings without the mediation of rabbis. Readers now encountered the text of scripture embedded, on every page, in a forest of commentaries—among them Rashi's explication of the Pentateuch. Rashi's commentary, along with the Targum (the early Aramaic translation of the Hebrew) and the Masorah (the brilliant critical notes on the text developed by scholars in the central Middle Ages), as well as the works of other medieval exegetes such as Abraham ibn Ezra, aimed to help the faithful guide themselves through the text. The Talmud—both the Babylonian and the Jerusalem versions—had also been a triumph of commentaries since it was first written down and collated by rabbis across several centuries in late antiquity. But Bomberg's edition was the first to present a unified version of this work and to enable the rabbis and their students to read it with attention to the variety of commentaries on individual passages. To be sure, only the brightest or most intrepid students would pass from the study of Torah to the study of the Talmud. Nonetheless, as most Jews recognized, the reading of the written Torah required deep learning in oral Torah (Jewish traditions), precisely what the Talmud provided.[36] As Solomon ben Isaac—also known as Shlomo Yitzhaki or the Rashi—had demonstrated in the eleventh century, the two Torahs, the written and the oral, were interdependent and mutually illuminating.

The opening page of the Torah. The Mikra'ot Gedolot or Rabbinic Bible (1524–25). *Bereishit,* the Hebrew word framed at the top of the page in the panel, means "in the beginning." Designed as a study bible, this and subsequent pages include not only the text of scripture but also the Aramaic *Targum,* the *Masorah,* and great rabbinical commentaries, here those of Ibn Ezra and Rashi.

While these and other great works and commentaries on Jewish teachings and law had circulated in the late Middle Ages, the great success of Bomberg and other printers in the early modern period was the increase in access that Jews throughout Europe and the Mediterranean—for the Jewish press was active also in the Ottoman Empire—had to this enormous corpus of Jewish texts. It was a momentous development. Traditionally rabbis had derived their authority from their ability to interpret the Torah and the Talmud, both of which were notoriously difficult texts. Not only did the circulation of these texts at times challenge the authority of rabbis and teachers in many Jewish communities, they also gave an increasing number of readers the opportunity to study for themselves texts that spoke of the End of History.[37] Many Jews, that is, could now read not only the Book of Daniel, to which they apparently gave considerable attention as the most apocalyptic of the Hebrew Bible texts, but also extensive commentaries on it. In 1525 printers brought out the Hebrew edition of Daniel in both Basel and Wittenberg, and in 1540 Robert Estienne, a Christian printer, published the *Sefer Daniyel*.[38] Commentaries on the book were also popular. Levi ben Gershom's *Commentary on Daniel* was published in the late fifteenth century and Isaac Abravanel's commentary on Daniel, composed around 1500, remained popular throughout the sixteenth and seventeenth centuries.[39] In Jerusalem the rabbi and kabbalist Abraham ben Eliezar Halevi, also drew on Daniel to prove that the End was near.[40]

Yet it was not only scripture that focused the minds of contemporaries on Last Things. In the early modern world men and women also looked to the stars in their efforts to know the future. Of course, astrology was an ancient science. Moreover—since all could observe the heavens—it was a science open to everyone, from highly educated mathematicians who studied the stars and predicted their movements with great precision to ordinary men and women who, in a highly unpredictable world, turned to astrology to determine the most auspicious timing for everything, from planting their crops to popular cures such as bleeding or purging, while princes relied on their court astrologers for advice on when to go to war. Crucially, astrology knew no religious boundaries. Earlier in the fifteenth century, a Jewish scholar had interpreted the appearance of Halley's Comet in 1456 as "a sign of the Messiah and the approach of the time of perfection and redemption."[41] At the end of the century the court of the Ottoman Sultan Bayezid II was teeming with astrologers.[42] And Columbus, as we shall see, embraced this science not only for practical purposes but also to deepen his grasp of history's course, from Creation to the End of Time.

Yet never had this science been so popular as it was in the early modern world, as a plethora of astrological works—almanacs, calendars, *practica,* and prognostications—which earlier had circulated primarily in manuscripts among courtiers, clerics, notaries, jurists, and schoolmasters, began to reach a broad public of merchants, artisans, and even workers and peasants.[43] Printed on broadsides or in pamphlets—short, cheap, quarto-sized volumes—these texts had a broad appeal. In Italy, they proved especially popular in the late fifteenth and early sixteenth centuries, as people struggled to make sense of the wars and other disasters that overshadowed much of the peninsula and thus welcomed the promise that the tribulations would soon end and a golden age would come. But such texts also exploded in France, Germany, and England.[44] Both Jews and Christians, who were often in contract with one another, published such texts.[45] Many of the Jewish texts, such as Abraham Zacuto's *Almanach perpetuum,* a copy of which Columbus owned and which he used during an eclipse to calculate the distance of Jamaica from Spain during his final voyage to the New World, were sophisticated works of mathematical astronomy.[46] In addition, the Hebrew press printed hundreds of calendars. These *sifrei evronot* often engaged eschatological themes and looked forward to the eventual triumph of the Jews. A map printed in Constantinople in 1510 held forth the promise that God "will gather the dispersed of Israel, and a redeemer shall come to Zion."[47] No one has calculated the number of Christian and Jewish books—pamphlets, calendars, broadsides—on this theme, but there is no doubt that, if a full census were possible, there would be hundreds of thousands if not millions of them.

While much of the appeal of these works lay in their practical applications, they often combined the astrologer's skill in interpreting the stars with an array of popular medieval prophecies and thus contributed decisively to the apocalyptic imagination of the period. To be sure, not everyone believed that the study of the stars could predict the End of History. Savonarola, for example, had expressed his disdain for this science in his *Treatise against the Astrologers,* which he published one year before his execution.[48] And, later on, Martin Luther—though he was never consistent on this point—asserted in a sermon of 1518 that putting one's faith in the stars was a violation of the First Commandment.[49] While this opposition was significant, it did little to undermine the widespread belief among early modern people that God spoke to them not only through the scriptures but also through the stars.

An early important text in this genre was Johannes Lichtenberger's *Prognosticatio*, first published in 1488.[50] Lichtenberger, who appears to have enjoyed some years as a court astrologer, spent the last twenty-two years of his life as a parish priest in Brambach, a town just to the west of Wittenberg. But, despite his decidedly humble office, his book was well known, appearing in some seventy editions over the course of the late fifteenth century and the first half of the sixteenth century. The text's explicit marriage of astrology with divine revelation, along with Lichtenberger's own claim to prophetic inspiration, struck a chord, enabling readers to find in Lichtenberger's prognostications an ultimately reassuring guide in an uncertain world. Lichtenberger's deeply conservative cast of mind may well have calmed readers. For his was a predominantly medieval vision of the three social orders—the clergy, the nobility, and the peasantry—and his insistence that each should continue to carry out its traditional function. At the same time, he predicted that the Church of St. Peter would "suffer in the storms and troubles of the world." But he was hopeful. At first he placed his hopes in the Emperor Frederick III, but, in subsequent editions to this work, Lichtenberger and then later revisers would transfer these hopes to different figures.

Lichtenberger placed his greatest hopes on Frederick's immediate descendants: the Emperor Maximilian and then his son Philip by Mary of Burgundy—and it would be Philip "who will be the prince and monarch of all Europe and who will reform the churches and the clergy, and after him, no one else will rule."[51] Finally, Lichtenberger—who, after all, was drawing on Joachim of Fiore—offered hope for a Beautiful Ending. The success of the Habsburgs would lead to a purification of the Church and then an era of peace.

In this same period, Lichtenberger's younger contemporary Johann Virdung also produced an extremely popular *practica*. Like Lichtenberger, Virdung derived his ideas from a family of medieval prophets—Joachim, Bridget of Sweden, pseudo-Methodius, and Cyril of Alexander—and offered hope for a Beautiful Ending.[52] These texts and others like them would be widely read throughout the sixteenth century. And they were not only Christian. Several works of this nature showed up in the inventories of Jewish households of late sixteenth- and early seventeenth-century Mantua.[53]

In 1537–38 the Venetian publishing house of Paganino Paganini printed the Qur'an in Arabic—a remarkable feat that had required the making of nearly six hundred Arabic characters and diacritical marks and that Paganino printed

A chorus of astrologers and prophets (1488). Though drawing on different forms of prognostication, Aristotle, Ptolemy, a Sibyl, St. Bridget, and Brother Reinhardt are all divinely inspired.

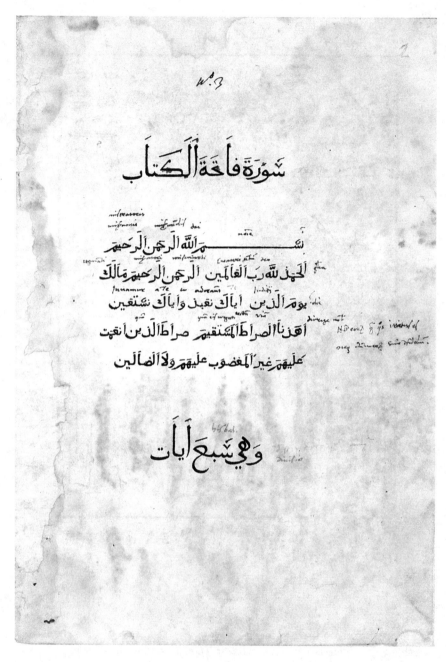

The *Sura al-Fātiḥah*, the opening verses of the Qur'an, from the *Corano in arabo* (1537–38). These seven verses are the most frequently recited chapter of the Qur'an. They are recited in each cycle of ritual prayers and throughout the social life of the faithful as an auspicious beginning and supplication, rooted in the worship of and reliance on God.

on especially fine paper. The orientalist Teseo Ambrogio degli Albonesi and the polymath Guillaume Postel, who were both in Venice at this time, knew of this work, as did Daniel Bomberg, whose printing of the Hebrew Bible may have inspired Paganino. Yet less than a century later no one could locate the book. The great Dutch orientalist Thomas Erpenius simply assumed that all copies had been burned, and in Protestant circles it was widely believed that a pope had ordered their destruction. In fact, however, the disappearance or near disappearance of the text—a copy finally emerged in a Franciscan library in 1987—was not a result of papal repression. For Paganino had not printed the text for either Christians or Jews but rather for the Muslim market.[54] Through his business contacts, he had hoped to sell the Qur'an in the Ottoman Empire—a hope that was based on a colossal misunderstanding. For while it may have seemed plausible to a Christian printer that he would find a sizable market for his book in Istanbul and other Ottoman cities, the Muslim approach to scripture differed radically from both Christian and Jewish understandings of their sacred writings. For, within Islam, the Qur'an had a sacramental quality, deriving its significance not from being written down or printed but rather from living in the hearts of believers who aspired to memorize the whole of the text. Indeed the language of the text, which is richly poetic (rather like the Book of Psalms), came to life through recitation, uniting believers with Allah and therefore to one another.[55] To be sure, manuscripts of the Qur'an circulated, but Muslims viewed the production of a sacred text in a fundamentally different way than did their Christian and Jewish contemporaries. It could never be seen as a business activity or a process involving mechanical reproduction. To the contrary, writing the Qur'an by hand was "regarded as a quasi-divine and mystical activity."[56] Thus, the calligraphers and illuminators who created copies of it were engaged in a holy act rather than a pragmatic one, producing more than enough copies to meet the needs of the religious communities in the empire. Accordingly, not only was there no need for printed Qur'ans, but the very idea of printing the Qur'an was viewed as blasphemous. We don't know what happened to the copies that Paganino more than likely shipped to Constantinople, but in the *Colloquium of the Seven about the Secrets of the Sublime*—a text traditionally attributed to the late sixteenth-century French political theorist and demonologist Jean Bodin but now believed to have been written or at least updated in the early seventeenth century—one of the interlocutors tells the story of a Venetian merchant, who, having "arranged for the Koran to be printed in this city and carried to Constantinople," was sentenced when he

Prophecy as art. Seven hadiths, written in diagonal lines in the middle, form a major part of this page from this Ottoman *muraqqa* or album created by the famous calligrapher Shaikh Hamdullah ibn Mustafa Dede about 1500.

arrived in the city with his copies for sale, but—through the good graces of the Venetian ambassador—had his sentence commuted. His right hand was cut off and all the copies of his book burned.[57]

Whether the story in the *Colloquium* is true or not, no one knows, but we do know that the calligraphers in the Ottoman Empire continued to produce and to circulate not only copies of the Qur'an but also other religious texts at an astonishing rate. Certain individual calligraphers produced as many as a thousand Qur'ans in their lifetimes. Indeed, there was considerable emphasis on making the Qur'an available on a much broader scale, and a large number of pocket-sized Qur'ans were produced in the seventeenth century.[58] Significantly, however, other texts with apocalyptic ideas also circulated widely. The early fourteenth-century Sufi Hibetullah b. Ibrahim's *Sa'atnāme* (On the Hour)—a Turkish work that, drawing on the Qur'an and the hadith, focused on preparing its readers for the Hour—encouraged its readers to have a copy of the book made and to share it, since doing so would ensure their salvation. Rich and poor alike produced copies, some in exquisite calligraphy and preserved on expensive paper, and others written out in a barely legible hand. Moreover, many of those who commissioned copies were women.[59] Other late medieval works, such as the jurist and Sufi Jalal al-Din al-Suyūtī's *Exposé concerning This Community Passing the Year 1000* enjoyed a robust afterlife throughout the sixteenth century, when many copied his manuscript.[60] But in the Ottoman Empire, it was books of omens—known as Falnama—that were especially popular. In these works—those produced for the sultans were especially beautifully illustrated—the believer could glimpse the future and catch sight of the return of the Mahdi or other Islamic visions of the End of Time.[61] In Spain a late fifteenth-century prophecy that foretold the coming of the Mahdi and a golden age for Islam, in a text that transformed John of Rupescissa's *Guide for the Tribulation* into a Muslim vision of a Beautiful Ending, circulated in the sixteenth century in *aljamiado* (Spanish in Arabic script).[62] Thus, even without the printing press, the libraries of the Islamic world were teeming with manuscripts.

Despite the proliferation of books, whether printed or in manuscript, most people heard and did not read the Apocalypse. We must not, however, draw too sharp a distinction between a textual and an oral culture. For one of the prime vectors in the spread of apocalyptic teachings were preachers—whether Muslim, Christian, or Jewish—who played a major role in transmitting their traditions, which they themselves were able to study and learn about through their readings, to the vast publics who crowded into mosques, churches, and synagogues to hear their sermons.

The Ottoman world was full of preachers, many of them celebrated mosque preachers, imams deeply learned in the sacred books and traditions of Islam and closely tied to aristocratic patrons. In Istanbul, these preachers were often closely associated with the sultan. The most eloquent of them could fill the great interior of the Süleymaniye Mosque.[63] The Ottomans often relied on such preachers to instill a clearly orthodox message in the minds of the faithful. Yet, in the Ottoman Empire, as throughout the Islamic world, members of the Sufi orders, drawing on mystical tradition, also preached. Many of them, drawing on hadith and often inspired and instructed by Ibn 'Arabi, frequently spoke in apocalyptic terms, promising the faithful that the time had come for a great renewal and that the Hour was near when the Mahdī would come.[64] They were especially influential in times of crisis. As news of Charles VIII's invasion of Italy in the 1490s triggered fears of a new crusade throughout the Ottoman Empire, preachers in the Balkans warned that the End of History was at hand.[65] In 1527, against the backdrop of uprisings in Anatolia, Molla Kâbiz, a Sufi in Istanbul, attracted large crowds with his promise that 'Isa would soon come and restore justice to the world. Judging his sermons to be politically dangerous, the authorities put him on trial and beheaded him. Throughout the sixteenth century, other Sufis, whose sermons struck some of the more orthodox as too free in their praise of 'Isa and the End Times—during a period in which the Ottomans were at war with the Christians of western Europe—were also condemned by the authorities.[66]

In Christian Europe, too, preachers and prophets, drawing on scriptures and more recent prognostications, drew large audiences.[67] Of course preaching had also been widespread in the Middle Ages. But, with the growth of towns and trades, ever greater numbers of friars and other itinerant preachers were on the move. Across the late Middle Ages and well into the early modern period, Franciscan friars were especially receptive to the teachings of Joachim of Fiore.[68] In the late fifteenth century, the Franciscan Giovanni da Capistrano, in his peregrinations across Germany, railed against worldly pride and explicitly drew on the teachings of Joachim, calling on those who gathered around him to repent and to prepare for the Millennium, the final stage of history.[69] And preachers such as Savonarola in late fifteenth-century Florence and the Augustinian Silvestro Meuccio—the early sixteenth-century editor of Savonarola's works and himself a great preacher—were only the most visible of the orators who presented apocalyptic ideas to the large crowds who flocked to listen to their sermons.[70]

Yet it would be the Reformation that would most emphasize preaching. Both Luther and the reformers in Switzerland had made the Word of God increasingly central not only to the faith of individuals but also—through the sermon—critical to expounding the meaning of scripture to the community. Sermons from the pulpits of Reformed and Lutheran churches drew large crowds, but there were also more and more popular preachers, whose independence and occasional radicalism worried Church elders. And sermons frequently unfolded new interpretations of the Apocalypse. Reformation-era preachers emphasized prophetic themes, pointing to the newly felt historical significance of such biblical figures as the prophet Elijah as a model for those seeking to establish a new church, if not a new society.[71] And Catholics responded in kind. Indeed, throughout the sixteenth century, Catholics came to see their own sermons as a critical bulwark against Protestant teachings, as they too came to place more and more emphasis on preaching. While the Roman Church had cautioned against prophetic teachings at the beginning of the sixteenth century, many of its preachers—especially Franciscans—continued to expound apocalyptic themes from the pulpit.

Finally, Jewish sermons also stressed messianic themes. The Spanish rabbi Shem Tob ibn Shem Tob placed great emphasis on the messianic in the late fifteenth century, as did Shlomo Molkho in the early sixteenth.[72] But it was in the late sixteenth and seventeenth centuries that Jewish preachers drew their largest crowds, throughout Europe and the Mediterranean, as Jews flocked to their synagogues from Venice to Istanbul. And many of these preachers, such as Solomon ben Isaac Levi in Salonika and Judah Moscato in Mantua, continued to develop messianic themes.[73] Across the early modern period, then, the pulpit, like the book, played a major role not only in Islam but also in Christianity and Judaism in the circulation and intensification of apocalyptic ideas.

CHAPTER 3

Christopher Columbus

Our Lord made me the messenger of the new heaven and earth which he revealed through St. John in the Apocalypse after having revealed it through the mouth of Isaiah, and he showed me its location.

—CHRISTOPHER COLUMBUS, letter to Doña Juana de la Torre (1500)

The discovery of these Indies was a marvelous event fashioned by God.... And now it would seem that Divine Providence customarily endows those it selects for a particular task with the necessary virtues and qualities.

—BARTOLOMÉ DE LAS CASAS, *Historia de las Indias* (1527–61)

Often they would hear the galloping and occasional shouts first. Then, looking up, they would see the men and the horses, and at times a carriage. In the quiet of the late medieval world, such riding parties—the parties of nobles and their retinues—captured the attention of peasants and townspeople. Those near the road could hear them, see them, smell them.

But no one had ever seen a cavalcade quite like the one that wound its way in the spring of 1493 along the ancient route—first built by the Romans centuries earlier—that stretched from Seville east to Valencia and then north, along the Mediterranean coast, to Barcelona. For, in addition to the expected nobleman and servants, the party included seven young men with coarse, jet-black hair. Their dress also drew attention. As Bartolomé de Las Casas—the friar who would create both the myth and the antimyth of Columbus—reported, "they wore masks, coverings made from fishbones and worked with pearls and gold, and similar belts, admirably fashioned with a great quantity and display of the finest gold, and many other things never seen or heard of in Spain."[1] And glorious birds—colorful green parrots in cages hanging from the carriages—made this riding party stand out even more. Something unknown, perhaps something from another world, was making its way across Castile and into the Kingdom of Aragon.

So unusual were these young men—the first men indigenous to the Caribbean to reach Europe—that word of their progress through Spain often preceded them. In towns and villages along the way crowds formed to see the captives and to catch a glimpse of the man who had seized them. "The news began to fly throughout Castile," Las Casas wrote, "that new lands that were called the Indies along with many, diverse peoples, and the newest of things had been discovered and that the one who had made this discovery was coming along the route and he was bringing along some of these people. Not only did people along the way by which he was passing come out of their houses to see him, even many peasants from distant villages crowded onto the roads to see him and hurried to the towns to greet him."[2] Writing of this same journey, Fernando Columbus noted that his father "would stop from time to time for a little while along the way, with such admiration of the people, for they rushed from all the neighboring places to see him, the Indians, and the new things that he had brought along."[3] Las Casas, who recalled seeing the captives as a boy in Seville where Columbus had briefly lodged them, would later condemn Columbus for seizing them, separating fathers from their sons and wives from their husbands.

But for most in Spain the captives were evidence of a great discovery. And nowhere was this more evident than in Columbus's reception in Barcelona. There, in this Catalan city, throngs of people, rich and poor, pushed up against one another to see Columbus and "the Indians and the parrots and many objects of art and jewels, and other things he had discovered, all made of gold, and which had never before been seen or heard of."[4] Then Columbus—about whom all were saying that he had discovered "another world"—was ushered into the Salón del Tinell, the grand Gothic room in the Royal Palace generally at the disposition of the Inquisition. There he made his way to the dais where the monarchs Ferdinand and Isabella, along with the prince, were seated. Then "having knelt down, he asked for their hands. They accepted and held them out before him, and, when he had kissed them, they asked him, with a most joyful countenance, to stand, and then they commanded that a chair be brought and they had him sit in their royal presence—the highest honor and grace which their Highnesses generally reserved for certain Grandees."[5]

Columbus proudly related his discovery to the monarchs, showed them samples of gold, and praised his captives, stressing their "simplicity, docility, and nakedness" along with "their most suitable disposition and aptitude" that made them ready to be converted to the Catholic faith.[6] The young men he brought with him, he noted, were proof of how easily they could be made Christian. Upon hearing this news, there was, again according to Las Casas, not a dry eye in the hall. Tears flowed everywhere. For the monarchs had "made known to all that the greatest joy and delight of their souls derived from seeing that they had been judged worthy, in God's judgment, to have enabled by their leave and by their expenditures . . . the discovery of so many infidel nations and that these were so well disposed that, in their lifetimes, they might know their Creator and be reunited to the flock of his holy and universal Church, and that the Catholic faith and the Christian religion might be thus immediately propagated."[7]

Las Casas, of course, had not been present in the royal hall in 1493. But even though this is a secondhand account of Columbus's reception, Las Casas's description of the powerful emotions felt by the monarchs upon learning that they could play a leading role in converting the infidel to their faith is likely accurate. If so, it underscores how central the underlying feelings of joy in doing God's work were to the formation of Spain's colonial enterprise. The colonization of the Americas was always—in Spain and beyond—more than a quest for gold and other resources and more than a quest for political power. It was also always a providential undertaking. But this was not, at least for many, an abstract or rationalized framework for the colonial enterprise. To some at least—and Columbus among them—faith in Providence was at the core of their inner life and a springboard both for moments of otherwise inexplicable courage and for moments of the most extreme cruelty.

Just a year earlier, in the first part of 1492, events had moved at lightning speed. In Granada on January 2, Boabdil, the last ruler of the Nasrid dynasty in Spain, had capitulated to Ferdinand and Isabella. Columbus had been present. "I saw the royal banners of Your Highnesses placed by force of arms on the towers of the Alhambra, which is the fortress of the said city," he wrote in a letter to the monarchs later that year, adding, "and I saw the Moorish King come out to the gates of the city and kiss the royal hands of Your Highnesses and of the Prince, my Lord."[8] Then, on March 31, the king and queen issued the decree—though it was not published until the end of May—expelling the Jews from Spain. Columbus also made a brief allusion to the expulsion in his letter.[9] Because he had been so close to the royal court over the previous several years, he was certainly aware that conditions in Spain for both Muslims and Jews had reached a tragic turning point. To be sure, Christian hostility to both Islam and Judaism had been growing over the previous century. In 1391 Christians in Seville had massacred more than two thousand Jews. A century later the Spanish Inquisition had begun systematically to persecute new Christians (converts from Judaism) who, many believed, continued to practice their faith secretly at home. In this same period, the Christian monarchs had also stepped up their efforts to eradicate the last Muslim powers in Spain. Fourteen ninety-two, therefore, proved to be an *annus terribilis* for non-Christians in Aragón and Castile. To be sure, the Muslims who stayed on after the fall of Granada enjoyed a modicum of religious freedom, at least in the short term. But Jewish families faced a wrenching decision. Either they had to leave Spain altogether or submit to baptism.

What did Columbus make of this religious suppression? While he does not comment directly on royal policy, it is certain that he not only supported the suppression of Judaism and Islam within Spain but also viewed it as part of his own hope to see Christianity triumph throughout the entire globe. Thus, in his letter of 1492 he noted that his monarchs—"as princes devoted to the holy Christian faith and propagators thereof, and enemies of the sect of Mahomet and of all idolatries"—had sent him "to the said parts of India, to see those princes and peoples and lands and the character of them all and of all else, and the approach one should take to bring about their conversion to our holy faith."[10] Moreover, Columbus would not only convert the heathen, he would also bring unprecedented wealth to the Spanish crown. The gold Columbus would find, so he assured his monarchs, would enable them to mount a crusade against the Muslims occupying Jerusalem—and, indeed, following his conversion of the peoples of Asia, the Great Khan of China, now a Christian, would attack the Muslims from the east, ensuring a Christian victory. Columbus, in short, inscribed his voyage to the Indies within a larger spiritual vision that animated the Spanish monarchy in just these years.[11] A Christian victory over the Muslims in Jerusalem would signal the Last Days, the Second Coming, and Eternal Peace.

While Columbus's design for a Christian victory in Jerusalem was as old as the Crusades themselves, his plan was nonetheless based on a radically new idea: he would sail west across the uncharted ocean to the Indies. To a large degree his years of experience at sea played a role in convincing him that such a voyage would be possible. Columbus had begun sailing at a young age. At first he accompanied merchants who put into port at Genoa, where he grew up, on relatively short runs to Provence and Corsica. Then, in 1474, when he was in his early twenties, he made his first long voyage, traveling to the eastern Mediterranean island of Chios, a major market for eastern silks and spices.[12] But it was his move to Portugal, where he settled in 1476, that brought him into an even wider, rapidly expanding world. From Lisbon, where he made his home, he sailed south as far as São Jorge da Mina, also known as Elmina, a Portuguese fortress on Africa's Gold Coast, and west to the Madeira islands, to the Canaries, and to the Azores. These voyages often arose from economic opportunity. In 1478, for example, he entered into a contract to become a sugar buyer for a wealthy Genoese firm. In 1479 he married Filipa Moniz Perestrelo, a noblewoman upon whose father the Portuguese crown had bestowed the governorship of Porto Santo, an island in the Madeira archipelago. The marriage may have brought Columbus more than a connection to the Portuguese nobility. In his account of his father's life, Fernando Columbus recalled—though we don't know if the story is true—that Filipa's mother had given Columbus "the writings and portolans which her husband had left her." And he added, "from these the Admiral grew all the more excited, learned of other voyages and navigations that the Portuguese were making to Elmina and along the coast of Guinea."[13] And although we know far less about Columbus's years in Portugal than we would like—the Lisbon earthquake of 1755 destroyed the city's archives—we do know that the decade he spent there played a formative role in shaping his desire to find a westerly route to Asia. From his experiences at sea, Columbus became convinced that there were many worlds that remained to be discovered and that the inhabitable world was much larger than either the ancient authorities or their medieval commentators had ever dreamed.

It is not only our experiences in the external world, though important, that make us who we are; we are also shaped by the texts we read, especially those that we study with care. And this was certainly the case with Columbus. Moreover, we know a lot about Columbus's readings. Las Casas, for example, noted that Columbus devoted the majority of his studies to "geometry, geography, cosmography, astrology or astronomy, and the nautical sciences."[14] Columbus himself provides a more expansive list. "I have," he wrote, "studied and made an effort to see all kinds of writings: geographies, histories, chronologies, works of philosophy, and other subjects."[15] Yet a stroke of good fortune makes it possible for us to know not only what Columbus read but how he read. Columbus's son Fernando—a bibliophile—preserved his father's books. Several of them are located today in the library Fernando founded in Seville. And these also include books that Columbus himself annotated. In these texts we can follow Columbus almost in real time as he read and reacted to many of the most significant geographical, astrological, and ethnographic texts of his time.

Columbus read to learn everything he could about the world he hoped to discover. Two of the texts he annotated were from antiquity. He owned both Plutarch's *Lives* and Pliny's *Natural History*, a well-known encyclopedia of classical knowledge of the various peoples of the world.[16] But Columbus was also drawn to more recent texts. He owned a Latin translation of the *Travels of Marco Polo,* and his extensive annotations in the margins of this book make it clear that he was fascinated not only by its description of the riches of Asia but also by its clues about this unfamiliar continent's geography.[17] He also read, in a similar way, *A Description of Asia and of Europe, and Their Histories,* by Aeneas Silvius Piccolomini, later Pope Pius II, with its vast learning on Asia, consistently annotating those passages that provided a mental map of a distant, largely unknown world.[18]

Yet Columbus also studied more theoretical works, and he did so at a propitious moment. It was, after all, in just these years that the *Geography* of Claudius Ptolemy, a Greek astronomer, cartographer, and mathematician who had lived in Alexandria in the second century, had begun to circulate in western Europe, often in beautiful illustrated editions, after having been virtually forgotten throughout much of the Middle Ages. The rediscovery of Ptolemy and his detailed description of the globe immediately gripped the imaginations of humanists, cartographers, and sailors, sparking debates and discussions about the structure of the world and providing new methods—above all in the presentation of the usefulness of coordinates, longitude and latitude, for accurately plotting any spot on earth. Columbus undoubtedly studied Ptolemy but he did so with some skepticism.[19] In the back of his copy of Piccolomini's *Description of Asia,* Columbus had bound two letters the Florentine mathematician Paolo Toscanelli had sent to Fernão Martins, a canon in Lisbon, and which Columbus had copied out in his own hand.[20] Like Ptolemy, Toscanelli demonstrated how by using a simple grid one could locate any place on earth in terms of latitude and longitude. But Columbus also found in Toscanelli the view that the Indies were closer to Spain than Ptolemy had assumed. Toscanelli, it seems, played a role in convincing Columbus that he could cross over to the Indies by sailing west.

The theoretical works that Columbus studied with the greatest care, however, were a series of treatises written by the French cardinal Pierre d'Ailly that had been published in Louvain in the early 1480s.[21] Scholars have long given attention to Columbus's reading of the first treatise in the volume—*On the Image of the World*. In this text, which d'Ailly had composed in the early fifteenth century in part on the basis of his reading of an early Latin translation of Ptolemy, Columbus read what was for him and his contemporaries a modern description of the globe. For Columbus this proved decisive. Las Casas confirms this, noting "that this doctor—that is, d'Ailly—much more than the ancients—moved Columbus to his enterprise [the Enterprise of the Indies]; his book was so familiar to Christopher Columbus that he annotated all of it and made red marks in the margins, in his own hand and in Latin."[22] One annotation is especially striking. Most of Columbus's *postille* are brief, often mere recapitulations of a central idea in the text. But in this case Columbus's annotations turned into a running commentary that filled not only the left-hand column and much of the right-hand column but also spilled out across the bottom of the page. Here, in short, the annotation is nothing less than the reflection of a "eureka" moment in which, while reading d'Ailly, Columbus conjoined the news of Bartolomeu Dias's discovery of a new route to the Indian Ocean with his growing grasp of world geography: "Note that in this year 1488 Bartolomeu Dias . . . reported to the Most Serene King [of Portugal] that he had sailed beyond what had already been navigated, as far as possible . . . and that he had reached a promontory to which he gave the name 'Cape of Good Hope' . . . and he described his voyage and provided an account of each place he saw on a portolan in order to put it under the eyes of the king. And I myself was involved in all this."[23] These are only the first words of the annotation—one that filled both margins and the bottom of the page—that Columbus made in Latin at the beginning of a chapter that d'Ailly had devoted to "the extent of the inhabitable earth." In the fifteenth century the question of how much of the earth was inhabited was disputed. Many believed, for example, that people were found only on a small part of the globe. But in d'Ailly's highly abbreviated prose, Columbus found several compelling arguments not only for the view that the earth was largely inhabited and that the various regions of the globe were connected by water but also—again in contrast to Ptolemy—that the Ocean could be crossed without great difficulty. One of the claims for the possibility of reaching Asia in this way, mistakenly attributed to Aristotle, was that "between the tip of Spain and the beginning of India, the sea is small and navigable in a few days."[24] But

The scholarly Columbus. While reading Pierre d'Ailly's *Imago mundi*, Columbus responds with an extensive annotation.

Columbus seized in particular on the Book of Esdras where the scripture maintained that five-sixths of the earth was inhabited and only a sixth was covered with water. Columbus then noted in the margins that Augustine, Ambrose, and Peter Comestor—all three great medieval authorities—had accepted this view, which again contradicted Ptolemy and many others who had held a much more limited sense of the extent of the inhabited world. Above all, d'Ailly's text gave him confidence that it would be possible to reach the Indies by sailing west from Spain.

Yet Columbus was equally drawn to d'Ailly's treatises on astrology, bound in the same volume as *On the Image of the World*. This might strike us as curious. After all, we easily understand and accept Columbus's fascination with geography. In the Renaissance this discipline played a foundational role in the development of a new cartography that is strikingly "modern." By contrast, astrology seems to represent an earlier, superstitious, palpably nonscientific view of the natural world, and some scholars have been reluctant to accept Columbus's intense interest in this discipline. Yet we must remember that to Columbus and his contemporaries both "sciences of the stars"—not only astronomy but also astrology—constituted essential ways of knowing and interpreting the world. No sailor could navigate without a knowledge of astronomy. The stars were the surest guide to location at sea.[25] But sailors and others in this period also believed that the heavens exercised a constant influence over affairs on earth. As Columbus noted in his annotations to an astrological treatise by d'Ailly: "This lower orb beneath the sphere of the moon is subject to celestial bodies," adding, "the form of heaven is a kind of natural book written by the hand of God; and when read, just like letters or like scriptures, it brings illumination."[26] Moreover, as a science, astrology could explain past events—from the Creation to the Flood and then to the Birth of Christ and beyond. But it also opened a window onto the present and future and could even help its students predict the fortunes of individuals, states, religions, and empires. When combined with scripture in which the arc of sacred history from the Creation to the Last Judgment could be discerned, astrology took on particular value, making it possible to predict when history itself would end.

Ultimately, it is in Columbus's quest to make sense of his place both in space and time that we can most easily see him as a Renaissance figure, endlessly curious about his own individual relation and the relation of his society to the cosmos. At the very time he was working through his own understanding of this question, humanists such as Giovanni Pico della Mirandola and Marsilio Ficino and the artist-inventor Leonardo da Vinci were also grappling with the question of man's place in the world—a world that they viewed, much as Columbus did, as only comprehensible through a grasp of its extensions in space and time and the relation of the individual to both.[27] For them and many thinkers of the Renaissance, man—and here they drew upon an ancient term—was a "microcosm." Columbus did not use such an elevated term, but his desire to connect his actions on earth to a deep understanding of the stars and of scripture reveal that he shared many of the same principles as some of the leading philosophers of his time. Thus d'Ailly's astrology was no less valuable to Columbus than was his cosmography. Certainly, Columbus annotated the astrological treatises with the same attentiveness that he had annotated *On the Image of the World.*

Columbus had been interested in astrology and indeed in matters of sacred history long before he read d'Ailly. As early as 1481, when still in Lisbon, he had written out on a loose sheet a chronology of the world's history "from the Creation according to the Jews."[28] Here he calculated the age of the world as 5,241 years. D'Ailly, therefore, was not his first introduction to astrology, but Columbus found in d'Ailly's treatises an especially compelling theory of time and eschatology. In *The Treatise concerning Laws and Sects,* for example, d'Ailly, closely following the teachings of the thirteenth-century English theologian Roger Bacon, had linked the movements of the planets to the rise and fall of world religions.[29] In this treatise, d'Ailly laid out the relationship of Jupiter's alignments with other planets, and how each major conjunction was associated with the rise of a particular religion. When Jupiter aligned with Saturn, Judaism was ascendant; when with Mars, the Chaldeans; when with the sun, the Egyptians; and when with Venus, the Muslims. D'Ailly also made it clear that the fall of Islam was coming soon. When Jupiter aligned with Mercury, the Christians would prevail; and then, when aligned with the moon, the Antichrist would reign. Columbus absorbed the teachings, making a brief table of these planetary changes and the past and future history of world religions in the margin.[30]

Then in his marginal notations to d'Ailly's *Treatise on the Agreement of the Truth of Astronomy with the Story of History,* Columbus filled the margins of his text with a brief outline of world history reaching from the Flood and the Fall of Troy down to the rise of the Romans, the birth of Christ, and the Crusades. But Columbus also found in d'Ailly a version of one of the most powerful prophecies of the Middle Ages: the amazing revelations attributed to the fourth-century martyr Bishop Methodius of Patara, who had foreseen the rise of Islam, the Crusades, and the reconquest of Jerusalem by the Muslims.[31] But now, as the prophecy made clear—it was actually a seventh-century forgery and the text has become known as the Pseudo-Methodius—there was a role for the emperor, in Columbus's mind the house of Ferdinand and Isabella, who were to conquer Jerusalem again. And this would lead to a period of profound and lasting peace, unlike any other, marking the end of the world. But then the "gates of the North will open and the people who were held back by Alexander" will burst forth and overcome Jerusalem. Yet a final battle remained. The prophecy continued: "the king of the Romans will ascend to the place where the Lord underwent death for us. The king will take the crown from his head and place it on the cross and will hand over the kingdom of the Christians to God the Father." And then it added: "When the Cross has been lifted up on high to heaven, the king of the Romans will directly give up his spirit. Then every principality and power will be destroyed that the Son of Perdition may be manifest." The prophecy concluded: "Then the sign of the coming of the Son of Man will appear, and he will come on clouds in the sky in glory, and he will kill him with the breath of his mouth according to the Apostles. And, then, the just will shine like stars, but the impious are tarred in hell."[32]

Apocalyptic desire was at the core of Columbus's enterprise. From his very first landfall in the Indies he believed that he had discovered, not a "New World," but rather a "New Heaven and Earth." His journal entries from 1492 reveal his admiration both for the exquisite beauty of the new landscapes—the enormous trees, the remarkable birds, the lush greenery—and the innocence of the peoples he encountered. He had not simply crossed an ocean; he had approached Eden. Moreover, he would play an active role in evangelizing the natives—in his view a necessary precondition for the End of Time. The Gospel of Matthew, after all, had stated that the "end will come," when "this good news of the kingdom will be proclaimed throughout the world" (Matthew 24:14). And as soon as Columbus returned to Spain in 1493, he devised a new signature that underscored the role he would play in bringing about the End of Time. Columbus fashioned himself in his elaborate signet as "Christoferens," the "Christ-bearer," the one who would bring Christianity to the Indies and beyond.

But it was on his third crossing to the New World, in the summer of 1498, that he came closest to touching heaven. As he recorded in his log, he believed that the further west he sailed, the higher the seas lifted his ships. He jotted down a novel idea: the earth is not—as Ptolemy had taught and most Renais-

Columbus's cryptic signature. Columbus fashions himself as "Xpo Ferens" or "Christoferens," "the Bearer of Christ." One possible interpretation of the letters is "Servus / Sum Altimissimi Salvatoris / Xriste, Maria, Yesu" or "I am the servant of the most exalted Savior, Christ, Mary, and Jesus."

sance astronomers believed—a perfect sphere. To the contrary the earth seemed to him "like half of a pear with a tall stem, or like a woman's nipple on a round ball."[33] And he thought that the earthly paradise could be situated on that stem. A few weeks later, when he encountered a large quantity of fresh water flowing from the powerful Orinoco River along the northern coast of what we now call South America, he believed that the waters likely came from the spring of the terrestrial paradise. Columbus, that is, experienced his discovery as part of an apocalyptic drama, and at the same time interpreted the lands he discovered in light of medieval maps that had often depicted the earthly paradise as located in the East, on the eastern fringes of Asia, precisely where Columbus believed he was.

Columbus nourished apocalyptic dreams throughout his adult life. But it was in the aftermath of a personal crisis that he offered his most sustained reflection on the End of History. In the fall of 1500, after facing dissent and even rebellion among the colonists, he had been arrested by the Spanish crown for having broken royal law in his repression of the rebellion, and he had been brought back to Spain in chains at the end of that year.[34] Although he was absolved by the crown, he lost his governorship of Hispaniola. Shaken, he spent much of the next year in refuge in the Carthusian monastery of Nuestra

Señora Santa Maria de las Cuevas, just outside of Seville, where his friend and fellow Italian Friar Gaspar Gorricio had invited him to stay. There, together with Gorricio, Columbus devoted much of his time to reexamining his own role within sacred history. He turned with particular intensity to the study of scripture. This, after all, was the ultimate source of apocalyptic faith. And he and Brother Gorricio assembled a collection of prophecies that spoke to Columbus's calling and faith. "The Holy Scriptures testify in the Old Testament by the mouth of the prophets, and in the New Testament through our redeemer Jesus Christ that this world must come to an end."[35] But Columbus did not draw only from the Bible. To the contrary, he found prophetic truths in a wide variety of authors from St. Augustine to the Sibyl, Merlin, Aquila, Joachim, and others, especially Nicolas of Lyra.[36] Nor did he draw from exclusively Christian traditions, even relying at one point on the Qur'an.[37] As he would write his monarchs: "I believe that the Holy Spirit operates in Christians, Jews, Moors, and in all others no matter what their faith."[38] He entitled the compendium *The Book or Manual of Authorities, Sayings, Pronouncements and Prophecies concerning the Matter of the Recovery of the Holy City and of Zion, the Mount of God, and concerning the Discovery and Conversion of the Islands of India and of All Their Peoples and Nations for Our Spanish Monarchs.*[39]

While this work—most often known as the *Book of Prophecies*—was a compendium, it is nonetheless possible to discern a common thread. Above all, as the text makes clear, Columbus was convinced that his crossings to the Indies had already fulfilled or begun to fulfil one of the major preconditions for the Apocalypse. Here, the Book of Isaiah seemed to address him directly. In that book, God had promised that He would "send some who have been saved to the peoples of the sea, to Africa and Lydia, to the ones who draw the bow, to the distant island, to those who have not heard me and have not seen my glory." Surely Columbus had gone as "Christo-ferens" to the distant island. And then there would be a great in-gathering. "They will declare My glory to the peoples," God spoke to Isaiah, adding: "They will bring all your brothers from all the peoples to the house of the Lord, on horseback, in chariots, in litters, on muleback, and in coaches to my holy mountain of Jerusalem, says the Lord, just as the Israelites bring a gift in a clean container into the house of the Lord. And I will accept some of them as priests and Levites, says the Lord. For like the new heaven and new earth, which I make to endure before me, says the Lord God, so will your seed and your name endure."[40] But in his discussions with Friar Gaspar, Columbus found other ways in which his voyages were fulfillments of the scriptures. In the Book of

Kings and in Chronicles, Columbus had read the account of Solomon's voyage to the islands of Tarshish and Ophir from which he had brought back "gold, silver, ivory, apes, and peacocks" (2 Chronicles 9:21). Now, in his *Book of Prophecies*, Columbus reiterated his view that Hispaniola was the Tarshish and Ophir of scripture. It was the gold he would find there, he believed, that would make it possible for the king and queen to conquer Jerusalem and to hasten the End of Time.[41] But above all, the *Book of Prophecies* provides us with a sense of joy that Columbus awaited in the reconquered Jerusalem. Again Columbus draws largely on the Book of Isaiah, where God had not only promised "a new heaven and a new earth," but also a New Jerusalem—a place of delight where "the sound of weeping will no longer be heard within her, nor the cry of distress."[42] And Jerusalem, in turn, would serve—and this was an ancient hope—as the site for the Second Coming. Here Columbus— inspired by his reading the Revelations of Methodius in d'Ailly—imagined that the conquest of the city by his rulers would lead to a period of peace for ten and a half years. The Heavenly City would descend. And for a fleeting moment there would be a time of abundance, justice, and peace. No one would suffer, no one would die. And then—as promised—time itself would end and the saved and redeemed would enter into the Eternal Kingdom. Columbus did not fear but rather longed for the Apocalypse.

Finally, the End Time was coming soon. Columbus's calculations convinced him of this. The clearest evidence for the imminence of the Second Coming was derived from St. Augustine's argument that the world would end after the conclusion of the sixth millennium.[43] But for his more precise calculations Columbus drew on both his careful study of d'Ailly's astrology and the famous Alphonsine Tables (a thirteenth-century work that enabled astronomers to determine the future positions of the planets). This allowed him to offer the calculation for the End of History. Thus Columbus argued that since, according to the Alphonsine Tables, 6,845 years had already passed since the Creation, with just under 5,344 years having passed from the Creation to the Birth of Christ and another 1,501 years since the birth of Jesus, it was clear that the end was near. "According to this calculation," Columbus concluded, "only one hundred and fifty-five years are lacking for the completion of the seven thousand years which, according to the learned opinions I have cited above, would be the end of the world."[44]

Columbus's vision of the End was never a purely intellectual matter. To the contrary, he saw himself as embedded in the drama of sacred history and fulfilling a prophetic role. His letter to Ferdinand and Isabella of 1501–2 underscored his conviction that God was playing a major part in his enterprise. To be sure, Columbus does not discount his own experience. He reminds his monarchs of his vast experience at sea, of his studies, of his efforts to convince them to underwrite his voyages. But ultimately his emphasis falls on Providence. "Our Lord has favored my wishes and given me an intelligent mind," he wrote, adding, "He has endowed me with a great talent for seamanship; sufficient ability in astrology, geometry, and arithmetic; and the mental and physical dexterity required to draw globes along with cities, rivers and mountains, islands and ports, with everything in its proper place." Then Columbus continued:

> With a hand that could be felt, the Lord opened my mind to the fact that it would be possible to sail from here to the Indies, and he opened my will to desire to accomplish the project. And with this burning desire I came to Your Highnesses.... Who could doubt that this fire was not merely mine, but also of the Holy Spirit who encouraged me with a radiance of marvelous illumination from his sacred Holy Scriptures, by a most clear and powerful testimony from the forty-four books of the Old Testament, from the four Gospels, from the twenty-three Epistles of the blessed Apostles—urging me to press forward?[45]

"Continually, without ceasing a moment, they insisted," Columbus added, "that I go on. Our Lord wished to make something most clearly miraculous in the matter of this voyage to the Indies in order to encourage me and others in this other matter concerning the Holy Temple about the holy spirit.... And so one should believe that this other matter will also happen; and as evidence of this, if what has already been said is not enough, I offer the Holy Gospel in which Jesus Christ said that all things would pass away, but not his miraculous word. He also said that everything that must happen has been spoken by Him and written by the prophets."[46]

God, Columbus believed, had chosen him for a prophetic role. He cited a passage attributed to Augustine. "Before you formed me in the uterus, you knew me; and before I left the womb, you preordained my life. Whatever things are written about me in your book, in the privacy of your dwelling place, etc."[47] And he accepted a popular prophecy that he was to play this role: "The Abbot Joachim of Fiore said that he who was to rebuild the house of Zion will come from Spain."[48]

In May 1502 Columbus left Spain for what would be his final voyage to the New World. He hoped to find a route to the Great Khan, and for the first time he explored the coasts of Honduras and Panama. The expeditions along the coasts proved disastrous, as the ships, many of them barely seaworthy, encountered ferocious storms. When finally the crews put to shore—in search of gold—they faced fierce opposition from the indigenous peoples and were attacked. At one point, Columbus, who was already ill, found himself alone on his own ship, the *Capitana.* Terrified, he cried out to God for help before exhaustion overcame him and he fell asleep. Prostrate from fear and exhaustion, at a point when he felt entirely abandoned, he heard a voice that reminded him of how much God had favored him and would continue to do so. Columbus believed his role was no less significant than those of Abraham and Moses within history.[49] No doubt these beliefs comforted Columbus down to his death in 1506, and similar beliefs may have comforted his son. In the *Book of Prophecies,* Columbus had included, first in the original Latin and then paraphrased in Castilian, a brief passage of Seneca's *Medea:* "The years will come, in the succession of the ages, when the Ocean will loose the bonds to which we have been confined, when an immense land shall lie revealed, and Tethys shall disclose new worlds, and Thule will no longer be the most remote of countries."[50] Shortly after his father's death, Fernando annotated this passage from Seneca in his own copy of the tragedies. "My father, the Admiral Christopher Columbus," Fernando wrote, "fulfilled this prophecy in the year 1492."[51]

The internal sense of destiny, calling, ambition was religious. Columbus desired the Beautiful Ending, the Second Coming, and the overflowing gift of salvation for all who had embraced the Gospels and repented. This is what motivated him, what pushed both him and countless others of his generation, including his monarchs who were also in the grip of an apocalyptic dream. The vision proved tragic. Out of it flowed a series of decisions by the Spanish to take possession of the islands and the peoples that Columbus had encountered on his first voyage. Thus, even if the prime motivation—the origin of the conquest—was a desire to evangelize, it is clear that it was Spain's desire for gold and power that underlay and sustained this missionary goal. Moreover, his voyages and actions—Columbus was convinced—would play a decisive role in accelerating the Heavenly Jerusalem for which he and so many of his contemporaries longed. But, like others before and after him, Columbus never found paradise. To the contrary—again like many others under the sway of an apocalyptic dream—he had discovered an imperfect world. Then, in a matter of a decade, he transformed it into a hell.

Ultimately, however, it is the Taínos—the indigenous people whose persons and lands Columbus and the Spanish came to possess—who provide the clearest sense of the contradictions embedded in Columbus's apocalyptic dream. To be sure, we know very little about the reactions of the natives to the conquest. Not only are the written sources from this period exclusively European, they are also deeply limited, for Columbus and the colonists not only didn't understand the languages of the peoples they encountered, they also tended to read Taíno culture either as purely innocent or purely barbarian. On their side, the Taínos—from their first glimpse of Columbus's caravels and then of the strangely-costumed men wading ashore—must have been both amazed and fearful. There may have been some joy in the exchange of gifts but there was also anger as the Spanish took several captives. The young men whom Columbus captured to serve as guides and translators—and eventually as trophies—frequently would have been terrified.[52]

In their resistance Taínos drew in part on their religious traditions. In his *relación* prepared for his sovereigns toward the end of February 1495, Columbus himself offered a glimpse of the native spiritual perspective. On July 7 of the previous year—Columbus reported—he and his crew had come ashore on the southeastern coast of Cuba after several weeks of difficulties in navigating along the island's shore. It was a Sunday, and the Spanish sailors erected a wooden cross and heard mass right on the beach. Several Taínos, all "naked as the day they were born" and some of them bearing gifts, stood close by. As the service ended an elder took Columbus by the hand and told him that, even though the Europeans had made the natives very fearful, he shouldn't let his exploits go to his head, for you are "a mortal just like everyone else." Then the elder continued, telling Columbus that "we are all born with and have a soul," adding "at the time of our death the soul separates from the body with great pain, and then the soul goes either to the King of Heaven or into the abyss of the earth, according to the good or for evil that one has done in the world."[53]

Of course, it's possible that Columbus himself invented the colloquy, making it up out of whole cloth. And we should be suspicious. At the very least it seems perfectly plausible that Columbus simply put Christian ideas about the afterlife into an invented speech by a tribal elder. Moreover, an invented speech, as one historian has suggested, might have helped Columbus convince Ferdinand and Isabella that it would not be overly difficult to evangelize the Taínos.⁵⁴ But, invented or not, Columbus's account of this extraordinary colloquy immediately captured the imagination of his contemporaries. Fernando Columbus alluded to the elder's speech in his life of his father; and Las Casas wove a version of it into his *History of the Indies*. In fact, Las Casas, who came to know the Taínos well, believed the sentiments of the elder's speech were accurate. "It is no wonder," Las Casas wrote, "that this elder spoke to the Admiral of another life, since collectively all the Indians of these Indies hold the opinion that souls do not die."⁵⁵ But the most compelling version appeared in Peter Martyr of Anghiera's *On the New World*, first published in 1516. "I warn you then to be aware that souls have two paths when they break forth from the body: one gloomy and hideous, prepared for those who cause trouble and are hostile to the human race; the other delightful and pleasant, appointed for those who in their lives have loved peace and quiet among nations," Peter Martyr has the elder tell Columbus, before adding: "If therefore you remember you are mortal and that rewards will be duly assigned to each in accordance with his present actions you will disturb no one."⁵⁶

Finally, the speech certainly fitted the context. For by the time Columbus reported it, it had become clear to the Taínos that the Spanish posed a genuine threat to their society. Indeed, from the very beginning of the arrival of the Spanish, the natives had fought back. Caonabo, one of the leading caciques of Hispaniola, and his forces killed the thirty-nine men Columbus had left behind on the island before returning to Spain on his first voyage. Inevitably, native alarm grew even more intense when Columbus returned in early 1494 with seventeen ships and over twelve hundred men. As the Spanish established a colony at La Isabela on the island's north shore and then began to march into the island's interior in search of food, slaves, and, above all, gold, the first major conflict was triggered when a group of Taínos "stole" some clothing off the Spanish. Alonso de Hojeda, a military leader whom Columbus had appointed to explore the interior, arrested the thief, cut off his ears, and delivered him along with his cacique and several of their relatives to La Isabela in chains. "This was the first injustice," Las Casas observed in his *History of the Indies*, "and the beginning of the shedding of blood, which afterward was so abundant on the island."⁵⁷

Then, as Spanish efforts to extract wealth from the island intensified, so too did Taíno resistance. A tipping point came in February 1495 when the Spanish took over fifteen hundred native men, women, and children captive, putting the fittest five hundred of them on ships back to Spain where they would be sold as slaves. In response to this and other Spanish attacks on the native culture, the cacique Guatiguaná organized his forces in the Vega Real, a valley to the south of La Isabela. But, as the battle was joined, the Spanish—with their horses, dogs, and guns—prevailed. They captured Caonabo, who had carried out the massacre at La Navidad, and several other caciques and began their pacification of the island. With an emerging network of coastal and inland forts, they were now able to control the native population, extracting their labor. The Taínos continued to resist, but a system of dominance and exploitation had been put into place. In the meantime, as Michele da Cuneo, who accompanied Columbus on this second voyage, revealed, not all the original five hundred captives survived the voyage to Spain. "By the time we had reached Spanish waters," he wrote, "approximately 200 of the men the Spanish had captured had died—I believe it was because they were unaccustomed to the air which is colder than theirs—and we threw them into the sea."[58]

From the beginning Columbus too had been callous. Even on his first voyage he had seized several captives, young men he sought to use as translators and guides. But he not only took them away from their families and familiar surroundings but treated them as objects. He brought them to Spain as trophies or tokens verifying that he had discovered something new. We don't know the details but it is certain that some of the captives died on the crossing to Europe and likely that Columbus would have cast their bodies into the Atlantic. A history of violence lay behind Columbus's decision to parade the seven young men across Iberia. "The Admiral and his Christians," Las Casas wrote, "and later all those who entered and stayed in these lands and kingdoms always made it their first task, as something they deemed necessary to carry out their intentions, to instill fear and dread in the hearts of all these peoples in such a way that, upon hearing the word 'Christians,' their very flesh would tremble."[59]

CHAPTER 4

Conquest and Utopia

Gold is most excellent. Gold constitutes treasure, and anyone who has it can do whatever he likes in the world, if God does not oppose him, and he can succeed in bringing souls to Paradise.

—CHRISTOPHER COLUMBUS, *Relación del cuarto viaje* (1503)

And they do not have private property but all things are held in common.

—AMERIGO VESPUCCI (1504)

The Arawak men and women who witnessed Columbus's landfall on October 12, 1492, sought safety in humanity's most archaic form of diplomacy: the exchange of gifts. In his account of the first contact Columbus emphasized this: "I gave some of them colored caps and glass beads, which they wore on their chests, as well as other items of little value in which they took so much pleasure that they became ours, which was a marvel." "Later," Columbus continued, "they came swimming out to us in our launches, and they brought us parrots and balls of cotton thread as well as spears and many other things which they traded for other items such as small glass beads and bells that we gave them. In the end they took and gave of what they had with good will."[1]

Yet Columbus was anything but interested in the free exchange of goods. To the contrary, his goal was to seize, to possess as much as possible. Even prior to the first voyage the crown had promised him a tenth of the profits of his enterprise, whether in "pearls, precious stones, gold, silver, spices, and other items and trade goods of whatever kind, name, or sort they may be."[2] His very first acts, not only upon his first landfall but also as he reached other shores in the Antilles, were to name the islands he found in honor of his faith or the monarchy, and then—coming ashore, unfurling flags, and uttering legal formulas—to take possession of them in the name of Ferdinand and

Isabella. Above all, mesmerized by the small golden ornaments that many of the natives wore on their bodies, he became obsessed with the pursuit of this precious metal. "Where is the gold?" was the question he most often directed to the natives. And it was the pursuit of gold that led Columbus and the hundreds of soldiers with whom he returned to Hispaniola in 1494 to push into the island's interior, raping, seizing, and murdering the natives and setting in place a pattern of colonization that would result in the torture, enslavement, and deaths of millions of people (at first the indigenous peoples of the New World and then African captives) over the course of the next several centuries.

Yet, even in the midst of his unbridled seizure of the natives and their land in the Caribbean, Columbus recognized that not only the material conditions but also the values of the indigenous peoples differed in striking ways from his own, especially in their seeming indifference to the accumulation of material possessions. "The people of this island and of all the others who, even without any earlier information, I happened upon and possessed," he reported in an account of this first voyage written during his return to Spain, "go about completely naked, both men and women, just as their mothers bore them." But this nakedness—a sign of innocence—was conjoined to their generosity, an absence of greed. After also noting that the natives had no weapons of steel and stressing their timidity, Columbus observed that "once they feel secure and their fear has subsided, they are so open and free with what they have that you have to see it to believe it. When they are asked for the things that they have, they never say 'no.' Rather, showing much love and opening their hearts, they entrust these items to others." Columbus then continued his account. He informed his readers that the peoples of these islands had a common language and that the men "are content with one wife, though to their chief or their king they provide up to twenty." In closing his account of the people he came back to their generosity. "Nor was I able to understand if they have private property, but it seemed to me that that all share in what each has, especially in food."[3]

Columbus's report—published almost immediately not only in Castilian but also in Latin, Italian, and German—reached a broad readership.[4] Crucially, in drawing on apocalyptic hopes and expectations, Columbus offered his readers a glimpse of paradise. He was especially struck by his encounter with a naked, generous people whose innocence contrasted so sharply with the social lives of Europeans, enmeshed in endless wars and the pursuit of wealth and power. Columbus's letter would play a role in stimulating the hope for a better world. Across the Atlantic lay Eden. Columbus had discovered, as he himself would write, "a New Heaven and a New Earth." Throughout his life, he would continue to believe that he had come, if not to paradise, at least to its very edges.

Reports of previously unknown peoples who lived in a state of innocence and without private property proved inevitably stirring in a Europe wracked by endemic warfare and unbridled greed. While Columbus offered and would continue to offer an interpretation of the New World that was largely apocalyptic, others believed the cultures and societies of the peoples he had described were more evocative of the ancient idea of a golden age, an imagined early period in history when humanity had lived without sorrow or pain, when aging brought no misery, and when nature provided, without their labor, all that they needed for nourishment and pleasure.[5] In the late fifteenth century it was only natural for many humanists, who had studied the ancient poets, to view the New World in this light.[6] Peter Martyr d'Anghiera—an Italian in the court of Ferdinand and Isabella—was certainly inclined to interpret Columbus's discoveries through this lens. "But I feel that our natives of Hispaniola are happier—more so were they converted to the true religion—because naked, without burdens, limits or death-inducing currency, living in a golden age, free without fraudulent judges, books, and content of their natural state, they live with no worries about the future," he wrote in his *De orbe novo*, or *On the New World*.[7] As he had also observed of the natives, "it was discovered that for them the earth, like the sun and the water, is held in common; nor does 'this is mine' and 'this is thine,' sentiments that are the seed of all evils, occur among them." And he then added: "For they are content with little, since in this spacious land there are more arable fields than needed. For them this is a golden age (*aetas aurea*), and they surround their farms neither with ditches nor walls nor hedges. They live in the open, without laws, without books, without judges, and their conduct is equitable in accordance with their own nature. They judge evil and wicked whoever takes pleasure in harming others."[8] In the social world of the indigenous peoples of the New World, that is, Peter Martyr found neither paradise nor Eden—as Columbus believed he had discovered—but rather nothing less than a golden age, a world in which men and women lived in plenty and in peace, uncorrupted by civilization. The ancient myth of the golden age, on the one hand, and the Christian myth of the messianic age, on the other, differed from one another in their underlying conceptions of time. The ancient authors had largely embraced a cyclical view of history; Jewish and Christian writers, by contrast, viewed history as linear. Yet, despite these tensions, these two different notions of historical time each offered a model of a more perfect world, and often in both the Middle Ages and the early modern period these two visions would converge, with the image of the golden age frequently grafted onto apocalyptic hopes for a Beautiful Ending. Indeed, it is no exaggeration to say that these two visions of history nourished and reinforced one another.[9]

Peter Martyr may have been the first to have characterized the lands Columbus had discovered as a "new world," but other texts soon reinforced this view. The most significant of these was the *Mundus novus,* or *The New World,* the title of Amerigo Vespucci's first published letter on his explorations of the Brazilian coast in 1501–2.[10] In this work, a gripping account of his observations of the coast of Brazil, Vespucci—a Florentine merchant and cosmographer—stressed the novelty of what he had witnessed. "In recent days," so he opened his letter to his patron Lorenzo di Pierfrancesco de' Medici, "I have written you quite amply about my return from those new regions . . . which we have sought out and found and which we might call a new world."[11] Then, after briefly reporting that what he and his crew had found was not merely another island but rather a continent unknown both to the ancients and to the Europeans of his own day, Vespucci turned to describe the peoples he had encountered. "Everyone of both sexes," he writes, "struts about naked, covering no parts of the body. Just as they come forth from their mothers' wombs, so they continue to the time of their deaths."[12] He was also struck, as were both Columbus and Peter Martyr, by their relation to property. "They have no cloth or wool or linen or cotton, since they do not require them nor do they have private property but hold all things in common."[13] Furthermore, "they live together without a king and without authorities, each man his own master. They take as many wives as they wish, and son may couple with mother, brother with sister, cousin with cousin, and in general men with women as they chance to meet. They dissolve marriage as often as they please, observing no order in any of these matters. Moreover, they have no temple and no religion, nor do they worship idols. What more can I say? They live according to nature, and might be called Epicureans rather than Stoics. There are no merchants among them, nor is there any commerce. The peoples are among themselves without art or order."[14] The language here is almost certainly Vespucci's, but the version of the *Mundus novus* that circulated incorporated a brief evocation, added by the text's editor, of the Apocalypse. Vespucci had not simply encountered new peoples; rather he had found—so this version of his text makes clear—the very "multitude that no one could count" who, as the Book of Revelation promised were those "who will hunger no more and thirst no more . . . . and God will wipe away every tear from their eyes."[15]

The *Mundus novus* circulated widely, but it was the longer *Letter on the Newly-Found Islands*—a compilation, attributed to Vespucci, of an account of Vespucci's crossings first published in 1504—that brought the news of his discoveries to a wide public. Though Vespucci himself was not the author of this work, it was nonetheless largely based on his private letters and the *Mundus novus* itself. Moreover, when a Latin version was published under the title *Quattuor navigationes,* or *The Four Voyages,* the work achieved unanticipated fame. This Latin version, after all, appeared in Martin Waldseemüller's *Cosmographiae introductio* of 1507, the publication that famously gave the name of America to the New World. As Waldseemüller wrote in his preface, in addition to the three parts of the earth previously known, "a fourth part has been discovered by Amerigo Vespucci." And then he honors the explorer: "Since both Europe and Asia are named for women, I see no reason why anyone would protest that this be called Amerige (or the land of Amerigo) or America after its discoverer, a man of wise ability."[16] On the planisphere that accompanied the text, Waldseemüller labels the New World "America" for the first time.

Thus, within two decades of Columbus's first reports of his discoveries, several other texts—Vespucci's *Mundus novus* and his *Quattuor navigationes,* along with Peter Martyr's letters and his *De orbe novo*—introduced European readers to social arrangements profoundly different from their own. In these texts—at times evocative of a golden age, at times apocalyptic—readers encountered the articulation of a new vision for human society. To be sure, both ancient and medieval authors had, in various ways, imagined ideal polities. And in the generation before the discovery of America a large number of Renaissance humanists, often drawing upon Aristotle, Plato, and Cicero, tried to do much the same, laying out the principles—just laws above all— that would produce the best states.[17] But one of the most significant, far-reaching consequences of the "discoveries" was the way in which ancient texts—Plato's *Republic,* Cicero's *On the Ends of Good and Evil,* even the monastic *Rule of St. Benedict*—now intersected with accounts of societies that offered visions of alternative social and political arrangements. The accounts of Columbus, Peter Martyr, Vespucci and others, in short, made it clear that it should be possible to realize a more just, more peaceful society. Columbus's vision of paradise had made utopia possible.[18]

At their core the reports of the discoveries opened up new spaces for the imagination. The very nature of these writings—especially in their incompleteness and in their inevitable imprecisions and moments of vagueness—compelled readers to imagine what it was that the writer was attempting to describe. This was especially the case in the account of Vespucci's purported "fourth voyage"—a voyage he in fact never made.[19] In what was meant to be the most ambitious crossing yet, a fleet of six vessels, one of them captained by Vespucci, had embarked from Lisbon—so the story went—on a voyage intended to carry them by a westerly route to Malacca. They never reached their destination. In the mid-Atlantic the fleet's commander struck a rock just off the coast of an "island in the midst of the sea." As the ship sank, men from the other ships sought to rescue the crew and to salvage whatever they could, while Vespucci, on the commander's orders, sought harbor alongside a nearby island. His understanding was that the fleet would soon join him, but, for reasons the story fails to make clear, the fleet's commander never came to find him but rather set out with the rest of the fleet to return to Portugal. Abandoned, Vespucci and his crew—along with another twenty-four sailors he had rescued from the sinking boat—decided to sail on to the west, managing to reach the coast of Brazil. There they explored the territory, loaded their ship with brazilwood, and built a fort where the twenty-four men Vespucci had rescued were to stay, and to protect for future trading activity. Leaving these men behind, Vespucci and his crew returned to Lisbon, where they learned the tragic news that the other ships of the fleet had been lost at sea.[20]

Slightly over a decade later, Vespucci's story of his "fourth voyage" would inspire another work of the imagination: Thomas More's dialogue *The Best State of a Commonwealth and the New Island of Utopia*, first published in 1516.[21] In this imaginative work, More, who casts himself as the character Morus, meets a sailor—an aging, shoddily dressed Portuguese philosopher by the name of Raphael Hythloday—in the Flemish port city of Antwerp where Morus was visiting his friend Peter Giles. After explaining that he had been one of the sailors who had not returned with Vespucci to Portugal—and here More was expanding Vespucci's fiction for his own ends—Hythloday described his further journeys. Along with a group of five companions, assisted by a local guide, he had set out to continue their explorations. At first they traveled across land, often by wagon. As they moved further and further south of the equator, they encountered towns and cities "in which there is a continuous commerce by land and sea not only among themselves and their neighbors but even with distant peoples."[22] Traveling, now on ships,

A map of Utopia. One of the several devices included in More's work, lending credence to the fiction. From the first edition published in Louvain in 1516.

throughout the southern hemisphere, Hythloday visited many nations, observing wise and prudent arrangements that could be useful to readers, but he was especially struck by the peoples of an island nation called "Utopia"—More coined this name from the Greek *ou-topos* ("no place")—which Hythloday visited before finally making his way back to Europe.[23] He had done so by continuing to sail west, passing through Ceylon (Sri Lanka) and then reaching Calicut where, encountering Portuguese sailors, he found a ship that brought him home.

It was through Hythloday's description of this imaginary island that More conveyed to his readers a vision of the ideal society. In contrast to early modern England, where many lived in poverty and were forced to work from dawn to night merely to scrape by, Utopia was an island republic with no beggars and where no one worked for more than six hours a day, leaving ample time for leisure. The society More envisioned contrasted with England in other fundamental ways. In Utopia, everyone enjoyed access to education and to medical care. Moreover, not only were there no arranged marriages but prospective partners were expected to see each other naked before agreeing to be wed, and, should the marriage not work out, there was the option of divorce. Furthermore, in Utopia there was no capital punishment, and there were no wars fought for the aggrandizement of the king. Finally, again in contrast to England and the rest of Europe, Utopia was a place of both religious tolerance and, at the same time, of a shared faith that encouraged virtue and belief in a Supreme Being. For readers, then, More's *Utopia* was an invitation to consider the possibility of an alternate set of social and political arrangements that would, ideally, favor the happiness of all.

Yet what set Utopia apart not only from More's England but also from all other European states was, above all, the fact that the Utopians had no private property but rather—as Columbus, Vespucci, and Peter Martyr had described them—held all things in common. Raphael Hythloday makes this point forcefully at the end of book 1: "My dear Morus—and here I am telling you all those things I have in mind—it seems to me that wherever there is private property and all things are measured in monetary terms, that it hardly ever happens that a commonwealth is either just or prosperous, unless you think that it is just when all the best things belong to the worst men or that it is a happy state when everything is divided among very few."[24] Hythloday continues, "Thus I am completely persuaded that no equitable and just distribution of goods can exist and no happiness in human affairs is possible unless private property is utterly abolished."[25]

This was a remarkably radical critique of the early capitalist society that was developing in Europe at this time—though, within the dialogue, the *character* Morus offers the classic defense of the benefits of private property. "But, to me, the opposite seems to be the case," Morus insists, "for it is never possible to live well when all things are held in common. For in what way can there be a sufficient quantity of goods when everyone avoids labor? Or when the desire for one's own gain does not urge one on and the faith in the labor of others renders one slothful? Moreover, when one is stimulated by poverty and is not able to keep for oneself what one has gained, is it not certain that one will be troubled by perpetual upheavals and seditions? And, I cannot imagine, how there will be a place for either the authority of or respect for magistrates, when we do not discriminate among men."[26]

Many of More's humanist readers—he was writing in Latin—would have recognized his presentation of the debate over private property as a recapitulation of the contrasting views of Plato and Aristotle. More had studied both of these ancient authors with great affection, but he was particularly inspired by Plato's writings about the perfect republic and his critique of private property. At the same time, More would have also been deeply familiar with Aristotle, whom the character Morus follows rather closely here, and his protest against Plato's support for the abolition of private property.[27] But, despite this classical antecedent, More's text was nonetheless radical. Under the veil of dialogue he offered his readers—at least for consideration—a powerful criticism of his own society and the hierarchical principles upon which it was based. But this is not merely a consideration in book 1. Book 2 also places great emphasis on how deeply the Utopians' view of property and private wealth differs from these institutions in other societies.[28] To further illustrate this, Hythloday underscores the use of gold and silver in the imaginary republic of Utopia, noting at the outset that it will not be easy to grasp, but as Hythloday notes, their use of these precious metals corresponds "more to their cultural logic (*moris rationem*) than to ours."[29] The result is that the Utopians use these metals as marks of shame. They fashion chamber pots and shackles for prisoners out of gold and silver; and they require that "those bearing the stigma of a crime wear gold in their ears, encircle their fingers with gold rings, their necks with gold bracelets, and crown their heads with golden crowns."[30] All of this, Hythloday makes clear, is part of a culture profoundly different from that of Europe. The Utopians, he continues, "marvel whenever a mortal takes delight in the uncertain sparkle of a little jewel or of a gem, when they can admire any star or even the sun itself."[31]

Yet More's vision of Utopia was not merely a legacy of classical Greek philosophy. Equally decisive was his deep Christian faith. More, who always struggled to balance his religious convictions with his sense of political duty, would end up paying the ultimate price for his religious commitments when, some twenty years after completing *Utopia,* he refused to take an oath to uphold the king's supremacy over that of the pope, and was beheaded.[32] Yet, as a work of imagination, *Utopia* enabled him to unify his religious and political views—at least on the level of the vision of a perfect society. And here he drew not only on Plato but also on the Gospels. Particularly close to his heart was St. Paul's description of the Apostles in the Book of Acts. "Now the whole group of those who believed were of one heart and soul, and no one claimed private ownership of any possessions, but everything they owned was held in common" (Acts 4:32). But closer to home as a resource for reflection on holding all things in common was the *Adages,* a book in which More's close friend Erasmus deciphered to the benefit and delight of his early modern readers a wide swath of enigmatic ancient proverbs or sayings. And the very first adage in his third edition of this work, published in 1515, was *Amicorum communia omnia* ("Between friends all is common"). On this one proverb Erasmus offered his readers a compendium of its use in antiquity. He found it in Plato and, before him, in Pythagoras. But he also found it in scripture, above all in the Acts of the Apostles. But Erasmus did not merely explicate, he also moralized. "Anyone who deeply and diligently considers that remark of Pythagoras, 'Between friends all is common,' " Erasmus wrote, "will certainly find the whole of human happiness included in that brief saying. What other purpose has Plato in so many volumes except to urge a community of living, and the factor which creates it, namely friendship? If only he could persuade mortals of these things, war, envy and fraud would at once vanish from our midst; in short a whole regiment of woes would depart from life once and for all. What other purpose had Christ, the prince of our religion? One precept and one alone He gave to the world, and that was love; on that alone, He taught, hang all the laws and the prophets. Or what else does love teach us, except that all things should be common to all."[33] Like Erasmus, More looked upon the ideal of "all things in common" through both classical and Christian lenses. It would be on the basis of these principles, as the historian David Wootton has demonstrated, that More would elaborate his vision of an ideal society.[34]

More was not merely a dreamer. In *Utopia,* after all, More managed to shift the focus from the widespread hope for a Beautiful Ending to a more earthly aspiration: to the possibility of reforming the economic and social arrangements that served as the foundation for political and ethical life. In part, More's approach reflected his own cast of mind—his immersion in classical works and his training as a lawyer. More's vision also reflected a current of Christian spirituality that rejected not only any human attempt to predict the End Times but also any vision of collective salvation or of a peaceful Millennium. To the contrary, More's focus on the Last Things stressed individual salvation.[35] Yet More's *Utopia* does offer his readers something not entirely unlike the Millennium. In his closing remarks on the imagined isle, Hythloday states, "So I am glad that this form of a commonwealth, which I would hope for all, has at the least favored the Utopians. They followed those habits of life by which they established the foundations of the commonwealth not only most felicitously but also, insofar as it is possible to foresee through human conjecture, that shall last for Eternity."[36] To be sure the basis of this hope was not the divine rescue of humanity but rather the moral reform of humanity through the effort of men and women, through giving up property, to abandon their pride, the basis of all sin. Nonetheless it is impossible not to sense that there are echoes of an age that was intent on finding a Beautiful Ending.[37] More, after all, lived in a time when many dreamed intensely of paradise. His genius lay in his translation of this otherworldly dream onto a more human plane. But we shouldn't underestimate his own spiritual longing. On a personal level, writing *Utopia* offered More a brief moment of believing that it might be possible—though he would be proved wrong—to unite his deepest religious convictions with service to a king.

Dreams of both God's millennial kingdom and utopia were by no means purely theoretical. Dreams, after all, are powerful things; especially when shared, they can motivate men and women to try to bring about fundamental changes. Indeed, the sixteenth century opened with a widespread sense that it would be possible to renew the world. In Europe utopian and apocalyptic dreams would lead to some of the most revolutionary social and political experiments of the era. But in the New World, too, utopian and millenarian dreams had an impact, influencing many aspects of the colonial societies that were beginning to take shape in just this period.

While dreams of a Beautiful Ending had done much to animate Columbus even on his first crossing, it was with the arrival of the first Franciscan mission in Mexico that millenarian aspirations first took deep root in the New World. Spain's success—in colonizing the Caribbean and Mexico under Hernán Cortés and Peru under Francisco Pizarro—not only raised enormously complex challenges about the administration and exploitation of vast new lands, but also urgent questions about the salvation of souls. Christianity, after all, had always been deeply proselytizing. Columbus had not brought a priest with him on his first voyage, but he did on his second. And soon a concerted effort at evangelization emerged. The first Franciscans had arrived in the Caribbean in 1500, and the first Dominicans followed them in 1510. But it was with the arrival of the first Franciscan mission to Mexico in 1524 that the missionary efforts took a decidedly millenarian turn. Many members of this religious order had long seen themselves as a spiritual elite whose vows of poverty merged with deep apocalyptic hopes. Indeed when the twelve Franciscan missionaries departed Spain for Mexico in 1524, their superior, Francisco de los Angeles, framed their mission as apocalyptical. "The day of the world," he told them, "is already reaching the eleventh hour; you . . . are called to go to the vineyard, not for wages as some do, but as true sons of your so great Father, seeking not your personal gain but only the advantage of Jesus Christ."[38]

The twelve friars took their mission and their vow of poverty seriously. Diametrically opposed to the brutal and greedy actions of the conquistadors, the Franciscans not only stressed their own poverty but also believed that their example would prove appealing to a native population that—as far as they could tell—did not have private property and that did not show any of the signs of avarice that were so widespread in European culture. In their actions they sought to show natives that their values more closely aligned with them than did the values of the Spanish who had come purely to conquer. Accordingly, shortly after arriving in Veracruz, the friars walked barefoot to Mexico City—a powerfully symbolic procession that a new type of Christian had arrived in the New World. But the most compelling example of their desire to align themselves with the natives was the ease with which Toribio de Benavente, one of the twelve, adopted Motolinía, the Nahuatl word for "poor," as his name.[39] Above all, despite the hardships the friars faced in their crossing and the living conditions they encountered in Mexico, they saw their work as part of a providential plan. Working with the natives this original group of twelve Franciscan missionaries believed that they would create a virtually utopian society and ultimately a millennial kingdom that would come at the End of Time.[40] "And as the church flourished in the east in the beginning of the world," Motolinía wrote, "so now, at the end of time, it must flourish in the west, which is the end of the world."[41] Converting the Indians to Christianity was, in their view, an essential prelude to the Millennium. The new virgin land of the Americas inspired both these original Franciscan missionaries and those who followed them across the Atlantic. In this new world they would see the establishment of Joachim's third age.[42] Or, as Motolinía put it in a letter to Charles V, in which he drew on the Book of Daniel, they would see "the Fifth Monarchy of Jesus Christ, which is to expand and embrace the whole earth and of which Your Majesty is to be the leader and the captain."[43]

Yet the impact of the Spanish led to a different transformation of the society from the one for which the Franciscans hoped. While some conquistadors may have dreamed of instant riches through the discovery of deposits of gold or silver, most recognized that the process of growing rich would depend primarily on developing agricultural and other natural resources. The Europeans early on became aware of the value of many New World crops that could be shipped back to Europe. Some of the products they could harvest, such as cochineal and brazilwood (both of which were used in the making of rich red dyes), chocolate, and tobacco, were native to the Americas. Sugar, by contrast, was an import, but the climates of the Caribbean and Brazil proved very favorable to the production of sugarcane, and sugar rapidly became one of the New World's most significant exports.

The exploitation of these crops and resources led to a system of *encomiendas*, or grants, to settlers that awarded them the rights to the labor of the indigenous peoples. Those who were granted such rights were generally relatively wealthy individuals who had already invested in the colonization of new territories or conquistadores whose exploits on the battlefield won the appreciation of their commanders. In exchange for the encomiendas, they were expected to provide the native Americans who labored for them with instruction in at least the rudiments of Christianity. This pattern of settlement in the Caribbean was followed closely in the settlements established by Cortés in Mexico and by Pizarro in Peru.[44]

To Bartolomé de Las Casas, the development of the settler economy, the theft of native lands, the exploitation of native labor, and the brutal massacres of tens of thousands of the indigenous peoples comprised the great Spanish crime against which he would fight most of his life. To be sure, there was virtually no hint at first that Las Casas would play this role. To the contrary, Las Casas, like many other settlers, was at first eager to make his fortune in the New World. At the age of eighteen he made his first voyage—in the same vessel as Pizzaro—to Hispaniola. And, as soon as he arrived, he not only went about the business of helping to manage one of his father's encomiendas on the island but also took part in slaving expeditions in order to increase the labor supply on his family's estate. In sum, while we know very little about his early years in the Caribbean, he appears to have only gradually become disillusioned with the violence and the brutality of the Spanish.[45]

An early turning point was likely the scandalous sermon preached by the Dominican Antonio de Montesinos in Santo Domingo on Christmas morning 1511. Montesinos excoriated the colonialists for their brutality and mistreatment of the natives. While Las Casas, who heard the sermon, did not immediately reform his own behavior, he must have nonetheless been haunted by the preacher's powerful questions about the indigenous peoples: "Are these not men? Do they not have natural souls? Are you not obliged to love them as yourselves?" Finally, it was in 1514 that he turned away from his earlier life. The final push came when Las Casas found himself struggling with a passage from Ecclesiasticus: "The bread of the needy is the life of the poor; whoever deprives them of it is a murderer. To take away a neighbor's living is to commit murder; to deprive an employee of wages is to shed blood" (Ecclesiasticus 34:25–27).[46] Here were words of God, Las Casas believed, that struck at the heart of the Spanish excesses in the New World. For what were the encomiendas but institutions that defrauded the laborer of his hire? Las Casas surely had read this passage before. On this occasion the reading proved transformative. Las Casas emerged as a critic of the very system he had helped establish. And over the course of the rest of his life—he died in 1566—he would devote most of his energies to defending the indigenous peoples of the New World.

Like many others who had come relatively early to America, Las Casas viewed the native populations there as largely innocent, living without signs of greed or the desire for material possessions, and open to hearing the Gospels. But the Spanish fell upon them "like the most cruel wolves, tigers, and lions."[47] And they were so greedy for plunder that—according to Las Casas in his vivid account of the Spanish massacres in Cuba—Hatuey, one of the local caciques, when asked by his fellow tribesmen why the Spanish were so cruel, explained that "they have a god whom they worship and love greatly, and they strive to subjugate us and kill us in order to seize this god from us in order that they might worship him." And then, holding up a basket of jewelry and gold, he said, "see, here is the God of the Christians."[48] Against this greed and barbarous cruelty, Las Casas argued again and again—before his contemporaries and especially the Spanish monarch—that it was necessary to abolish the institution of the encomienda and to do all that was possible to end the enslavement and exploitation of the labor of the natives. In its place he envisioned a world of free farmers among whom it would be possible for clergy and friars, living an apostolic life of poverty, to bring the True Gospel. In this sense, he shared much of the vision, though not the apocalypticism of the Franciscans. Las Casas did, however, understand mission in a providential key. Only in this way, he believed, would it be possible to bring to them the True Gospel.

While Las Casas was not an apocalyptic thinker, his hopes were nonetheless deeply rooted in his sense that God's Providence guided his work and that of other religious in the New World. Above all, he believed it would be possible to tear down the encomienda system and to develop a far more human way to create peace between the natives and the settlers in the New World. In 1515 he composed a fascinating treatise entitled *A Proposal on Remedies for the Indies,* in which he laid out his plans for a more peaceful colony.[49] Curiously many of Las Casas's proposals seem to anticipate those that More presented in his *Utopia*. Yet a direct connection between the texts seems unlikely. Rather—in a cultural environment increasingly intent on improving society—Las Casas and More shared certain visions for a better world. There is, however, a fundamental difference between the *Proposal* and *Utopia*. While More wrote from a position of privilege and leisure, with little sense of urgency, Las Casas wrote against the backdrop of an unfolding human catastrophe. All about him—as Las Casas and many others observed—the Indians were dying at a horrifying rate. Overcome by a variety of diseases and subject to the most appalling conditions, it was imperative to do all that was possible to create a safer and healthier environment for them.

Accordingly, Las Casas envisioned a fundamental restructuring of society. In place of the encomienda system, in which Indians worked directly under the supervision of Spanish colonialists, he envisioned a loose network of *pueblos* (villages), each with a population of about a thousand souls.[50] Above all, Las Casas—who was clearly horrified by the incalculable number of deaths that he had seen taking place around him in the Antilles—viewed these communities as making it possible for the indigenous peoples to live in health. First, he insisted that they not be overworked, whether in the fields or in the mines. Thus, Las Casas proposed a system in which those deemed capable of labor would alternate every two months between labor and rest. Moreover, their work days too should be regulated; and he proposed that "every day they should be given, during the time they have for meals, four hours of leisure . . . and for this to be done better, let them have hourglasses so that they neither work nor rest too much."[51] Second, he focused on ensuring medical care, and here he proposed a network of hospitals, each capable of treating up to two hundred patients at a time and well-provisioned with "a doctor and a surgeon and a pharmacist with a well-stocked apothecary," along with appropriate bedding, supplies of food, and a cooking staff—all with the expectation that, whenever a native should begin to feel ill, he should be sent to the hospital without any delay.[52] Furthermore, he stressed the importance of spiritual care. There were to be ten clerics in each community and instruction in the faith, noting that "the ultimate goal for which all has been ordered is the salvation of these Indians."[53] In closing his memorial, in an address to his monarch, Las Casas noted that, "if such remedies were not carried out, you could lose the oversight of the Indies as if they had never existed—and not only those discovered which have already been destroyed, but even the ones yet to be discovered, which likewise will be destroyed. . . . I implore your most reverend Lordship to order likewise to be considered—and I know without doubt it will be considered—that the first and ultimate goal that ought to compel the remedy for those sad souls should be God and the effort to bring them to heaven; because He did not redeem them nor did he discover them so that they would be thrown into hell."[54]

Las Casas did not only imagine new, utopian communities, he also sought to bring them into existence. In 1520 he won royal support for his proposal to establish what he hoped would prove to be a model and humane community for the evangelization of the natives in Cumaná in Venezuela. This community failed. At first Las Casas was crushed. He fell into a deep depression, but he did not give up hope, and ultimately he launched an even more ambitious plan for a new kind of native community. He and his fellow Dominicans—he had joined the order shortly after the failure of Cumaná—would take the Gospel into one of the most forbidding and untouched places then known in the Spanish territories of Central America: the province of Tezulutlán in the rough highlands of Guatemala. The Spaniards called this area the *Tierra de Guerra* (Land of War). But Las Casas viewed the remoteness of this territory as an advantage. If he and his fellow Dominicans could succeed in reaching and evangelizing the indigenous peoples there, they would be less likely to be disturbed by the colonists who so often treated them horribly and thus kept the cycle of violence going. Remarkably, the Dominicans enjoyed some initial success. Their music and their gentle approach so impressed a local chieftain that he allowed them to enter his territory. Then, for a few years, Las Casas could believe that the peaceful community of which he had long dreamed would survive. But local politics—both within and beyond the indigenous community—made this impossible, and the community was torn apart by violence in 1556.[55]

Verapaz, or the "Land of True Peace"—as the Dominicans came to call what had once been the "Land of War"—may have been Las Casas's last utopia but it was far from the last of the utopias either imagined in the New World or the Old over the course of the sixteenth and early seventeenth centuries. For throughout the "age of discovery," millenarian and other apocalyptic dreams would continue to contribute significantly to the utopian imagination—a source of hope for those who longed for a better world. Stumbling upon the earthly paradise, Columbus had launched a new way of thinking about the future.

CHAPTER 5

The Last World Emperor

And God wishes that under this emperor there be only one flock, only one pastor.

—ARIOSTO, *Orlando furioso* (1532), on Emperor Charles V

It is clear now that the ghazi Süleymān,
The sultan, son of Sultan Selīm Han
Is the epitome of bravery

. . .

This one is either the *mahdī* or his commanding general.
He is the paragon of all sultans.

—MEVLĀNĀ 'ISĀ, *Compendium of Hidden Things* (ca. 1543)

On September 4, 1538, an Ottoman fleet, with seventy-two ships and sixty-five hundred men, attacked the Portuguese garrison on the island of Diu, just off the coast of the Kingdom of Gujarat. The Portuguese had long recognized the strategic importance of this site, a major transit point along the spice routes that reached from the Moluccas (Spice Islands) through the Near East and into the Mediterranean. Initially their efforts to fortify Diu had been frustrated. Then, in the mid-1530s, the Sultan of Gujarat, fearing the expansionist ambitions of the Mughal emperor Humayun, turned to the Portuguese for help. In 1535 the construction of a massive fort was underway; and by 1538 its walls were raised when the sultan, in a sudden change of heart, now sought to expel the Portuguese from the island and turned to the Ottomans for assistance.

The Ottomans were only too willing to comply. They too had their eyes on gaining greater power in the Indian Ocean. Indeed, in 1509, in alliance

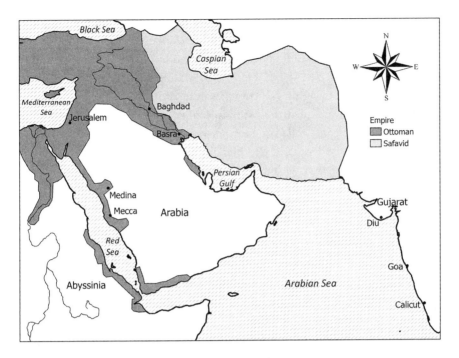

The Ottomans and the Indian Ocean. After their conquest of the Mamluks in 1517, the Ottomans sought to strengthen their naval presence in the Arabian Sea, as the two battles of Diu make clear.

with the Mamluk rulers of Egypt and critical assistance from Venetian engineers, they had routed the Portuguese at the First Battle of Diu. Now, following their conquest of Syria in 1516 and Egypt in 1517, they were even more prepared to project their power past the Red Sea and the Gulf of Aden toward India itself, especially since they feared Portuguese domination of the spice routes would lower the profits they received from their taxation of the trade. Yet, despite their overwhelming numbers and twenty days of battering the fort with their cannons, the Ottoman fleet abandoned the siege. Worrisome rumors that a Portuguese flotilla was arriving from Goa had reached them. The Ottomans withdrew to Suez, while the Portuguese victory at Diu consolidated their position in the Arabian Sea.[1]

In the early sixteenth century, the Ottomans, much like the Spanish and the Portuguese, who were also extending their global reach in these very years, proved capable of projecting their power far from their heartlands—in this case, into the Indian Ocean. Nor were the Ottomans, Spanish, and Portuguese the only new monster states of this period. To the contrary, the sixteenth century witnessed the emergence of a large number of "gunpowder empires" capable of bringing vast stretches of territory under their control. In Asia these empires included the Ming in China, the Mughals in India, and the Safavids in Persia, while, in Europe, not only the Portuguese and the Spanish but also the Russians, the French, and the Habsburgs constructed empires in this age.[2] But, within this emerging choreography of great states, it was the Habsburgs and the Ottomans whose growth would do the most to shape Europe at the dawn of the early modern era.[3]

In the first half of the sixteenth century the rapid expansion of Ottoman power astonished contemporaries. Bayezid II's son Selim I, who had replaced him as sultan in 1512, seemed unstoppable. In 1514 he scored a major victory against the Safavids at the Battle of Chaldiran. Then, in 1516 and 1517, he defeated the Mamluks, bringing Egypt along with Mecca and Medina under his control. Now "Protector of the Holy Cities," Selim was invested with the title of "caliph."[4] At the very beginning of the sixteenth century the Ottomans were now able to proclaim themselves preeminent among all the Muslim states, confirming in the eyes of many that they had a special place in history.

Upon Selim's death in 1520, these hopes were projected onto his son Süleyman, who not only consolidated these gains but also extended his empire into central Europe through the Danube valley. His first major victory came with his conquest of Belgrade in 1521. Then, in 1526, Süleyman dealt a major defeat to the Hungarian forces at the Battle of Mohács and, less than two weeks later, occupied Buda, the Hungarian capital. But his most ambitious strike into the heart of "Christendom" came in 1529 when, in a bid to defeat the Habsburgs, he mounted a siege of Vienna which, while it ultimately failed, made it clear that the Ottomans were a major European power.[5] In these same years the Ottomans pushed north as well into Wallachia, Moldavia, and Transylvania. Süleyman was also aggressive in the Mediterranean. In 1522 he had seized the island of Rhodes. In 1537, he opened a war on a new front against the Venetians in the Adriatic; and he continued to support the Islamic statelets in north Africa. In these same years, states from Libya to Tunis and Algiers either became subject to, or protectorates of, the Ottoman Empire.[6]

Contemporaries may have interpreted this expansion as evidence of divine favor, but Süleyman was bolstered by an expanding economy that produced greater and greater wealth, much of which the state was able to capture through its taxes.[7] And nowhere was the prosperity of the empire more visible than in Istanbul, which, by the mid-sixteenth century, counted more than five hundred thousand inhabitants, making it the most populous city in Europe. A major port, with a diverse population, the city bustled with trade while serving also as the home to the court of the sultan and to an expansive government bureaucracy. Splendid villas lined the Bosphorus, and grand mosques, with their exquisite domes and elegant minarets, transformed the urban landscape.[8] But the Ottomans—like their contemporaries throughout both the Islamic world and Christian Europe—also benefited from the new technologies of warfare. The musket and the cannon, which the Turks adapted from French and Hungarian prototypes, gave them greater power than ever before. They proved decisive, for example, in their early sixteenth-century victories over the Mamluks, who rejected the use of gunpowder and artillery, viewing these new inventions as an insult to the honor of warriors.[9] But perhaps most decisively, the Ottomans proved especially adept in the recruitment of their military forces.

The lynchpin of the Ottoman military was the corps of janissaries who served not only as the sultan's guard but also as the most disciplined and highly trained warriors in the empire. These men, levied from subject Christian communities as boys, were quickly converted to Islam. They were slaves of the sultan. This corps both manned the key garrisons of the empire and accompanied the sultan on his campaigns. They were so successful that they became an object of awe among many western observers. In his early sixteenth-century *Commentary on the Affairs of the Turks,* for example, the Italian humanist Paolo Giovio praised their valor and discipline, even maintaining that they were superior to Christian forces.[10] Yet in reality the janissaries, of whom there were some ten to thirteen thousand under the sultan's command, constituted only one of the several components of the Ottoman forces. Far more numerous were the some fifty thousand knights or *sipahis* who offered military service to the sultan in exchange for the great landed estates—*timars* or fiefs—they were awarded for their loyalties. Finally, the state levied *azabs* (conscripts), who served as infantrymen on the battlefield and as sailors on the Ottoman fleet.[11]

Despite the unprecedented wealth and military power of the Ottomans in the sixteenth century, Süleyman's power was limited. First, the very extent of empires imposed constant challenges on those who sought to rule them. In the early modern world, information traveled slowly; even when couriers were arranged, speeds were extremely irregular, since the poor conditions of most roads, which were often washed out by storms or blocked by fallen trees, made the movement not only of troops but also a ruler's representatives continuously challenging. With Süleyman's domains reaching from Egypt and the Middle East through Anatolia, Greece, and the Balkans into Hungary and northward into Wallachia (Romania), Transylvania, and Moldavia, distance itself was a great enemy. So too was the landscape. Certain regions such as the deserts of North Africa and the mountains of the Balkans were virtually unreachable. To be sure, in Anatolia and other low-lying regions of the Balkans, Süleyman was able to exercise relatively direct forms of authority. It was from these regions, for example, that the government drew the *sipahis*. Further away from these traditional homelands the Ottomans exercised a much less direct form of rule. In Egypt and Yemen, for example, the Mamluks and sharifs enjoyed considerable autonomy, while the corsair states of North Africa as well as the "principalities" of Transylvania, Wallachia, and Moldavia, though required to pay tribute to the sultan, were essentially self-governing.[12] Ottoman law was also flexible. Throughout this diverse empire, jurists were careful, whenever possible, to synthesize sharia, or Islamic law, with local customs, as they grappled with the conflicts arising from every aspect of life in the empire, from contracts and divorce to robbery and violent crimes.[13]

This flexibility was equally evident in Ottoman policy on religious matters. Within the Islamic world, the Ottoman rulers saw themselves as fierce defenders of Sunni teachings and, largely on this account, they were implacable enemies of the Shi'ite rulers of Iran. At the same time, the Ottoman state did not see either its Christian or its Jewish subjects as a threat to its authority. To the contrary, in the early 1490s Sultan Bayezid II had actually encouraged the settlement of Jews in his empire following their expulsion from Spain, and the Ottomans allowed both Christians and Jews to live as *dhimmis* or protected minorities, and thus to continue to practice their own religions and to regulate themselves in accordance with their own laws. In exchange for this relative religious freedom, Jews and Christians paid a special poll tax and the Christians, as we have seen, were required to give up some of their sons as janissaries—an often painful sacrifice but, at times, one some families embraced as they viewed it as a form of upward mobility. The result was an imperial system radically different from that of the Habsburgs, where the drive for religious unity—and intolerance toward both Jews and Muslims—was one of the defining traits of their reign.[14]

But to many western observers the most startling aspect of the Ottoman system lay in the sexual politics of the sultan's court. The Ottoman emperors did not marry into other aristocratic families but rather reproduced through a system of slave concubines. The mothers of their children were women who had been taken into captivity and brought into the sultan's harem. When one of these women delivered a son, she could never again return to the sultan's bed but would be sent off with her newborn to raise him in the provinces, where he would eventually become governor. And the sultan, meanwhile, would choose another slave as the mother to his next child. If there were multiple sons, all but one (there were exceptions) would be killed to avoid struggles over succession. This was how the Ottomans solved the riddle of reproduction and ensured the continuity of the rule of the House of Osman. It was in this institution, above all, that the Ottomans differed most clearly from the West, where rulers and their dynasties intermarried.[15] Yet Süleyman famously broke from the Ottoman system with his concubine Roxelana or Hürrem, originally taken into captivity as a young girl during a raid on Ruthenia (Ukraine). Süleyman had taken her as his consort immediately after succeeding his father as sultan. By the time she bore him their first child, the following year, he had fallen in love with her. He did not have her leave for the provinces but insisted that she remain by his side. Altogether she bore him five sons and one daughter. In the early 1530s—we don't know the exact date—he married her. It was an extraordinary break from tradition and it invested Roxelana with immense prestige and power. She bore her responsibilities with grace and an admixture of steel. She played a part—though her precise role is unclear—in effecting the execution of Süleyman's first-born son, Mustafa, who had been born to an earlier concubine when he was still prince. With Roxelana, the Ottoman Empire, like its great western rivals, witnessed the ascendancy of a new form of female power that an earlier system had discouraged.[16]

In their drive into the west, the Ottomans were propelled also by powerful religious traditions. The call to jihad, or holy war, had long animated Ottoman policy toward the infidel, and had invested warriors with a religious calling. These warriors, known as early as the fourteenth century as *ghazis* (fighters for the faith), played a major role in the early expansion of the empire. But, by the time of Süleyman, the dream of world empire and the hope for the End of History had grafted itself onto this earlier spirit of holy warfare. Indeed, many came to view Süleyman as the world emperor who would lay the foundations for a millennial age of peace, justice, and prosperity if not as the Mahdi-Messiah himself.

It was a stunning development. At the start of the early fifteenth century an Ottoman dream of world domination would have seemed absurd. Rather, this role appeared to belong to Timur (or Tamerlane), who had awed his contemporaries with his command over vast stretches of Central Asia, India, Russia, Iran, and Iraq—lands that he had conquered, often with brutal measures, over the last three decades of the fourteenth century.[17] Then, in the summer of 1402, he defeated and captured the Ottoman sultan Bayezid I at the Battle of Ankara, plunging the Ottomans into a protracted war of succession and reducing their territories to a fraction of what they had been thirty years earlier.

A year before Ankara, during Timur's siege of Damascus, the great jurist and historian Ibn Khaldūn, now an aging scholar, had had himself lowered from the city walls in a basket for a colloquy with Timur, whose encampment was nearby. While Ibn Khaldūn had been skeptical of many of the millenarian movements of his time, he viewed Timur's lightning-like rise to world domination as evidence that this central Asian ruler was destined by the stars to play a transformative role in history. Ibn Khaldūn, that is, rooted this view not in the hadith but in astrology; in particular, in the science of conjunctions, the belief that the great periods of historical change corresponded with specific alignments of the planets. Especially decisive, the astrologers agreed, were conjunctions of Jupiter and Saturn. Writing at about this same time as the Christian scholar Pierre d'Ailly, who had drawn on this science to explain the rise and fall of world religions, Ibn Khaldūn emphasized the correspondences of the conjunction of these two planets with political events. The "small" conjunctions that occurred every 20 years signaled the appearance of rebels; the "medium" conjunctions, that occurred every 240 years indicated the appearance of persons seeking royal authority; and the "great" conjunctions that took place every 960 years pointed to great events "such as a change in royal authority or dynasties, or of a transfer of royal authority from one people to another."[18] To Ibn Khaldūn, Timur's great power was evidence that he merited the title *ṣāḥib-qirān*, or "Master of the Auspicious Conjunction." Timur himself craftily resisted this title. Nonetheless, within a few years after his death, the designation gained traction. On the one hand, it offered a cosmic explanation for Timur's ability to amass a great empire. A conjunction in the year 1364—about the time Timur came to power—had predicted "a powerful one who would arise in the northeast region of a desert people, tent dwellers, who will triumph over kingdoms, overturn governments, and become masters of most of the inhabited world." On the other, the title gained traction among Timur's successors throughout the Islamic world—in India, in Iran, in the Ottoman Empire—who would increasingly use the title "Master of the Auspicious Conjunction" to enhance their own sovereignty.[19]

The view of the Ottoman sultan as Master of the Auspicious Conjunction first appeared in the court of Bayezid II, whose courtiers Uzan Firdevsi and Mirim Çelebi had viewed the sultan as *ṣāḥib-qirān*. The court of Bayezid's son Selim also made use of this title.[20] Not surprisingly, these traditions continued to resonate in the court of Süleyman. Based not only on hadith predicting the End of Time but also on an array of complex sciences—astrology, calendrics, geomancy, and hurufism—early sixteenth-century court seers crafted a particular image of Süleyman. The *qadi* or judge Mevlānā 'Isa, for example, represented Süleyman's reign as the culmination of sacred history. Drawing on the writings of Ibn 'Arabi and other earlier prognostications of the End of Time, Mevlānā 'Isa portrayed Süleyman not only as the *ṣāḥib-qirān* but also—with the approach of the conjunction of Saturn and Jupiter in 1552–53 (on the Islamic calendar exactly 960 years after Muhammad's *hejira*, or migration to Medina)—as the forerunner of the Madhi if not the Madhi himself. Meanwhile the courtier Haydar-ı Remmal also described Süleyman as the *ṣāḥib-qirān*. Haydar, a geomancer who had come to Istanbul from afar (likely from Iran) and quickly found favor with the sultan, established himself as one of the most powerful figures in the palace. And, like Mevlānā 'Isa, Haydar too drew on Ibn 'Arabi. But he also found evidence of Süleyman's status in his victories. "All signs, including the unprecedented scope and magnificence of your regime and the approach of the Lunar Age . . . show . . . that this is the millennial age to precede immediately the appearance of the Mahdī and the descent of Jesus."[21] Süleyman was to defeat his Christian enemies and ensure the victory of Islam as the world's dominant faith.[22]

These and indeed other apocalyptic beliefs that Süleyman was destined to play such an important role in the final drama of history only intensified the sultan's efforts to push westward with the goal of conquering Rome and ultimately bringing the inhabited world under the sway of Islam. To be sure, Süleyman was fighting for wealth and for power, but he was also driven by a sense of religious mission; in particular, the belief that his victories would eventually lay the foundations for a golden age within the Last Days of historical time. But this was not merely an elite dream. Apocalyptic hopes and expectations pulsed throughout much of the population, not only in Istanbul but also throughout the provinces.[23] At times, such hopes inspired movements—such as the series of Turkoman rebellions that swept through Anatolia between 1511 and 1537—that opposed the sultan.[24] And Ottoman authorities also occasionally cracked down on apocalyptic visions that put Jesus rather than Muhammad at the center of the divine drama.[25] But, in general, the vision of Süleyman as the Last World Emperor aligned with popular hopes for the End of History.[26] In the early modern world, religious longings for a Beautiful Ending did much to fuel the territorial ambitions of emperors.[27]

Like the Ottomans, the Habsburgs witnessed a meteoric rise to political and military prominence in the early sixteenth century, forging an empire that would ultimately include not only the Germanies, the Low Countries, Castile, Aragón, Navarra, Sicily, Naples, and Milan but also much of the Caribbean, Mexico, and Peru. If the Ottomans dominated the eastern Mediterranean and southeast Europe, with their power extending eastward through the Red Sea and the Persian Gulf into the Indian Ocean, the Habsburgs were their counterpart in central and western Europe, with their power extending westward across the Atlantic to the New World. The Ottomans and the Habsburgs ruled mirror empires.

Bella gerant alii, tu felix Austria nube! "Let others wage war, but you, happy Austria, marry!" Traditionally the story historians tell of Habsburg success is a story the Habsburgs told of themselves—of brilliant dynastic marriages. Even before he was crowned emperor in 1493, Maximilian I had implemented a strategy of family alliances that, in the near term, led to a remarkable consolidation of territories. In 1477 his own marriage to Mary, Duchess of Burgundy, had brought some of the richest lands in Europe into the Habsburg realm. Then the marriage of their son Philip the Handsome to the princess Joanna, the daughter of Ferdinand of Aragón and Isabella of Castile, combined the Habsburg fortunes with those of Spain. In 1506 these territories passed to Charles, son of Philip and Joanna. In 1516 Charles became King of Spain; and then in October 1519 he was crowned emperor at Aachen. There, in the city's ancient cathedral, he was anointed and invested with symbols of ancient authority: Charlemagne's ring and sword, Otto I's crown, imperial insignia that included a scepter, an orb, and a mantle, and even one of the most holy relics, the lance that had pierced Christ's side.[28] Charles was not, therefore, just any ruler. Rather he embodied all the power and authority of the *Sacrum Romanum Imperium*, the *Heiliges Römisches Reich*, the Holy

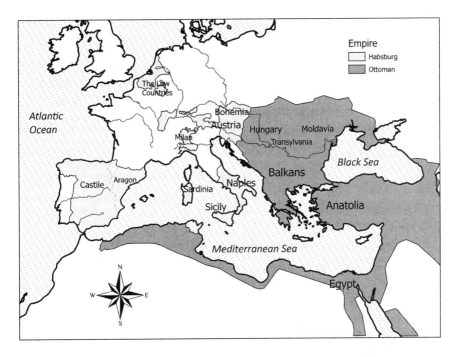

The Habsburg and the Ottoman Empires in the age of Charles V and Süleyman the Magnificent.

Roman Empire—holy, because it fell to the emperor to protect the pope and Christianity; Roman, because it was a revival of a glorious past; and empire because it aspired to bring together all of Europe.

But the Habsburg ascendency, no less than that of the Ottomans, was also the result of structural changes. The Habsburgs too benefited from the growth of population and the quickening of trade across the second half of the fifteenth century, which had enabled them to amass greater wealth and power than ever before. And, like the Ottomans, the Habsburgs also benefited from the evolving technology of warfare. It was in this period, for the first time, that infantrymen, whose services were not costly, had begun to prove stunningly effective. A single soldier now could use a matchlock to strike the weapon's powder while he took aim at his enemies with both hands steadying his gun. And Charles held the advantage of having Spanish regiments—many of them Castilians—in his employ. Well-armed, these highly disciplined infantrymen had, for example, managed to repel an attack by King Francis I of

France on the Italian town of Pavia, just outside Milan, in February 1525. For the Spanish marksmen outflanked the French cavalry, picking off the knights as they charged. More French nobles died that day than in any battle since Agincourt a century earlier. Francis himself was taken prisoner. As in the Ottoman Empire, the musket and the arquebus had trumped the lance and the sword. The Gascon nobleman Blaise de Monluc, who had fought alongside Francis at Pavia, expressed his dismay that the new firearms were replacing knighthood. "I wish to God," Monluc wrote, "this miserable engine [the musket] had never been invented." From his privileged perspective as a European nobleman, proud of the traditions of aristocratic warfare, it was an outrage that mere infantrymen—"cowards," he called them—could now defeat the most valiant knights.[29]

Like the Ottomans, the Habsburgs faced considerable limitations on their power. Again, given the vast extent of their empire, distance proved a constant obstacle to centralized control. Also like the Ottomans, the nature of their rule varied from region to region. Even within Spain, where Charles was not only emperor but king, he did not so much rule a unified realm as a mosaic of kingdoms, and he was expected to honor the diverse laws and privileges not only of Castile but also of Aragón, Valencia, and Catalonia.[30] He faced even greater limitations in territories beyond Iberia: in Sicily, Naples, Milan, and the Low Countries. Finally, Charles's power within the Holy Roman Empire was far from "absolute." The Holy Roman Empire had always been and would remain much more a federation than a centralized territorial state. This was clear in the constitutional requirement that the emperor be elected—a procedure that had been formalized in 1356 when the *Reichstag* or Imperial Diet issued the so-called *Golden Bull,* a decree stipulating that seven of the most powerful princes—from among the most influential in the empire—would share the duty of choosing the emperor.[31] Moreover, under Maximilian, Charles's grandfather and predecessor, the various constituencies of the empire had gained a greater say in the financial and judicial administration of the Empire in exchange for their support of the emperor's wars in Italy. As a result, when Charles became its ruler, the empire was not a unified but rather a "mixed" monarchy with its various estates—the electors, the princes, and the cities—"collectively constituting the Empire with the emperor."[32] Finally, the Protestant Reformation, which broke out in 1517 and would roil the empire throughout the rest of Charles's reign, did much to undercut his authority in the Germanies.

Early modern empires, in short, were largely patchwork. Machiavelli succinctly captured this aspect of empire in his *Discourses,* written in the early sixteenth century, where he underscored the indirect nature of the authority of the ruler of the Germanies, noting that he was a "sign of the emperor," who, "should he happen not to have forces, nonetheless has so much reputation among them that he is a conciliator for them, and eliminates every scandal with his authority by interposing himself as a mediator."[33] And indeed most early modern people would have understood the nature of empire in terms not unlike those of Machiavelli. To be sure, humanists in this era had encountered a more centralized notion of empire in their study of the history of Rome, many of whose "princes" had overwhelmed the traditional constitutional institutions of the former republic and had managed to rule autocratically. And some in this period would have thought of certain smaller states, such as Venice, as "empires," though in a loose sense, since the Venetians too granted the territories over which they ruled considerable autonomy.[34] But what is critical is that no one at the start of the sixteenth century understood empire in the sense that would begin to take shape—under the influence of Charles in the west and Süleyman in the east—as a strong center intent on accumulating resources from afar, establishing colonies, and ruling its subject populations with scripture and the sword.

Yet, despite the fact that the imperial title was formally restricted to Charles's rule over the Germanies, his status as the king of Aragón and Castile, as well as duke of Burgundy, impressed many contemporaries. Like other western European rulers in this period, Charles's political authority, while vast, was dynastic, and cobbled together on the basis of inheritance and conquest. In keeping with the political logic of Europe's first modernity, its individual parts were not bound to one another, but each was bound to Charles. "The factor binding all his territories together was," one historian has written, "dynastic right."[35] But Charles's vast territories did impress his contemporaries. Some spoke of this conglomeration of territories as *la monarquía española*.

Finally, as in the Ottoman world, religious expectations invested Charles with a decisive role in the sacred drama of history. In part, recent events—the defeat of the Muslims at Granada, the expulsion of the Jews from Spain, and Columbus's discovery of the New World—had cast the Spanish monarchs Ferdinand and Isabella in a providential role—one that Charles would inherit. But, unlike the Ottomans whose prognosticators had placed considerable emphasis on astrology, the Christians in Charles's circle, while not ignoring astrology, fashioned his millennial role, above all, from several centuries-old prophecies. In the *Apocalypse of Saint Methodius*—composed shortly after the first Islamic conquests in the Mediterranean in the seventh century—they found prophecies foretelling the emergence of a "Last World Emperor" who would overcome the Muslims and lay the foundation of the victory of the Antichrist and the Second Coming of Jesus.[36] And they also found support for his special role in the Second Charlemagne Prophecy, which had identified the Last World Emperor with a ruler by the name of Charles who would unite the world under a Christian emperor. It had been this prophecy, for example, upon which Savonarola had drawn in his sermons about the French king Charles VIII in the 1490s.[37] And now many in Spain and other Habsburg territories transferred these hopes to the new Spanish king Charles. But even these prophecies were nestled with others that, drawing on the Book of Daniel, foresaw a Fifth Monarchy that would bring about Christian unity before the End. All would convert to Christianity and there would be one shepherd and one sheepfold. This, at least, was the dream of the Dominican Giovanni Annio da Viterbo, who had foreseen the unification of Christianity and the destruction of the Turks in the late fifteenth century.[38]

These ideas circulated broadly, but they would come to play a major role in Charles's court through the agency of Mercurino Arborio di Gattinara. In late 1515 or early 1516, Gattinara, a native of Piedmont and a brilliant jurist, had retired to the Carthusian monastery at Scheut just to the west of Brussels soon after losing his position as an advisor to Margaret of Austria, governor of the Spanish Low Countries. His autobiography presents this as a period of considerable inner turmoil and anxiety—concerns that led him to immerse himself in the monastery's collection of apocalyptic writings.[39] Then, as it became clear that Charles was about to be named king of Spain, he composed a work of millenarian flattery in which—clearly he was in search of a royal appointment—he wove together the various apocalypses of Methodius, the Second Charlemagne Prophecy, and Giovanni Annio da Viterbo, to portray Charles as the king destined to be the Last World Emperor. Charles, Gattinara wrote, would become "the king of the Romans" who "will ascend Golgotha where the wood of the holy cross was fastened, and the Lord endureth death for us. And the king will take the crown from his head and put it on the cross. Stretching his hands to the sky, he will hand the kingdom to God the Father."[40] Gattinara certainly attracted the attention of Charles, who appointed him chancellor in 1518. And, when Charles was made emperor the following year, Gattinara doubled down on his messianic aspirations. "God the Creator has given you," he wrote to the emperor, "this grace of raising you in dignity above all Christian kings and princes by constituting you the greatest emperor and king who has been since the division of the empire, which was realized in the person of Charlemagne your predecessor, and by drawing you to the right path of monarchy in order to lead back the entire world to a single shepherd."[41]

Nor was Gattinara alone in expressing these messianic hopes. As the elector of Brandenburg put it at the time of Charles's election: "the oracles in this time promise that the emperor, in the breadth of his power, will surpass his ancestors. And they add another praise, much more pleasing, for I prophesy that it will be Charles and Charles alone who will exceed all other rulers in goodness and mercy. Therefore, I wish that God commit the *imperium* to him and that he will govern for the salvation of the world."[42] Then, shortly after Charles's troops sacked Rome in 1527, Giles of Viterbo—Augustinian friar, humanist, and kabbalist—appealed to Charles as a New Moses and a new Caesar whose victories, Giles predicted, would gather "all sheep into one fold."[43] In this same period, the poet Ariosto expressed this sentiment in his great poem *Orlando furioso:* "And God wishes that under this emperor there be only one flock, only one pastor."[44] And there were similar effusions of messianic hope following Charles's coronation as emperor by Pope Clement VII in Bologna in 1530. During this ceremony—which took place only one year after Süleyman's attempted siege of Vienna—many of Charles's courtiers were focused on the Ottoman threat, and Charles vowed to fight the Turks and to do all in his power to ensure the triumph of Christianity.[45]

For the Habsburgs, therefore, as for the Ottomans, apocalyptic ideas animated many of the aspirations of the emperor. They made it clear that he had a divine mandate to strive to impose universal rule over his territories and eventually over the globe. Moreover—given the widespread popular apocalyptic hopes of the era—such a vision of empire also did much to enlist support for his projects from his subjects not only in Spain but throughout much of Europe and the western Mediterranean. On the ground, empires were fragile, patchwork, centrifugal. Given these limits, the mystique of empire and the providential role of their rulers did much to cement them together, investing them with a common ambition for world domination.

That both Charles and Süleyman claimed to be the world emperor—and did so simultaneously—by no means escaped the attention of contemporaries. In 1531, as war loomed in the Mediterranean between the Ottoman Empire and the Habsburgs, the humanist Erasmus of Rotterdam observed that they were battling "for a great prize, whether Charles or the Turk be the monarch of the entire globe, for the world cannot any long bear to have two suns in the sky."[46] And at just this time the *qadi* Mevlānā 'Isa recognized that both Charles V and Süleyman laid claim to divine favor in "the last age of the world."[47] As Grand Vizier Ibrahim Pasha put it, mirroring Erasmus's concern, "just as there is only one God in heaven, there can be only one empire on earth."[48]

In 1518, almost immediately upon assuming the Spanish crown, Charles had led his first strike against the Ottomans. Over the previous several years the Ottomans had begun to consolidate their power across the North African Maghreb. From an archipelago of ports and coastal strongholds, several powerful Muslim rulers, now allied with the Ottomans who supplied them with resources and official titles, threatened Christians in the western Mediterranean, seizing ships, raiding coastal towns, and enslaving captives. Given these conditions and an atmosphere hypercharged with apocalyptic expectations, it was not difficult to persuade Charles to take decisive action, and he led a large force against Oruch—a legendary "pirate"—with success. The one-armed Oruch was killed and the threat the Muslims of Algeria posed to Spain was reduced, but only briefly—and it became almost immediately clear that Spain's ability to control North Africa would remain, at best, limited throughout the rest of the sixteenth century.[49]

Charles's ability to fight the Ottomans was hampered at almost every turn, for, like many emperors, he was forced to fight on multiple fronts. In the 1520s, for example, the Habsburgs were engaged above all in war against the kings of France, especially in Italy where both Aragón and the Valois (the ruling house of France) fought for control of the Kingdom of Naples and the Duchy of Milan, two of the richest states of the peninsula. Nonetheless, when possible, the Habsburgs did attempt to contain the growth of Ottoman power. They were not always successful. In 1529, for example, the Ottomans routed Charles's brother Ferdinand, archduke of Austria and king of Hungary and Bohemia. But in 1532 Charles himself took the helm of the empire's defense when Süleyman threatened a second attack on Vienna. Mobilizing a vast army of his own forces—in addition to Austrian and papal troops— Charles planned to intercept Süleyman on the Danube. The direct confrontation between the two emperors never occurred. When Süleyman learned of the massive opposition he was to encounter, he changed his plans, attacked and plundered territories in Hungary and Styria instead, and retreated to Istanbul.[50]

Charles also defended Habsburg interests in the western Mediterranean, especially against the corsair states of North Africa, whose pirates posed a constant threat to Spanish interests. In 1535 Charles personally led a flotilla to victory against the Ottoman captain Khair-ad-Din Barbarossa—Oruch's brother—in Tunis. But a second invasion of a North African port, this time of Algiers, in 1541, was hampered by violent rains and ended in a resounding defeat.[51] Effective opposition to the Turks in the Mediterranean often required a cooperative approach. Here the Venetians too played a key role, as they were willing to commit enormous resources to defending their trading interests in the Mediterranean. They had gone to war against the Turks as early as 1463, a decade after the fall of Constantinople, and were ensconced in their third major war against them from 1537 to 1540, this time in a "Holy League" with the Habsburgs. For the most part, however, theirs was a losing battle, as the Ottomans managed to pick off colony after colony from them, even besieging several towns along the Adriatic coast in nearby Albania Veneta.[52]

The clash between the Ottoman and Christian powers in the west also shaped Jewish messianism in this period, as was evident in the preaching and teaching—and perhaps especially the chutzpah—of two improbable, nearly inexplicable figures who captured the attention not only of Jews but also of Christians and Muslims in the early sixteenth century. One, a certain David ha-Reuveni, "dark in aspect, short in stature, gaunt, his language Hagarish and a little Jewish," showed up in Venice in December 1523. Claiming to be from a mysterious eastern kingdom and to have contact with the lost tribe of Reuben, this well-dressed interloper stayed on briefly in Venice before setting out to Rome where he met with Pope Clement VII, and then, in 1525, to Portugal where he met with King João. His goal—likely inspired by his careful reading of Abraham ben Eliezar Halevi's prophecy that the Christians would conquer Mecca—was to muster Portuguese naval assistance, a ship and some cannons, for an assault on Jeddah, a major Islamic port on the Red Sea and only two or three days west of Mecca.

In the early age of globalization, such a plan resonated with widespread dreams among Christians for the militant triumph of their faith—dreams that involved connecting the great Christian powers of Spain and Portugal with lost Christian communities in the East and then, having forged an alliance, striking a deadly blow against the Ottomans. Ha-Reuveni, however, harbored the belief that a Christian conquest of this magnitude would set the stage for the coming of the Jewish Messiah. Evidently the Portuguese king sensed a discrepancy in their goals.⁵³ And he was also worried that ha-Reuveni might induce some of his Marrano subjects to reconvert to Judaism. He wasn't wrong. One of his royal scribes, the converso Diego Pires, was inspired and converted to Judaism, circumcising himself and taking the name Shlomo Molkho—an act that turned the king instantly against ha-Reuveni, since he feared that ha-Reuveni's presence would lead to mass conversions or reconversions of the large Marrano or converso population to Judaism, the faith the Portuguese crown had sought to suppress in 1497 when King Manuel ordered the forced baptism of all Jews in his territories.

Ha-Reuveni fled to France, where—so he claims in his diary—he was imprisoned, but then spent the next two years in northern Italy, seeking to find a sympathetic audience wherever he could. Molkho, by contrast, moved to Salonika where he rapidly became a leading scholar of Kabbalah and increasingly came to see himself as a messianic figure. Then, following the example of ha-Reuveni, he too traveled to Rome to meet with Pope Clement. Again, while skeptical about Molkho's prophecies, Clement, who also had an interest in Kabbalah, nonetheless protected him when he fell afoul of the Inquisition. As a result Molkho was able to keep preaching, and later in 1532, after reconnecting with ha-Reuveni, the two men—not only united by their vision of a war against the Turk but also by their conviction that Jews themselves must take action in order to hasten the arrival of the Messiah—traveled to the imperial diet at Regensburg where they hoped not only to convince Emperor Charles V of their plans but even to convert him to Judaism. In this undertaking they were doubtless inspired by contemporary Jewish prophecies which, drawing on Christian visions of Charles as the Last World Emperor, viewed him as "the universal monarch who, having defeated the Muslims, would set the stage for the coming of the Jewish Messiah."⁵⁴ Charles found their ideas not only absurd but offensive, and had them handed over to the Inquisition. Molkho, whom the emperor had transported back to Italy in a cage, was burned at the stake in Mantua. Ha-Reuveni's sentence was commuted, but he was sent into exile in Spain where he languished in prison. Whether he died at the stake or of natural causes—even when he died—is unclear.⁵⁵

Charles's destiny was not limited to Europe. With the New World opening up, Charles, many believed, would bring about the Christianization of the globe, preparing the way for the Second Coming. The connection between conversion and the Second Coming was clear in the New Testament: "And this good news of the kingdom will be proclaimed throughout the world, as a testimony to all the nations; and then the end will come" (Matthew 24:14). Sixteenth-century writers built on this connection. In this same period global aspirations were certainly central to Giles of Viterbo, for whom the great voyages of his own age were one of the several signs—among them the reform of the Church and the flourishing of the arts and of learning—that presaged a golden age. In his *Scechina,* Giles concluded his argument with an address to his Charles V: "Therefore, magnanimous Caesar, it falls to you, as emperor, to discover, correct, and bring order into the world, and as advocate of the Church, to nourish it, raise it up, and extend it," adding later: "Here you see a new earth and a new heaven: a new Asia, a new Africa, a new Europe, and, in like manner, new seas and new islands which you alone, as their discoverer, are to know and investigate."[56]

While Spanish involvement in the New World had begun well before Charles took power, many in his court or under his command envisioned Charles as pivotal in the extension of his empire to the Americas. As early as 1523 the conquistador Hernán Cortés had put it in Charles's mind that he might well be on his way to building an empire in the New World. "There are so many [lands and peoples] and of such a kind that one might call oneself the emperor of this kingdom with no less glory than of Germany."[57] And in roughly this same period the friar Francisco de Ugalde described Charles's empire as one "on which the sun never sets."[58] Charles also took concrete steps to bolster his imperial rule. Over the course of the first half of the sixteenth century, an organizational framework for the administration of the colonies took shape. In particular, the Council of the Indies, under the supervision of a president and four to five councilors, played a major role in attempting to govern the Spanish colonies from afar. It appointed governors for New Spain and determined which expeditions to send to the New World, many of which were led by independent and largely private individuals. In 1522 it dispatched four governors to help Cortés in the administration of Mexico, though the real goal was to keep Cortés from becoming too independent. The council also sought frequently to ensure that proper measures were taken to protect the indigenous peoples. But Charles's major step in taking greater direction over the New World came in 1535 when he named Antonio de Mendoza viceroy to New Spain; and only seven years later, in 1542, appointed the first viceroy to Peru.[59] Furthermore, Charles supplemented the work of the viceroys with *audiencias,* judicial and legislative councils, that were designed to ensure that Spanish law was followed and Spanish interest

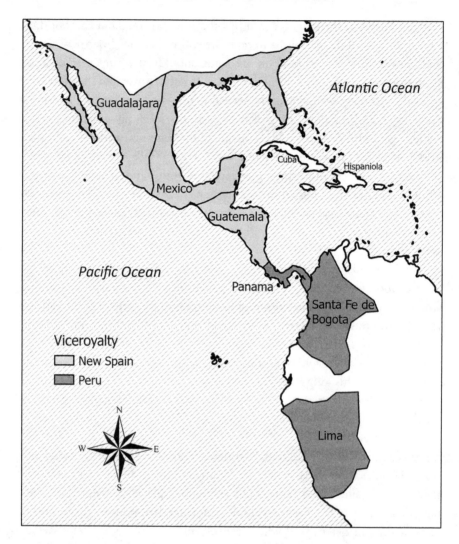

Spanish possessions in the New World in the age of Charles V.

served. None of these steps—given the great distance between the New World and Spain—brought the Spanish colonies under direct royal control. The settlers would do much to forge their own policies, and they did not always follow the directives from the crown. Even so, by the time of Charles's abdication of the throne in 1556, a new imperial system was in the making.[60] Charles had inherited the title of emperor from the most traditional of empires, but by the end of his life the transatlantic system that had developed around him constituted the foundation for a new form of empire: the seaborne colonial empire that would become one of the major features of the modern world.

Indeed, from the very beginning, some of Charles's contemporaries had sensed that he was going beyond the limits of an earlier civilization. This was clear in the ambitious iconography of his reign. When, in 1517, the seventeen-year-old Charles sailed from the Netherlands to Spain to claim his kingship, the mainsails of his ship bore his new royal device: the Pillars of Hercules with the revolutionary motto *Plus ultra* ("Further beyond") inscribed below. And, in 1519, to celebrate Charles's elevation to the imperial office, the German artist Hans Weiditz used the same device to commemorate the newly elected emperor. Whether or not the significance of these pillars pointed to Charles's future New World conquests in these early years of his rule is a matter of debate.[61] What is certain is that they did so by the time Charles abdicated in 1556. Charles had crossed through the Pillars of Hercules and created not only a new empire but a new form of empire at that. He had, in short, gone *plus ultra*—farther than any ruler had gone before.

CHAPTER 6

Antichrist and Reformation

> Lord, keep us steadfast in Thy Word
> And break the Pope's and Turk's fell sword
> Who fain would hurl from off Thy Throne
> Christ Jesus, Thy beloved Son.
>
> —MARTIN LUTHER, Hymn (1542)

In a revolutionary moment the woodcuts make the stakes clear. Believers must make a choice. Either they can embrace the True Gospel or they can follow the Antichrist. In the black-and-white images Jesus is depicted over and over again in great humility, while the pope—who is portrayed as the Antichrist—lives in luxury and splendor. In the first two pairings, we see Jesus as he rejects the royal diadem and wears a crown of thorns, a stark contrast to the pope who glories in all the trappings of secular authority and wears an elaborate tiara. Subsequent pairings make a similar point. In the next-to-last set, the twelfth, Jesus drives the money changers from the temple, while the pope uses his position as the head of the Catholic Church to enrich himself. And the cycle closes, in a thirteenth pairing, with Christ's ascension into heaven, while the pope is cast down into hell.[1] Brilliantly constructed and self-consciously drawing on popular apocalyptic imagery, these images—presented in the pamphlet *The Passion of Christ and the Antichrist,* printed in May 1521 and an immediate bestseller—was a work designed to fan the flames of antipapal feelings in early sixteenth-century Germany and, in so doing, to build support for the evangelical movement launched by Martin Luther.

In some respects, the pamphlet hardly exaggerated. Fifteen years before its publication, Pope Julius II had initiated one of the most ambitious and expensive architectural undertakings in European history: the building of the

Christ and Antichrist (1521). On the left Christ washes and kisses the feet of a humble man; on the right princes kiss the feet of the pope.

new St. Peter's Basilica—a project that would, in the end, consume the talents of dozens of architects and thousands of workmen over the course of the next 150 years. To meet these goals Julius's successor Leo X, a Medici, proved especially extravagant. He depleted the papal treasury so quickly that, only four years into his pontificate, he found himself scrambling for funds. To raise monies he sold ecclesiastical offices to the highest bidder, with little regard to the spiritual qualifications of those he appointed. He also sold indulgences. Originally granted to the crusaders as a means of shortening their time in Purgatory, they had become, in the increasingly commercial economy of the late Middle Ages, commodities, items the popes could put up for sale. Thus, in 1516, when one of his allies, Albrecht, the younger brother of the elector of Brandenburg, needed to borrow money to purchase the archdiocese of Mainz, the pope authorized the Dominican friar Johann Tetzel to sell indulgences throughout much of the Holy Roman Empire, a territory that had earlier been exempt from this activity. The mutual understanding was that the proceeds from the purchase of indulgences would go, in part, to pay off Albrecht's loan and, in part, to fill the papal coffers, providing Leo with the monies he needed for the ongoing rebuilding of St. Peter's.

While many Germans, swayed by Tetzel's preaching and promises, did purchase indulgences, many others were sharply critical of the practice. The monk and professor of biblical theology Martin Luther was among these critics. One tradition holds that, on All Hallows' Eve 1517, Luther nailed his objections—his famous *Ninety-five Theses*—to the door of the Castle Church in Wittenberg. Whether Luther actually posted his theses on the church door we don't know, though what is certain is that the theses were quickly printed, at first in Wittenberg itself and then in many other towns and cities throughout the empire. Many immediately recognized that Luther was not merely attacking a single instance of the abuse of papal power but in fact articulating a radically new understanding of salvation, one that would do much to undercut the very foundations of papal authority.

Ever since he had become a monk, Luther had been preoccupied with the matter of his own salvation. In the Augustinian house in Wittenberg, the young Luther struggled to make himself acceptable to God through frequent confession, seeking absolution for his sins from his father confessor, Johann von Staupitz, vicar-general of the Augustinians in Germany.[2] Yet no matter how hard he tried, Luther felt that he could not do enough to satisfy God, whom he tended to see as a strict, punitive deity. He worried, for example, that if he failed to confess a sin, even one that he had forgotten, he would never be forgiven. Exceptionally scrupulous, he worried above all that no matter how hard he tried, or how thorough he was in the confession of his sins, he would not be able to set himself right before God.

Staupitz patiently counseled Luther to have more faith in God and to place less burden on himself. Nonetheless, it was primarily Luther's study of the Bible that led him to a new understanding of God's Justice. Indeed, one of the most exciting discoveries in Reformation history came early in the twentieth century from the study of Luther's lectures delivered at the University of Wittenberg, especially his course of lectures on the Psalms and the Letters of St. Paul that he gave in 1513–15.[3] In his study of the Psalms, Luther wrestled with the word *justice,* which he had first understood to refer to a vengeful, punishing, unforgiving God: "When I first read in the *Psalms* and sang *'in iustitia tua libera me'* ['in your justice deliver me']," Luther told his companions years later, "I was horrified and hostile to these phrases 'the justice of God, the judgment of God, the work of God.' For I knew only that *iustitia Dei* meant his stern judgment. Now, was he supposed to save me according to his stern judgment? If so, I was lost forever."[4] And he encountered a similar obstacle in his reading of St. Paul's letter to the early Christian community in Rome. "For I am not ashamed of the gospel," Paul had written, adding, "it is the power of God for salvation to everyone who has faith, to the Jew first and also to the Greek. For in it the righteousness (*iustitia*) of God is revealed through faith for faith; as it is written, 'the one who is righteous (*iustus*) will live by faith' " (Romans 1:16–17). Again Luther stumbled over the word *justice*. As he wrote about his early interpretation of this passage in a brief autobiographical fragment many years later, "I hated that phrase 'justice of God' which, through the tradition and practice of all the doctors I had been taught to understand philosophically as His formal and active justice (as they call it), through which God, who is just, punishes unjust sinners."[5]

Yet Luther would come to a new understanding of God's Justice. In part it was through his continuing study of Paul's text, a study informed by his reading of St. Augustine. Gradually, as he recalled, "I began to understand the justice of God as a justice by which a just man lives as by a gift of God, that means by faith."[6] And he provided a similar account of his new interpretation of the Psalms. In his reflection on God's mercy he had a new insight. "When I then understood the substance of the matter and knew that *iustitia Dei* meant the righteousness by which He justifies us through the righteousness given in Jesus Christ," he told his companions, "then I understood the *grammatica*, and finally the *Psalter* tasted sweet to me."[7]

Luther's new understanding that God's justice was not punitive but rather liberating brought him a sense of inner peace and a confidence that he had earlier lacked. Luther no longer looked on God as a harsh father who judges each individual's actions but rather as a power in whom Christians, if they wished to be saved, merely had to place all their trust. Salvation, therefore, did not depend on works but instead on one's inward state and, in particular, on one's faith. From this vantage point, the Christian—so Luther argued in an outpouring of publications—was no longer saved by pilgrimages or penitential acts, by celibacy or making oneself a monk, but by faith alone. And, from this perspective, the sale of indulgences appeared especially abusive, for as Luther stressed in his *Ninety-five Theses*, it is the life of repentance, an inward state, that ensures salvation, not the purchase of a pardon from the papacy.

Like many young men Luther believed that his own ideas, which he viewed as firmly based in scripture, would prevail. He had expected reasoned debate and discussion of scripture with the leading theologians in the Church. What he had not been prepared for was the virtually total opposition he received from Rome. As the papacy grew concerned over the speed with which

Luther's teachings were taking hold, Luther began to expand upon his theology. In his debate with the theologian Johann Eck at Leipzig in 1519, Luther stressed the primacy of scripture. Then, in a flurry of writing in 1520, Luther produced three powerful treatises, each of which opened up what would prove to be an unbridgeable divide between his own understanding of Christianity and that of Rome. The first of these three, his *Address to the Christian Nobility of the German Nation,* turned the traditional understanding of Church-State relations on its head. No longer would the Church dominate. To the contrary, Luther made it clear that the State should not only work to curtail the abuses of the existing Roman Church, "whether it strike at popes, bishops, priests, monks, nuns, or whoever it may be," but also oversee the establishment of an entirely new and reformed church in their territories.[8] Then, in his *Babylonian Captivity of the Church,* published in October that same year, he called for the dismantling of the walls that separated the clergy from the laity, insisting on the "priesthood of all believers," and pared down the sacraments from seven to two—preserving only baptism and communion.[9] Together these two treatises alone served as the foundation for a new Lutheran confession.

But by far his most revolutionary work of this year was the shortest of these three treatises, *On the Freedom of a Christian,* which Luther published in November. In this work faith became the transformative protagonist in the life of a Christian. Rejecting all notions that a believer could be saved through his or her outward acts—ceremonies or works—Luther maintained that it was faith and faith alone that saved. The Christian, Luther argued, was required only to accept God's gift of his Son and his Son's sacrifice. And by placing trust in Christ, faith itself would become transformative. Faith alone would bring individual salvation, a deep sense of inner spiritual freedom, while at the same time serving as the foundation for service to others. Again and again, Luther portrayed faith as dynamic, a power flowing within the heart of Christians. "From faith," Luther wrote, "flow forth love and desire for God, and from love a free, willing, and joyful life in unconditional service to neighbor."[10] "The good things we have from God ought to flow from one to another, and become common to all."[11] The metaphor of "flow" was a favorite of Luther's. Tellingly, he would even use it to explain his extraordinary productivity: "I have a quick hand and a swift memory. When I write, the words flow to me."[12] Faith, from his perspective, was the source of action in the world. And works, therefore, were no longer the means to salvation but the fruit of faith. "Good, righteous works do not make a man good and righteous," Luther wrote, "but a good, righteous man does good works."[13]

Here Luther was able finally to express for a broad readership the fundamental insight into the nature of salvation that he had begun to develop years earlier, first when lecturing on the Psalms and St. Paul and then when opposing the preaching of papal indulgences in 1517. The power of *On the Freedom of a Christian* derived above all from its simplicity, for, by making the faith of the individual Christian the gate to heaven, Luther made it clear that all the traditional trappings of the Roman Church—from its hierarchy of priests and bishops to its emphasis on works—were not only entirely unnecessary but also had, in Luther's view, led to a radical distortion of the Gospel message. The True Gospel, Luther maintained, demanded nothing more than faith.

Even before the Reformation, the Holy Roman Empire had been highly fragmented. It consisted of numerous semiautonomous principalities, bishoprics, and cities scattered across western and parts of central Europe from the Baltic and the North Sea south to Austria and east to Bohemia. The constituent states of the empire—over three hundred in all—enjoyed a wide array of privileges and liberties that rendered most of them virtually independent states in their own right. The emperor's powers over them were restricted, and individual princes and urban magistrates exercised a wide array of fiscal and legal powers on their own. Thus, while monarchs in France, Spain, England, and elsewhere largely managed to consolidate their power over the magnates of their realm, in Germany particularism triumphed, a major factor in frustrating the creation of a unified German state before the late nineteenth century.

Within this mosaic of larger and smaller states, it was above all in the urban settings of the German-speaking lands that Luther's ideas first began to take hold. But what predisposed these cities to embrace the new religious messages of Luther? It is almost certain that the appeal of the Protestant message varied from individual to individual, and from city to city. Some townsmen were fed up with what they saw as widespread abuses in the Church. Some, like Luther himself, had come to resent the burdens of late medieval piety. Still others, especially those artisans, merchants, and professionals who were themselves literate, readily embraced the Protestant emphasis on the reading of scripture and the erasure of the traditional divide between a literate clergy and an illiterate laity. Finally, some found the Protestant message, with its emphasis on Luther's ideals of "freedom of a Christian" and the "priesthood of all believers" to be fundamentally ennobling. Often literate and proud, merchants and artisans were particularly attracted to Luther's doctrine of "the priesthood of all believers" and its implicit acknowledgment of the equality of every citizen before the law, promising a more republican and less centralized or hierarchical form of urban government. Luther's teachings ennobled the lives of workers and artisans and offered a far-reaching critique of Catholic prelates who became, in the eyes of many, symbols of both the political and ecclesiastical hierarchies that had developed in the late Middle Ages. Whatever the causes, there is no doubt that German cities became the major arena for the spread of the new religious and political ideas. And the ideas spread quickly. Even while Luther was working on his translation of the Bible in the Wartburg Castle in 1521, Wittenberg reformers—many of them more radical than Luther—began to enact a series of liturgical reforms that would make Wittenberg, in a short period of time, the first Protestant city. Other cities in Germany also quickly went over to the Reformation. In 1524–25 Erfurt, Gotha, Magdeburg, Nuremberg, Bremen, and Altenberg all became Protestant. Most of the major cities in Germany—some two-thirds of its *Reichsstädte* or free imperial cities—would become Lutheran.[14]

From the very beginning of this movement, the Catholic Church had sought to stop it. Leo X and Emperor Charles V had hoped to arrest Luther, but Luther's own prince, Frederick the Wise of Saxony, offered him protection. Moreover, with Charles's attention largely focused on the Turks and on France, he was unable to do much to keep the burgeoning Lutheran movement in check. Nonetheless, Luther's supporters recognized that they were at risk. In 1531 the German princes who supported the Reform entered into a confederation, the Schmalkaldic League, aimed at protecting the Lutheran churches in their territories. For fifteen years, this system functioned in relative peace, enabling the new Lutheran confession to strike deeper and deeper roots. But in 1546 Charles V, at last freed from his commitments in the Mediterranean, launched an assault on the Protestant princes, roundly defeating the Schmalkaldic League at Mühlberg in April of the following year. But it was an empty victory. The princes rallied again, and dealt a defeat to Charles two years later. Ultimately the warring parties reached a peace at the Diet of Augsburg in 1555. The emperor would allow those territories whose princes and magistrates supported the Reformation to remain Lutheran, while the lands of those princes and magistrates who opposed it would remain Catholic. *Cuius regio, eius religio*—the prince's religion would be the religion of his territory.

The theological disputes that had opened up in the wake of Luther's teaching were never purely a matter of doctrinal debate. To the contrary, from the very beginning, the period's arguments over biblical interpretation, church organization, and ritual practices were fueled by the hopes and fears—the spiritual passions—of a society roiled by apocalyptic expectations. Indeed, apocalyptic anxieties and dreams of a Beautiful Ending were decisive in shaping the religious turmoil of the period. To be sure, such hopes had ancient roots and they had recurrently played an important role in medieval politics and culture, but they had become especially pronounced, as we have seen, over the course of the late fifteenth and the early sixteenth centuries. The Turkish expansion into both Europe and the Mediterranean, the discovery of a New World, the consolidation of both Ottoman and Habsburg imperial power, and the invention and then the success of the printing press all contributed to the growing appeal of apocalyptic ideas and images. What could not have been anticipated before Luther was the degree to which the Reformation would not only draw on apocalyptic ideas but also function as an accelerant, intensifying the hold of visions of the End Times over so many who were increasingly anxious not only about their collective safety and security but also about their own salvation.

And, indeed, in the Holy Roman Empire, where the printing press was most developed, apocalyptic ideas were especially broadly diffused. But print was not the only medium through which apocalyptic ideas reached the populace. Popular preaching in Germany had played a role. In the 1450s the Franciscan Observant Giovanni da Capistrano preached the Apocalypse in towns and villages throughout the empire. Like Savonarola, whose sermons would mesmerize Florence a generation later, Capistrano drew on the teachings of Joachim of Fiore, calling his fellow Christians to repentance in preparation for the End.[15] In the 1470s, Hans Behem (or Böhm), a young German shepherd with a fondness for drumming and piping, pushed Capistrano's ideas in a more radical direction, calling for an end to clerical privileges and for everyone to share all things in common. And while this shepherd who became known as the Drummer of Niklashausen was not explicitly millenarian, there is no question that his preaching struck an apocalyptic chord.[16] And in this same decade Johann Hilten, a Franciscan, lashed out against the corruptions of his day and prophesied the coming of a great reform of the Church—even predicting that the great reform would begin in 1516, only a year before Luther would launch the Reformation.[17] Then, in the 1490s, a writer—we don't know his identity—wrote out a manuscript which opens yet another window onto millenarian currents in this period. Like Hans Behem, the author of this text experienced a vision. The archangel Michael had appeared to him and commanded him to form a band and "to re-establish a firm Christian faith on earth so that the words of our Savior may be fulfilled: 'There shall be one flock, one shepherd.' "[18] It was in this context too that Dürer's *Apocalipsis* became a bestseller.

Given this atmosphere it is hardly surprising that Luther himself would embrace much of the apocalyptic imagery. To be sure, Luther never supported millenarian ideas. But he did believe history was racing to an end. And he believed that a great struggle was opening up between the Church of the True Gospel and that of the Antichrist—in the very terms that Cranach had made plain in his woodcuts for his *Passional* in 1521. To be sure, Luther's first identifications of the papacy with the Antichrist had been tentative. But by 1520 he was unequivocal. In his treatise *Address to the Christian Nobility of the German Nation,* published in this year, he maintained that the papacy was not only the major barrier to the teaching of the Gospel but was in fact a minion of Satan, leading believers away from the true teachings of Jesus, which called for faith, to the false doctrines of the Roman Church, which emphasized external acts. In his *Against the Execrable Bull of the Antichrist,* also of 1520, he attacked the pope as "God's enemy, Christ's persecutor, the destroyer of Christianity, and the true Antichrist."[19] In 1521, Luther again identified the pope as the Antichrist.[20] Thus the battle was joined. Luther had inscribed his teachings within an apocalyptic struggle and one, moreover, with which many were already familiar. For, as we have seen, in the late Middle Ages many had preached that the rule of the Antichrist would mark the Last Days, the rapid acceleration of the End of History. Not surprisingly, many Christians saw Luther as fulfilling an apocalyptic and prophetic role. Michael Stifel, who would eventually publish a celebrated calculation of the End of History based on his studies of mathematics and numerology, identified the pope as the Antichrist as early as 1520 and Luther as the angel in the Book of Revelation in 1522.[21]

In a quieter time such ideas about the End of History may have remained largely a matter of scholarly debate or, at most, public discussion. But this was far from a quiet time. Beginning in late 1524 a series of rebellions and protests broke out among peasants in many regions of the empire. The upheaval, which peaked in the first half of 1525 and was finally suppressed in 1526, has come to be known as the German Peasants' War.[22] Undoubtedly the causes for this conflict lay primarily in a series of economic, social, and political transformations. Over the previous decades bishops, abbots, and other feudal lords—united in their effort to consolidate their territories and to develop more efficient forms of taxation on rural populations—had tightened the screws on the peasants, provoking a series of sporadic rebellions over the course of the late fifteenth and early sixteenth centuries. But now, religious ideas were inflaming an already combustible situation. Some peasants and other poor groups of townsmen and miners, inspired by Luther's emphasis on "spiritual freedom," took his message as a legitimation of their economic and political demands. But apocalyptic ideas—above all the widespread belief that the Millennium was at hand—proved especially inflammatory.

Apocalyptic visions, after all, bridged heaven and earth, grounding the promise of a more just and even egalitarian society in the widespread longings of many peasants for the Second Coming and the establishment of a millenarian kingdom. It was precisely expectations and hopes of this sort that such figures as Niklas Storch, Thomas Stübner, Thomas Dreschel, and Thomas Müntzer offered in their sermons during the peasant wars.[23] But Müntzer, a former Catholic priest who had at first been inspired by Luther, was the most influential of these preachers. A visionary, Müntzer did not believe, as did Luther, that religious truth was located in scripture alone. To the contrary, in his view, priests and Lutheran pastors were distorting the original, pure message of Christ. "They teach and say that God no longer reveals his divine secrets to his beloved friends through proper visions or his spoken word," he preached in his famous sermon to the Saxon Princes in July of 1524, adding, "so they keep to their inexperienced ways and make fun of those who are familiar with the revelation of God."[24] But Müntzer himself did have such visions. "It is true—I know it so—that the spirit of God is revealing to many elect and pious men at this time the great need for a glorious and triumphant reformation in the near future."[25] In his interpretation of the Book of Daniel, he argued that the work of the ending of the Fifth Empire of the world was now in full swing. And in a letter from this period, he called upon his readers to "consider the transformation of the world that is at the door."[26]

But by the spring of the following year, Müntzer, frustrated by the lack of support from the princes, took a more radical turn, as he came not only to associate the Elect with the poor but also to believe that violence was the only solution. As he wrote to the people of Allstedt in April: "Strike, strike while the fire is hot. Don't let your sword grow cold, don't let it down. Let your hammers ring on the anvils of Nimrod; cast their tower to the ground. As long as they live, it will not be possible for you to empty yourselves of your mortal fears. One cannot say anything about God to you as long as they dominate you. Strike, strike, while you have time. God is leading you: follow,

follow!"[27] Müntzer explicitly identified the Elect with the peasants and with miners. "Daniel," he wrote in his letter to the people of Eisenach in early May, "says that power will be given to the common folk." It is a passage that he will cite on several occasions over the next few days. Then, alluding to what he saw as the inevitable rebellion, he warned the wealthy: "You would do well not to scorn the lowly (as is your habit) for, to overthrow the powerful, the Lord raises up the weak."[28] Equally important was his millenarianism. His sermons stressed that not only salvation but also an earthly kingdom of peace and justice awaited the Elect—now, as we have seen, identified with the poor—who suffered so much under the pressures from their lords. Repeatedly he appealed to Nebuchadnezzar's dream and its promise that "the kingdom and the dominion and the greatness of the kingdoms under the whole heaven shall be given to the people of the saints of the Most High; their kingdom shall be an everlasting kingdom, and all dominions shall serve and obey them" (Daniel 7:27).

It was with this utopian vision in sight that Müntzer led a band of some three hundred miners and peasants from Mühlhausen to Frankenhausen in Thuringia. There, on May 15, 1525, they joined a large force of peasants in rebellion against the princes. As they set out for battle they took heart in a rainbow—surely a sign from God that they would prevail—but they were no match for the forces of Philip of Hesse and the other magnates who overwhelmed them. Taken captive, Müntzer was interrogated under torture on May 16. The official transcript of his testimony, while it can't be entirely trusted, rings true. His aim, he told his captors, was equality for all in Christendom, adding that all things are to be held in common, and that goods "ought to be distributed to each according to his need, as occasion arises. Any prince, count, or gentleman who refused to do this would first be warned, and then would be beheaded or hanged."[29] The next day, May 17, his captors cut off his head. Somehow, just before his execution, he managed to write a final letter to the people of Mülhausen in which he cautioned them to be on good terms with every man and to "no longer embitter the authorities."[30] Whether these were his own words or words of his captors, undoubtedly concerned about further violence, it is impossible to know.

Müntzer was far from alone in shaping a utopian vision of society in the expectation of the Millennium.[31] Indeed, many peasants drew on precisely such ideas in articulating their more radical, utopian demands. Framed in a powerful denunciation of the Antichrist—here construed as those who opposed the peasants in their efforts to live by the True Gospel—the *Twelve*

Articles, the most widely circulated tract calling for social and religious reform during the Peasants' War, demanded that their traditional rights be restored; that tithes, rents, and labor services be just; and that the peasants continue to have rights to the commons in order to hunt, fish, and gather wood.[32] These ideas—which appear to have given expression to peasant resentment against an increasingly aggressive feudal regime—spread rapidly not only through South Germany (with the noted exception of Bavaria) but also into eastern France, parts of Switzerland, the Alto Adige, Austria, and Carinthia. At times their utopianism would lead to a communist dream. Nearly a year after the *Twelve Articles* were published, Michael Gaismair, a reformer who led a series of peasant rebellions in the Tyrol, spelled out a program for the religious, social, and economic transformation of society in his *Territorial Constitution for Tyrol* with the goal of abolishing nobles and their castles and establishing a democratic peasant republic.[33]

Ultimately, all the regions in which the outbreak of the revolution had occurred suffered a similar fate, and some seventy to a hundred thousand peasants and commoners were killed by the counterrevolutionary forces of the German princes. Yet, even after the defeat of the peasants, many men and women—the great majority from the poorer strata of society—continued to place their hopes in visions of the End. And, like Müntzer, many put their faith in a promised Millennium that would lead to a far more just society. Kings and lords were to be thrown into an abyss; property was to be held in common; and, for many, free love would reign. Often the proponents of these ideas also embraced teachings which emphasized that salvation could only be found in communities of those who rejected not only traditional Catholic teachings but also the Lutheran accommodations of secular power. Perhaps most significantly, these radicals rejected infant baptism, a fundamental sacrament in the Catholic community, and insisted instead that believers, as adults, make a conscious decision to accept and to be baptized into Christ. Indeed, it was for this reason that they came to be called—on the basis of the Greek for "re-baptizers" (*ana-baptizimi*)—Anabaptists.

One of the most compelling apocalyptic utopian visions was articulated in a short pamphlet entitled *On the New Transformation of the Christian Life,* printed in Leipzig in either late 1526 or early 1527—that is, almost immediately after the defeat of the peasants. We don't know the author of this work, even if many have attributed it to Hans Hergot, a Nuremberg printer who was arrested for distributing it. But what is certain is that the author outlined a vision for a reformed society shaped largely by his prophetic visions and his deep sense that the End of Time was at hand. Powerfully rejecting the "scribes" who denied that there were any living prophets, the author embraced his position as a prophetic, poor man to argue for a fundamental change in the social order. "I, a poor man," he writes, "know these things which are in the future: that God will humble all social estates, villages, castles, ecclesiastical foundations and cloisters. And he will institute a new way of life in which no one will say, "that is mine." "And the people," the text continues, "will all work in common, each according to his talents and his capacities. And all things will be used in common, so that no one is better off than another."[34] And, rather like Thomas More's *Utopia,* this work too emphasized opportunities for education and the need to create institutions to care for the ill. But the work went well beyond an emphasis on the social and economic order. Its vision offered a rearrangement of the political order, one that would be based on the village community but regulated ultimately through a series of elected officials that reached up to a global scale. At the summit of this hierarchy would be a chief lord, who would be confirmed by God, and who would also serve as a single shepherd who would gather all into one sheepfold. "All religious divisions," the author insisted, "will pass away together and be made into a single church."[35] Finally, it seems likely that the author had been inspired by Joachim of Fiore, whose ideas were circulating in Germany in these years. The text opens with a tripartite or trinitarian division of historical time. After the Age of the Father and the Age of the Son, so the text begins, "the holy spirit will bring about the third change with a future transformation of the bad situation in which people now find themselves."[36]

Similar ideas surfaced in the writings of Hans Hut, a book peddler and visionary who had fought alongside Müntzer at Frankenhausen.[37] Others, such as the one-time priest Oswald Leber, drew on a variety of radical Jewish teachings about the End of History. Indeed, after the defeat of the peasants, Leber, who had preached the Apocalypse to the peasants during their uprisings, had sought refuge—and anonymity—in the Jewish community in Worms where he spent several years studying Hebrew and Kabbalah. But it was not only individuals who gave expression to such hopes. In Esslingen on the Neckar, Augustin Bader, a journeyman weaver from Augsburg and several companions sought to establish God's Kingdom on Earth. Inspired in no small part by Leber's synthesis of Jewish and Christian messianic hopes, Bader came to believe that the prophet Elijah had come to dwell within him and that his own son would be the Messiah; and he was convinced that this messianic kingdom would begin in 1530—the date favored by many Jews. Like both Hut and Leber, who had influenced him, Bader's dream of the Millennium was decidedly inclusive. In the anticipated kingdom people of all faiths—Jews, Muslims, and Christians—would live together in peace. Bader had special regalia made to underscore his status as the messianic king and withdrew with his wife, who was then expecting their second child, his son, and several followers, to a remote mill near Ulm to await the End. The radical hopes of this group frightened the authorities. In March 1530 Bader was arrested, tried, and executed.[38]

Finally, similar visions pulsed among workers and artisans in the Low Countries and in cities along the Rhine, especially Strasbourg.[39] There, in the 1520s and early 1530s, Lienhard and Ursula Jost—a poor, illiterate couple—began to experience a series of powerful visions about the Last Days. Lienhard, who likely earned his living as a woodcutter, experienced his visions as a clarifying light that inscribed prophecies into his heart.[40] Convinced he was living in the Last Days, Lienhard preached the need for greater charity and a more just order in Strasbourg. His wife, Ursula, also experienced prophetic visions, though hers took on a more traditional and more enigmatic form. When, in 1529, this small group of worker prophets encountered Melchior Hoffman, a Swabian furrier who, after several years of preaching Luther's message in north Germany and in Sweden and Lithuania, had already begun, on his own, to preach that the Last Days were at hand. They quickly persuaded him—perhaps he had already begun to have this belief himself—that he was the reincarnation of the prophet Elijah.[41] Now Hoffman began to preach the Apocalypse. He published Ursula's and Jost's visions; like them he viewed both the emperor and the Lutherans as Antichrists; and he too proclaimed that Strasbourg would be the New Jerusalem. When the local authorities sought to arrest him, Hoffman fled into the Low Countries, where he continued his preaching among workers and artisans in Leiden. Eventually he returned to Strasbourg, continuing to find many followers there as well, especially among poor townsmen.

Yet it would be in the Westphalian city of Münster, just over the border from Holland, that the apocalyptic expectations inspired by Hoffman and others took their most radical turn. Like many other German cities, Münster, with a population of some nine thousand, had itself only recently passed over to the Reformation. But the population was spiritually restless and largely receptive to the prophecies of the End. The constant prognostications, astrological reckonings, and prophetic predictions had rendered many members in the public endlessly curious about the meaning of contemporary events, their relation to the secret meanings of scriptural passages, and the signs that the gifted could read in the stars. To be sure, Hoffman's predictions had failed to materialize, but others took up his message about the coming of the Millennium. Two of them—John of Leiden, a tailor, and Jan Matthijs, a baker—did so with irrepressible conviction. Certain that Jesus would come again in his full glory, in 1534, they became leaders of a revolutionary utopian experiment in Münster where many of the local population allowed Matthijs to rebaptize them, while those who rejected his prophecies found it wise to leave the city. At the same time, well over two thousand others, inspired by his teachings, poured into Münster, which quickly fell under Matthijs's authority. With a new council now governing the city, Matthijs transformed the city into a biblical utopia. The government outlawed not only private property but also money; apart from the Bible, all books were confiscated and burned; residents were executed for even minor transgressions, while, for the ruling elite, polygamy was instituted. Matthijs had twelve wives.

As word trickled out of the coming of the Kingdom of God in Münster, it quickly became clear that Matthijs and his followers posed a genuine threat to the social order. Neither neighboring Lutheran nor Catholic princes could tolerate the making of this radical eschatological state in their midst. On February 23 Prince-Bishop Count Franz von Waldeck began a siege of the city, while Jan Matthijs and his followers interpreted the presence of the prince-bishop's cavalry just outside the city walls less as a threat than as a confirmation of the Great Battle that must accompany the End of History. As the prince-bishop tightened his grip on the city and food was running in short supply, Jan continued to reassure his followers. Christ, he told them, would come on Easter morning.

When Christ failed to appear at the expected time, Matthijs took matters a step further. On or shortly after Easter, Matthijs along with some twenty of his followers rode out into the fields outside the city, convinced that their small band would repel the diabolical forces of the prince-bishop that, for the last two months, had held their city of saints under siege. This was a folly. Such a small group stood no chance against the forces of the prince-bishop. But it is by no means clear that Matthijs was, as many have believed, mad. What is clear is that he shared an apocalyptic faith that was widespread in European culture at this time. And clearly many of his fellow saints shared this faith. From inside the city hundreds scrambled up onto the ramparts to watch. At first they watched as Jan and his followers slowly made their way toward Miller's hill, a knoll just outside the city. Many hoped that, through some act of divine favor, Jan would triumph. Others feared that this small band of holy warriors was about to be slaughtered. Suddenly the appearance of a regiment of the prince-bishop's cavalry coming up over the hill broke the suspense. The onlookers saw Jan and his comrades surrounded. Shots were fired, but Jan and his band were outnumbered. When the onlookers saw a few of those who had gone out on this sortie with Jan escape, they were heartened. They watched in hope as Jan plied his axe. Then, in horror, they saw one of the Prince-Bishops's knights gore Jan with his pike. As the sounds of the slaughter began to fade, they witnessed a frenzied scrum of knights hack Jan and his comrades to pieces and then hoist up his severed head to celebrate their victory.[42]

Yet this was not the final victory for the prince-bishop. The faithful remnant was undeterred. In September John of Leiden was crowned "king of righteousness over all," "king of the people of God in the New Temple." He began his messianic reign by running naked through Münster in wild religious ecstasy. He appointed twelve men in charge of the affairs of the city, instigating a reign of terror and wild innovations including polygamy. He indulged himself in excesses while subjecting the citizens to austerity. Under growing frustrations with his leadership and especially under the hardship of the siege, the saints were unable to hold out. In June of the following year Protestant and Catholic forces combined, and under the leadership of Philip of Hesse, they besieged the city. Six months later, before a crowd of onlookers, executioners tortured John and two other leaders of the Anabaptist community in Münster before knifing the three men. They then placed the mangled bodies in three iron cages that they suspended from the steeple of St. Lambert's Church—"as a perpetual memorial and to warn and terrify restless spirits lest they attempt something similar in the future."[43]

Luther, who was kept abreast of the events unfolding in Münster, was dismayed by the violence. But he didn't believe a theological rebuttal of the radicals there was worth his time. Instead, he simply called upon the political authorities to restore order. On this matter Luther was consistent. He had never intended for his ideas to lead to social and political reforms. "Only through the Word," he had written in 1521, "is the world overcome and the church preserved and through the Word it will be made strong again. The Antichrist will be ground down through the Word without a hand even stirring itself."[44] Yet, despite his emphasis on spiritual freedom, he had made clear his support for secular authority. Luther consistently distinguished

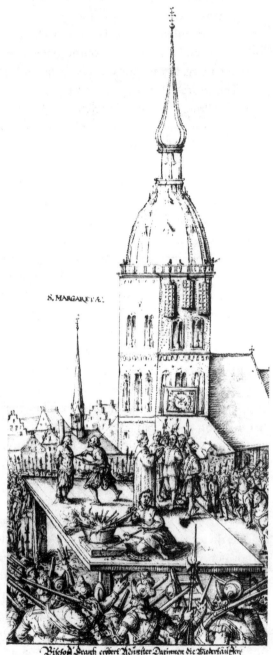

The end of the New Jerusalem. In 1536 the authorities in Münster tortured and executed John of Leiden, Bernhard Krechting, and Bernhard Knipperdolling, the leaders of the Münster rebellion. Afterward their bodies were hoisted above the city in three cages suspended from the spire of St. Lambert's Church.

clearly between what he called the *Zwei Reiche* ("the two kingdoms"): one of Christ, the other of the World. The first of these is spiritual, the second temporal. The Church, in Luther's view, had no temporal authority. Its concern was purely with the souls of believers—a sharp break with the traditions of Catholicism through which the papacy and the bishops had exercised significant political power. By contrast, Luther invested the state with enormous authority. Its function, divinely ordained in his view, was to preserve the social order, to support the church, and ensure the preaching of the Gospel. But the church the state supported was—and here Luther was emphatic—the visible church: the parishes, their pastors, and their properties. The true church was invisible: an *ecclesiola in ecclesia*—a gathering of true Christians within the larger institutional church, their identities known to God alone.[45] Within this framework, Luther was appalled by those of his contemporaries who drew on religious beliefs to challenge the political order. His tract *Against the Robbing and Murdering Hordes of Peasants* that he had written in 1525 had been part and parcel of this, as was his letter "A Shocking History and God's Judgment," in which he denounced Müntzer that same year.[46]

Yet, while renouncing the millenarianism and radicalism of the peasants and the Anabaptists, Luther himself believed that he was living in the Last Days and his sense of the growing struggle between the True Gospel and the Antichrist played itself out in this eschatological frame. If anything, his sense of this struggle would only intensify over the course of his life. We can see this, in part, in his shifting attitude toward the Book of Revelation. Early on, he had found this scripture baffling, even disconcerting. "My spirit cannot propel itself into this book," he had written in his brief preface to the text in 1522.[47] Yet, as he witnessed more and more obstacles to his message, he was ready by 1530 to offer a more charitable reading. The new preface he composed to this book in this year provided—with some caution that he couldn't be certain he was right—a reflection on the history of the Church down to his own days. Indeed, Luther believed that it would be only by examining the interconnections between the figures and the symbols in Revelation not only with the long arc of history but also with recent events that one could make sense of the book.[48] Yet Luther did not only dwell on the tribulation of his age. In his preface to the Book of Daniel, written that same year, Luther noted that "the prophecies of Daniel and others like them are written, not only in order that men may know the events and the tribulations that are to come, and satisfy their curiosity, but in order that the righteous shall be encouraged and made happy, and strengthened in faith and hope and patience, since they here see and hear that their misery has an end, that they are to be freed from sins, death, the devil, and all evil, and be brought into heaven, to Christ, into His blessed, everlasting kingdom. So Christ, too, in Luke 21:28, encourages His own with terrible news, and says, 'When ye shall see these things, then look up and lift up your heads, for your redemption is near.' So here, too, we see that Daniel always ends all his visions and dreams, however terrible, with joy."[49]

Moreover, even if Luther had been hesitant at first to embrace the Book of Revelation, he had been consistent throughout his ministry that the appearance of the Antichrist was a sign of the Last Days. On one level, there was little that was original about this view. The Antichrist was a central figure in popular spirituality, frequently represented in religious plays and in block books, and generally portrayed as a single figure, the spawn of the Devil, who would appear in the Last Days.[50] But Luther's understanding, which drew on a more learned tradition, was more abstract. As we have seen, in his early identification of the papacy with the Antichrist, he had come to understand the Antichrist not as a single figure but as something far more pernicious, what today we might call an underlying structure of beliefs and institutions that, from Luther's perspective, not only had misled Christians for well over a millennium about the true nature of their faith but now, as the Reformation took root, had done everything in its power to condemn Luther and his teachings. In his view, after all, the papacy had distorted the teachings of the Gospels from its very founding. And it was under the power of the papacy that "the world had been filled with all kinds of idolatry—monasteries, foundations, saints, pilgrimages, purgatory, indulgence, celibacy, and innumerable other creations of human doctrines and works."[51] In his *Against the Roman Papacy: an Institution of the Devil* of 1545 he attacked all the popes of history as "full of all the worst devils in hell—full, full, and so full that they can do nothing but vomit, throw, and blow out devils."[52]

Yet, for Luther, the Antichrist was present in other enemies of the Gospel as well. When, in 1529, the Ottomans laid siege to Vienna, Luther composed in short order three treatises against the Turks. In the second of these—his *Battle Sermon against the Turks*, which he published in 1530—he described the Turk as the Fourth Horn of the Apocalypse. Furthermore, blind to the violence embedded in his teachings, Luther came to embrace the idea of a holy war against the Ottomans, reassuring German Christians that their deaths in battle against the Turks would win them a place in heaven.[53] At the same time, Luther even believed it important that Christian scholars study Islam—in order to guard against its pernicious doctrines, of course. In 1542 he urged on the publication of an edition of the Qur'an and wrote a preface that both condemned its teachings and argued that the text needed to be read by those who opposed it, for, as he wrote, "you must open the injury if you want to heal it."[54] Finally, Luther also came to view the Jews as the Antichrist. When he was first preaching the True Gospel, Luther had believed that the Jews would convert to Christianity. This was the message of his treatise of 1523, *That Jesus Christ Was Born a Jew*. But over the next decades Luther shifted from a desire to convert the Jews to a call for their destruction, as he had come to view them as obstinate deniers of Christ. In his *On the Jews and Their Lies*, a work he wrote in 1542, he unleashed his most virulent and hateful attack upon them. Frustrated by their refusal to see the truth of the Gospel and to recognize Jesus as the Messiah and inspired by long traditions of anti-Jewish Christian writings, Luther called for their synagogues to be burned, their homes destroyed, and their sacred texts confiscated, with individual Jews, stripped of social status, forced into hard, demeaning physical labor.[55]

For Luther, therefore, the Antichrist was a powerful, many-horned beast. "How many different enemies have we seen in our time—the defenders of the pope's idols, the Jews, a multitude of Anabaptist monstrosities, the party of Servetus and others," he wrote in his preface to Bibliander's edition of the Qur'an.[56] Then, in a sermon before a number of his followers who had gathered into the Church of St. Andreas in Eisleben in February 1546, Luther made much the same point. He understood the anxieties they were feeling. The Lutheran church's future was still uncertain. The papacy and the Catholic princes of the empire, both with the support of the emperor, threatened its survival. And, even though the Turks had failed in their effort to capture Vienna in 1529, the Ottomans remained a threat to the peoples of Central Europe.

Yet Luther also expressed his own hopes about the End in this sermon. To be sure, unlike the more radical apocalyptic thinkers of the period, he never imagined an earthly kingdom of the saved. But he did offer his followers a consoling message that it would be possible, for those with faith, to experience—as individuals—a Beautiful Ending. "If you are facing oppression, death, or torture, because the pope, the Turk, and emperor are attacking you, do not be afraid," he assured them, adding:

> It will not be heavy for you, but light and easy to bear, for I give you the Spirit, so that the burden, which for the world would be unbearable,

becomes for you a light burden. For when you suffer for my sake—and here Luther appears to fuse his own identity with that of Christ—it is my yoke and my burden, which I lay upon you in grace, that you may know that this your suffering is well pleasing to God and to me and that I myself am helping you to carry it and giving you power and strength to do so. So also say Psalm 31 and Psalm 37: "Let your heart take courage, all you who wait for the Lord," that is, all you who suffer for his sake. Let misfortune, sin, death, and whatever the devil and the world load upon you assail and assault you, if only you remain confident and undismayed, waiting upon the Lord in faith, you have already won, you have already escaped death and far surpassed the devil and the world.[57]

Luther struggled as he spoke these words. Over the previous years his health had deteriorated. Now, once again not feeling well, he broke off his preaching, telling the congregation he was too weak to continue. Two days later, experiencing chest pains, he went to bed early. And later that evening his family and friends, fearing the worst, gathered around his bed and were with him when he died some two or three hours after midnight. Four days later, on February 26, Johannes Bugenhagen, Luther's pastor, preached at his funeral in Wittenberg. Even in this final moment, the Apocalypse pressed in on Luther. Drawing on the Book of Revelation, Bugenhagen compared Luther to the "angel" who had preached the Gospel, while Philip Melanchthon portrayed Luther as a Second Elijah.[58] Even before his death many had seen him as a prophet, associating him at times with Daniel, at times with John the Baptist, at times with the angel of the Apocalypse but most often with the prophet Elijah or Elias.[59]

CHAPTER 7

"No One Knows the Hour"

> So when they had come together, they asked him, "Lord, is this the time when you will restore the kingdom to Israel?" He replied, "It is not for you to know the times or the periods that the father has set by his own authority. But you will receive power when the Holy Spirit has come upon you; and you will be my witnesses in Jerusalem, in all Judaea and Samaria, and to the ends of the earth."
>
> —ACTS 1:6–8

In 1534, when John Calvin learned that Michael Servetus, whose *On the Errors of the Trinity* had been published three years earlier, was in Paris, he raced back to the city to debate him. The book had, after all, attacked one of the fundamental tenets of Christianity: the doctrine of the Trinity, the teaching that God, while one "substance," was somehow at the same time the three "persons" of Father, Son, and Holy Spirit. Allowing Servetus's attack to go unchecked, Calvin feared, would endanger Christianity's central tenet: the belief that Jesus was the Son of God. Servetus did not show up for the debate, and it would be another twenty years before the two men would meet face to face.[1]

In Servetus's mind, the doctrine of the Trinity, first embraced at the Council of Nicea in 325 and a central teaching of Christianity ever since, was not based on the Bible. Rather it was a construction—Servetus called it a "monstrosity"—that had transformed the figure of Jesus Christ into a philosophical abstraction. And, indeed, one of the primary goals of Servetus's text, like so many other reform works in this period, was to strip away the later philosophical accretions that had, so he believed, distorted the original message of the Gospels. God the Father, Jesus, and the Holy Spirit were not "persons" but rather—so Servetus argued, as he himself veered into a metaphysical account—"dispositions," different ways in which God manifested himself to humanity. God the Father alone was Yahweh and before him there could be no other gods. Jesus was not only the Messiah and the son of God the Father but also "a likeness, a face, an effigy, a sign, a character, a seal, a distinguishing mark, a kind of engraving" of the being of God.[2] And the Holy Spirit, which Servetus characterized as "a divine agitation in the spirit of man" was a similar emanation.[3] In Servetus's vision—and here Servetus appears to have been drawing on kabbalistic ideas—God remained the "source of being," a primal light from whom "multiple rays of divinity" flowed.[4] Thus Christ and the Holy Spirit were, above all, manifestations of the underlying reality of the divine through whom it became possible, for those with faith, to know God.

But the doctrine of the Trinity was not only, in Servetus's view, an obstacle to a clear understanding of the foundations of Christianity to Christians themselves; it also set up a stumbling block to Muslims and Jews. "How often this tradition of the Trinity has been an occasion for derision for the Mohammedans—what grief!—only God knows," Servetus wrote, adding, "even the Jews shrink from adhering to this fantasy of ours and make fun of our folly over a Trinity and, because of this blasphemy, do not believe Jesus to be the Messiah who was promised in their law."[5] Servetus based this critique not merely on hearsay but on a careful study of both Judaism and Islam.[6] He actually drew his views in part from the Qur'an. "Hear also what Mohammed says," he wrote, "Christ was the greatest of the prophets, the spirit of God, the virtue of God, the breath of God, the very soul of God, the Word born through the breath of God of a perpetual virgin. . . . He says, moreover, that the Apostles and Evangelists and the first Christians were the best of men, and wrote the truth, and did not hold the doctrine of the Trinity, or of three Persons in the Divine Being, but men in later times added this."[7] Ultimately, if the Jews and the Muslims could be converted, Servetus argued, the End Times would come.

For Servetus, who had grown up in Spain in the early sixteenth century, the Jewish and Muslim rejection of Jesus as the Messiah was not a purely abstract issue. To the contrary, he would have known from firsthand experience that many Jews and Muslims in Spain, although nominally converted to Christianity, continued to practice their ancestral religions in private. For many of these men and women, the doctrine of the Trinity did nothing less than render the Christian conception of God and the Messiah so alien to their conceptions of the divine that their true conversions seemed impossible.[8] It is even possible, though we can't be certain, that Servetus had been born into a *converso* family.[9] But whatever the details of his personal biography, Servetus was far from alone in attempting to find a bridge between

Christianity and Judaism. In fact, in the decade before he wrote his *On the Errors of the Trinity,* Cardinal Francisco Ximénes de Cisneros, the most influential prelate in Spain, had sponsored the publication of the most ambitious biblical project yet imagined: *The Complutensian Bible,* a massive, expensive polyglot edition that made it possible for readers to examine the Hebrew, Greek, Latin, and Aramaic versions of the holy scriptures on each page. But what was clear to many was that this was not merely a scholarly undertaking. It was also a millenarian project, since Cisneros, among other things an ardent student of Savonarola, and others in his circle hoped that such a text would help convert both Jews and Muslims to Christianity—conversions that many Christians saw as necessary preludes to the Second Coming.[10] Servetus greatly appreciated this effort, but was convinced, as his powerful reconsideration of the Trinity makes clear, that it had not gone far enough. As a young man—he was in his early twenties when he wrote his diatribe—he thought that he would win the day with his careful exegesis of the original Hebrew and Greek of scripture. What he could not have foreseen was the violent reaction his text produced. Indeed, many of his contemporaries, mistaking his attack on the Trinity as a denial of the divinity of Christ, saw him as an arch-heretic.

There is much we don't know about Servetus's movements in the decades following the publication of his *On the Errors of the Trinity.* But it is clear that, now living under the name of Michel de Villeneuve, he divided his time primarily between Paris and Lyons. In Paris he excelled in the study of medicine and was recognized as an outstanding anatomist, a skill that would lead to his discovery, a century before Harvey, of the pulmonary circulation of blood. In this same period he published works in pharmacology and in astrology. In Lyons, drawing on his formidable language skills, he served as editor for two important editions of Ptolemy's *Geography* as well as to the Santes Pagnini Polyglot Bible of 1542.[11]

Then, in the 1540s and early 1550s, Servetus fashioned a seemingly more settled life in Vienne, a town in the Rhône Valley. There, while employed as a physician to the archbishop, he came back to his religious studies and produced what would be his magnum opus: *The Restitution of Christianity.* As in his earlier writings, he once again offered a devastating critique of the doctrine of the Trinity. But in this text, he did more; he also made plain his belief that the End of History was at hand. He saw signs of the Last Days everywhere, and he was certain that a great apocalyptic struggle between the forces of good and evil had already begun. But, like so many of his contemporaries,

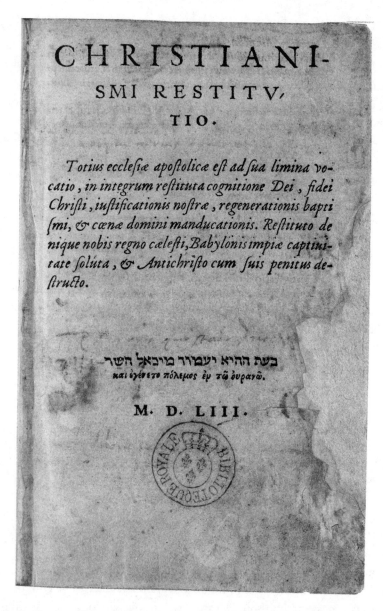

Predicting the Apocalypse. Title page of *Christianismi restitutio* (1553). In addition to the Latin title, Servetus included two scriptural passages. The Hebrew passage reads, "And at that time the Archangel Michael will arise"; the Greek, from the Book of Revelation, reads, "And war broke out in heaven."

Servetus was profoundly hopeful. He believed that the faithful would experience a Beautiful Ending, a thousand years of peace and blessedness on earth before the Last Judgment.[12]

In his reading of the opening of the seals in the Book of Revelation, for example, Servetus, here focusing on only the first six seals, associated the first of the Four Horsemen of the Apocalypse with the temporal power of the papacy; the second with the pope's bellicosity; the third with his gluttony; and the fourth with his support for the new religious orders. He then read the fifth seal as standing for the pseudo-prophets, and the sixth seal—with its allusions to earthquakes and storms—to our spiritual death itself. But, to Servetus, the victory of the Elect was at hand. *The Book of Revelation* promised that the Elect would rise again *ad mille annos*—for a period of one thousand years, the Millennium.[13] Finally, Servetus offered a tantalizing calculation. He took the Council of Nicea, which had embraced what was to become the Christian doctrine of the Trinity in 325, as the starting point for the reign of the Antichrist (the papacy) and then added Joachim's calculation of 1,260 years as the length of this reign. According to this calculation, history would end in 1585. Thus, within a generation of his writing, history would have its victorious culmination. With this divine victory, Servetus wrote, "We ourselves are truly made Gods. Through participation in the divinity of Christ we are indeed made participants in the divine nature."[14]

Servetus understood that his ideas were controversial. Nonetheless, he believed his arguments still had a chance of winning the day, and it was in this spirit that, in 1546, he had begun a correspondence with Calvin whom he believed he could convince of his position.[15] But when he shared a manuscript copy of *The Restitution of Christianity* with Calvin in early 1553, the latter was so offended that he alerted the Catholic authorities in Vienne, who in turn had him arrested. And they certainly would have executed him had he not had the luck to escape.

Servetus then inexplicably walked directly into Calvin's hands. We don't know why Servetus came to Geneva. Some have theorized that he was associated with a party there intent on overthrowing Calvin. Or, since his ultimate goal was to reach Naples where he hoped to practice medicine, perhaps he was simply trying to throw the French authorities off his trail by taking a circuitous route. But he never reached Italy. The morning after he arrived in Geneva, he was recognized and arrested. Calvin accused him not only of rejecting the doctrine of the Trinity but also of a kind of pantheism—of seeing God in everything—and even of drawing favorably upon the teachings of both the Jews and the Muslims. At the conclusion of his trial on October 27, 1553, Servetus was condemned to death. Guards then took him outside the city walls to Champel, a gentle field overlooking the town below. Along the way, Calvin's colleague, the theologian Guillaume Farel, attempted to convince the Spanish prisoner to renounce his teachings against the Holy Trinity. But Miguel refused. Brusquely, with the magistrates and other prominent citizens looking on, the guards tied him to a stake. They fastened a copy of *The Restitution of Christianity* to his body, and lit the kindling below. It is said the green wood burned slowly.[16]

Unlike Luther, we know very little of Calvin's spiritual awakening. Certainly, his early years unfolded in a traditional Catholic environment. Even as a boy he had held ecclesiastical benefices secured for him by his father; and he had started his university studies in theology in Paris. Then, at his father's urging, he changed his program to law, which he studied first in Orléans and then in Bourges. But he was restless. In 1531 he left the law behind to devote himself to humanistic studies and came to Paris where he hoped to master Greek and Hebrew, the languages of scripture.[17] It was in Paris in the early 1530s that Calvin first became involved in the evangelical movement that was intensifying in France in just these years, and by 1533 Calvin was decidedly in the Protestant camp. When King Francis I, reacting to the growth of Protestant activism in France, began to persecute evangelicals in 1534, Calvin, fearing for his safety, fled to Switzerland. He went first to Basel and later, inspired by Farel, to Geneva. And here, safe across the border, Calvin would continue to revise and expand his masterpiece, *The Institutes of the Christian Religion,* over the course of his life.[18]

Calvin's starkest and most distinctive innovation was his teaching on salvation. Even from before the birth of each individual, so Calvin famously argued in the *Institutes,* God had determined who would be saved and who would not. This belief, at the core of Calvin's theology, would come to be known as the doctrine of predestination. This was, of course, a major departure from both Catholic and Lutheran teachings. Catholics had taught and continued to maintain into the early modern period and beyond that, even though God's grace played a role in the salvation of an individual, good works were also necessary. And even though Luther had railed against the doctrine of works, his doctrine of solafideism—his insistence that the believer is saved by his faith alone—seemed to Calvin to place too much agency within the individual believer. Indeed, to Calvin, Luther's emphasis on the individual's faith constituted a kind of work, since salvation still seemed dependent on the individual Christian's will and acts of faith. By contrast, to Calvin, there was nothing—*absolutely* nothing—that individual men and women could do, whether through their acts or their faith, to earn salvation.

Calvin's argument was not entirely original. Zwingli had made a similar argument in his *Sermon on the Providence of God* in 1530.[19] Nonetheless, Calvin expressed his views in starker, more rigorous, more exacting terms. God and God alone, he argued, determined who was to be saved and, at the same time, who was to be damned. As Calvin wrote in blunt terms in the fifth edition of his *Institutes:* "God adopts some to hope of life and sentences others to eternal death," adding, "we call predestination God's eternal decree, by which he has determined within himself what he willed to become of each man. For all are not created in an equal condition, but eternal life is foreordained for some, eternal damnation for others."[20]

God's sovereignty, in short, was absolute. "Since the disposition of all things is in God's hand, since the judgment of salvation or of death resides in him," Calvin wrote, "he so ordains by his plan and pleasure that among men some are born cursed for certain death from the womb, who glorify his name by their own destruction."[21] And he illustrated this point vividly with the biblical story of Jacob and Esau, twins born to Rebekah, the wife of Isaac, Abraham's son. Even before their birth God determined that he would favor Jacob, the second born, and not Esau—predestining Jacob and not Esau to be the bearer of His people of the covenant.[22] Calvin, in short, invested Providence with a kind of judicial edge and finality.

Not surprisingly, to many who read Calvin, it seemed incomprehensible that God would foreordain some to eternal damnation and others to eternal salvation without any attention to their works or even intentions. They refused to believe that God would act in such a capricious manner. But to Calvin the idea of questioning God was itself absurd, for there could be, he argued, no human accounting for this divine decision. For in attempting to comprehend it, the human mind fails. Our understandings are simply too limited. Calvin pushed the contrast between God's omniscience and the frailty of the human understanding to the extreme. For the weakness of our intellectual capacities, Calvin insisted, was of a piece with our fallen state, a state that Calvin again read in contrast with the majesty of God. Therefore, in Calvin's view, there was absolutely no way for the human to comprehend the justice of God on his or her own. "The predestination of God," Calvin wrote in his commentary on St. Paul's Letter to the Romans, "is truly a labyrinth from which the human mind can by no means extricate itself."[23] "When they inquire into predestination," Calvin wrote in the *Institutes,* "they are penetrating the sanctuary of divine wisdom. If anyone securely and confidently breaks into this place, he will not succeed in having his curiosity satisfied and he will enter a labyrinth from which he can find no escape."[24] It is not, therefore, in the human being's capacity to understand predestination according to logic; rather it must be learned through the Bible where ultimately, Calvin argues, it is taught that God's election of those He chooses to save "is gratuitous; and in no way depends on men, so that in the salvation of the pious nothing higher must be sought than the goodness of God, and in the destruction of the reprobate nothing greater than his just severity."[25] To many this teaching appeared terrifying, the expression of a profoundly anxious soul, fearful before a judging father. Calvin himself thought it "a thing horrible to hear" that "of so large a multitude only a small number would obtain salvation."[26]

Equally important was Calvin's reconceptualization of divine providence and his ability to allay, in a reassuring way, the hypercharged anxieties of his readers about the End Times. This is not to deny that there were not elements of Calvin's own teachings that contributed to the apocalypticism of the era. In his *On the Necessity of Reforming the Church,* a treatise he addressed to Emperor Charles V at the Diet of Speyer in 1544, Calvin had equated the papacy with the Antichrist.[27] Moreover, in his *Institutes,* Calvin explicitly looked forward to the Second Coming. "The Kingdom of God," he wrote, "will be filled with splendor, joy, happiness, and glory."[28] Christ "will come down from heaven in the same visible form in which he was seen to ascend. . . . The sound of the trumpet will be heard from the far corners of the earth, and by it all will be summoned before his tribunal, both those still alive on that day and those whom death had previously taken from the company of the living."[29] At the time of the Last Judgment, the Elect were to pass immediately into the glories of paradise and the reprobate into the torments of hell. Thus, there was to be no Millennium of peace on earth. In a real sense the drama of history shifted from the social realm to that of the individual. Moreover, Calvin not only sought to attenuate the anxiety around the Last Days through his theology of the Last Judgment, he did so as well through his denunciation of the idea that men might know when the End would come. "Assuredly the day of the Lord will come suddenly and unexpectedly and will, like a thief in the night, take unbelievers by surprise," Calvin wrote in his commentary on St. Paul's First Letter to the Thessalonians, published in 1551, noting "it would be foolish to wish to have a fixed time on the basis of portents and signs."[30]

In developing his ideas of Providence, Calvin had offered a radically new conception of sacred history. The sixteenth century had opened in a period of intense apocalyptic hopes and fears. And many preachers and prophets in this period intensified these anxieties through their arguments predicting that the End—the Millennium and the Last Judgment—was at hand. To Calvin, such prognostications were baseless. The human mind, he argued, was incapable of knowing when the End Times would come. Moreover, he explicitly rejected the overly literal understanding of the Millennium of many of his contemporaries. "Those who assign the children of God a thousand years in which to enjoy the inheritance of the life to come do not realize how great an insult they are casting upon both Christ and his Kingdom," Calvin wrote in his *Institutes*, arguing that it was necessary to understand God's kingdom as eternal and, thus, in no way bound by time.[31] Assurance, in Calvin's view, came from the individual Christian's conviction that he or she was one of the Elect and thus belonged already to an invisible community of saints. That is, he restored the Augustinian idea of the two cities, with the promise that those who lived in faith and with absolute trust in God were already saints in the City of God. Sanctification, therefore, did not lie in an "unknowable" future but rather in the heart of the believer. Calvin, in short, effected a revolution in the understanding of time itself. It was no longer a sacred or predominantly a sacred category, and salvation was no longer a collective promise. This shift brought with it deep assurance. Above all, it provided what the historian Denis Crouzet has called a *désangoissement*—a release from the intense religious anxieties of the age.[32]

Calvin's attacks on prophecy did not only grow out of his concern for the contemporary atmosphere. They stemmed as well from his reconceptualization of the notion of Providence itself. And, yet, at the same time, Calvin and many others viewed his teaching on predestination as, ultimately, deeply reassuring. For Calvin's ideas about predestination were closely related to the promise of divine providence. "When that light of divine providence has once shown upon a godly man," Calvin wrote, "he is then relieved and set free not only from the extreme anxiety and fear that were pressing him before, but from every care. . . . This, I say, is his solace that he might know that his Heavenly Father so holds all things in his power, so rules by his authority and pleasure, so governs by his wisdom, that nothing can befall except he determine it. Moreover, it comforts him to know that he has been received into God's safekeeping and entrusted to the care of his angels, and that neither water, nor fire, nor iron can harm him, except insofar as it pleases God as governor to give them the occasion."³³ Calvin, in short, had relieved the individual Christian not only from the burden of obtaining salvation but also from the anxiety about the End of History.

Calvin was not the only one to react so forcefully against Servetus. Many Protestants shared Calvin's views. So did the Catholic authorities. In Vienne, the Inquisition, after Servetus's escape, burned his body in effigy. Protestants and Catholics held sharply opposing views on many matters—from the mass and the other "sacraments" to the nature of the church—but they both condemned antitrinitarian and millenarian teachings. For Calvin and for Luther these were largely matters of theology and biblical interpretation. The divinity of Jesus was central to their view of Christ as mediator between man and God. While they may have believed that they were living in the Last Days, they rejected the belief in a peaceful Millennium for the saved, holding instead that individual Christians would meet their Maker upon death. Many Catholic theologians shared these views. Moreover, the Catholic Church—which was itself undergoing far-reaching reforms in its institutions in this very period—continued to teach that individual Christians should put their faith in its teachings and sacraments for their salvation. Most crucially, following St. Augustine, the Catholic Church interpreted the Millennium as the era reaching from the conversion of Constantine down to the present and likely beyond—in short, the period that coincided with the Church's ascendancy.

This does not mean that there were not many millenarian currents within Catholicism. But it does mean that, on an official level at least, the Church acted both to repress and to contain them. This explains, in part, the Church's repression of Savonarola at the end of the fifteenth century and its condemnation of the teachings of Joachim of Fiore and other millenarian thinkers at the Fifth Lateran Council in the early sixteenth century.³⁴ Then, by the mid-sixteenth century, the Catholic Church in Italy was able to police prophetic claims that it found threatening. On July 4, 1540, Pope Paul III, at the urging of those members of the papal curia who had long hoped for a more aggressive stance by the papacy against heresy, appointed a commission of six cardinals to serve as Inquisitors General, and three weeks later, on July 21, he established, through the bull *Licet ab initio,* the Roman Congregation of the Holy Office, or the Inquisition.³⁵ This new Inquisition, which claimed authority throughout all Christendom on matters of the orthodoxy of religious doctrines (to whose oversight, in theory, even the Spanish Inquisition, was subject) centralized the authority to make judgments concerning the orthodoxy of religious beliefs and practices in Rome itself. Unlike the Spanish and Portuguese Inquisitions that had been established in reaction against concerns over crypto-Judaism, the Roman Inquisition was formed to counter the spread of Protestantism. And certainly most of those it brought to trial were seen as Protestants. But the Holy Office also went after prophets and messiahs. When a French prophet showed up in Venice in 1566 preaching that the End was near, he quickly landed before the Tribunal of the Holy Office in the city, as did a number of local artisans who, only a few years later, became convinced that a new era of peace was about to dawn.[36]

Even more significantly, in this same period, the Church reaffirmed its role as the conduit of divine grace and, in rearticulating its powers, did much to contain apocalyptic expectations. *Extra ecclesiam nullus salus* ("There is no salvation outside the Church")—this, ultimately, was the meaning of the Council of Trent. This gathering—which met in three long phases between 1545 and 1563 at Trent, a town in the foothills of the Alps—played a major role in giving doctrinal and institutional shape to early modern Catholicism. As a response to Protestantism, Trent enabled the Catholic Church to reiterate its fundamental principles. In its first session it rejected the Lutheran doctrine of "salvation by faith alone," a step that, along with its rejection of Luther's doctrine of the "priesthood of all believers," ensured that the Catholic Church would continue to play a central role in the lives of its followers in mediating the relationship between the faithful and God. The delegates also upheld, again in contrast to the Protestants, the validity of all seven sacraments and reasserted the truth of the doctrines of transubstantiation and of Purgatory. The Church also rejected the reading of the Bible in the vernacular, upholding the Vulgate Bible and its interpretation in light of the traditional teachings of the Church as the source of religious authority. Reading the Word of God in the vernacular, the Church fathers believed, could easily spill over into heresies and apocalyptic fantasies.

The clergy, which the council insisted needed to be better educated, was to remain the arbiter of doctrinal truth. But the delegates also grappled with matters of reform within the Church itself. They emphasized the importance of developing seminaries to foster a better-educated clergy; they encouraged bishops to play a more active pastoral role within their dioceses; and they sought reforms in the sacrament of marriage that would do much to protect women from being forced into unions they did not want. The Tametsi Decree of 1563, for example, stipulated that marriages were to be public and take place in churches—a measure aimed above all to protect women from the relatively widespread practice of clandestine marriages that resulted in their being abandoned by men who, falsely, promised that they would be their husbands. Perhaps most of all the council did much to instill a sense that the Roman Church was not merely reacting to the Protestant challenge but was also finding a way to reform itself from within with growing attention to the importance of meeting the needs of ordinary men and women.[37]

Above all, the Church would embrace all who sought salvation. Drawing on St. Augustine, sixteenth-century Catholic theologians viewed the Church itself, and not a future Golden Age, as the Millennium. Accordingly, it would be through the Church's institutions and practices that the faithful would find salvation. In short, the Church in the sixteenth century had come to function as a *katechon*—the "restraint" that, as St. Paul had described it in his Second Letter to the Thessalonians, was holding back the Second Coming. Until then the faithful were invited to seek salvation through the rituals—actions that, in theory, held time still and moved the matter of salvation, as had Calvin's teaching, from the collective to the individual plane.[38]

Yet Catholic reform was not merely a top-down process driven by the papacy and the leading members of the clergy, secular and regular. To the contrary, throughout the late fifteenth and sixteenth centuries many currents of reform pulsated through the Catholic world, many of them well outside the institutional church. Throughout Spain and Portugal as well as Italy and much of France and indeed those parts of the Holy Roman Empire that did not pass over to the Reformation, many Catholics sought to reinvigorate their faith, often through pragmatic engagement in the world. In late fifteenth century Genoa, Caterina Fieschi, a noblewoman, devoted her life to the care of the poor while, at the same time, emphasizing a quiet internal piety. Her example inspired Ettore Vernazza to establish what would come to be known as the Oratory of Divine Love in 1497. And this movement in turn inspired Gian Pietro Carafa (later Pope Paul IV) and Gaetano da Thiene to create a new religious order: the Theatines, who would also become known for their works of charity. And there were similar impulses outside of Italy. In Spain, in the early sixteenth century, John of Ávila sold his goods and devoted his life to the poor and stressed inward piety and prayer, while in France in this same period Renée de Bourbon continued the reform of Fontevrault, where males were under the authority of an abbess.[39]

But no religious reform movement within the Catholic world would have a greater impact on the development of the Church in this period than the

Jesuits. Founded by the Basque courtier and soldier Ignatius of Loyola, this order came to exercise a profound influence upon the Church and society not only within Europe but across much of the globe. Ignatius himself was a charismatic figure who had embarked on what proved to be a transformative spiritual journey not merely for himself but for many of his contemporaries after he was wounded in battle in 1521. Over the next two decades, especially in Paris where he was a student in the 1530s, he gathered around himself a number of loyal followers, attracted by both his intelligence and his deep spirituality. In 1540 Pope Paul III officially recognized the order, and by the time of his death in 1556 its membership had grown to over a thousand.[40] Within Europe, Jesuits would prove pivotal in pushing back against the advance of Protestantism in the Holy Roman Empire and they helped to re-Catholicize both Bavaria and Austria. But they also would come to play a major role in evangelizing Asia, Africa, and the Americas. Its members were expected to subordinate their own wills to that of the pope and to work tirelessly for the reform of the Church and the preaching of the Gospel abroad—"to go, without complaint," as the Jesuit vow read, "to any country whither [the pope] might send us, whether to the Turk or other infidels, in India or elsewhere, to any heretics or schismatics, as well as to the faithful, being subject only to the will of the pope and the general of the order."[41]

The rapid growth of the Jesuits struck many as providential. Perhaps inevitably apocalyptic hopes soon surfaced among many of Ignatius's followers. In 1550 an anonymous manuscript—almost certainly written by the Spanish Jesuit Andrés de Oviedo, later patriarch of Ethiopia—connected the origins of the order to the prodigious events of 1492. Not only had this been the year of Columbus's voyage, the defeat of the Muslims at Granada, and the expulsion of the Jews—all of which the author celebrated as providential—but also (Oviedo believed) of the birth of Ignatius of Loyola. And it was Ignatius's birth, so Oviedo maintained, that provided the promise for the fostering of a Beautiful Ending in the New World. Unlike the earlier Franciscan and Dominican missionaries who had not served God but rather aided the Spanish in their brutal conquests in the Caribbean, Mexico, and Peru, the Jesuits would be the "true spiritual men" who would usher in the Age of the Spirit. Nor was Oviedo alone in these views. To the contrary, he was a member of the household of Francisco Borja, duke of Gandia, who had—after an earlier life at the court of the Emperor Charles V—entered the Jesuit order in 1548. And millenarian expectations circulated among the Jesuits in Borja's circle, with many insisting that the duke was to be the angelic pope.[42]

When Ignatius learned of these expectations and prophecies, he reacted no less firmly to the apocalyptic currents within the Catholic world than Calvin had reacted to them among Protestants. In a long letter composed with the assistance of his secretary Juan Polanco, Ignatius cautioned Borja and his circle against such prophecies. "This spirit of prophecy or feelings, especially about the reformation of the Church and the angelic Pope, etc., which have been circulating here for many years, must be regarded with reason as highly suspect," Ignatius wrote. He then listed a broad slate of discredited prophets, some from the recent past such as the Portuguese João da Silva e Menezes and the Italian Girolamo Savonarola. Ignatius also noted that "it is a wondrous thing in our day how many have meddled in this business," and pointed to such well-known figures as Cardinal Pietro Galatino, papal chancellor Ambrogio Recalcati, and the sometime Jesuit and polymath Guillaume Postel, as well as to a number of self-anointed messiahs. In Urbino, one of these had begun dressing up as pope and had created cardinals; another had done much the same in Calabria; and one of these newly elected popular figures had even shown up in Rome to tell Ignatius that he had been elected pontiff and was the angelic pope whom the prophets were expecting. With this last figure, Ignatius was patient, telling him that papal elections only occur when the Holy See is vacant and that he should go and learn for himself whether Pope Paul III is "still living or not, in order to see if his election had really happened."[43]

But the conviction of Oviedo and others that Borja would be the long-awaited angelic pope was clearly a more serious matter. Ignatius did not deny that this could be possible—as all things were possible with God—but he disapproved strongly of those who prophesied about this matter. Above all, such prophecies risked causing harm to the papacy and the Society of Jesus, and, in the end, Ignatius sensed that their prophecies were prompted by the Devil rather than by God. Thus, it was essential that the Jesuit leaders use their discernment to discourage and to discredit this prophesying that had found such an important set of followers in the order, as it was better ultimately "to keep clear of such thoughts, leaving it to God's divine goodness to bring things to completion."[44]

Thus, like Calvin, Ignatius pushed back on those who prophesied about the End of Time—against the apocalyptic dreams of those who longed either for an angelic pope or for the Millennium. To be sure, reservations that Ignatius expressed on this front were motivated more by sweet reasonableness and pragmatism than were those expressed by Calvin, whose position drew more rigorously on his theological understanding of salvation. Crucially, moreover, for Ignatius the spiritual drama, in the end, was internal not institutional. Thus—in much the same way as the earlier followers of Luther had cast his struggles as part of a great battle between Christ and the Antichrist—Ignatius too was deeply aware of a great spiritual struggle between the followers of Christ and those who, by contrast, followed Lucifer. Accordingly, as Ignatius would make clear in his *Spiritual Exercises,* the believer should reflect not only on Christ and on the splendors of paradise but also on the Devil and the horrors of hell.[45] And in his meditation on the Two Standards, the one of Christ and the other of Lucifer, Ignatius offered his readers as vivid a contrast between the values and virtues of a Christ who was humble and committed to the poor, and a Lucifer who represented an indulgence in worldly honors and the pursuit of wealth, as Cranach had in his *Passion of Christ and the Antichrist,* but without, as the Protestant artist had done, associating the Antichrist with the papacy. To the contrary, Ignatius's spiritual teaching, far from rejecting the papacy, presented the papacy and the ecclesiastical hierarchy as central to the Christian life and, throughout his own life, Ignatius would insist that his followers not indulge their own wills but submit themselves entirely to the teachings of the Church. "What I see as white," he wrote in an appendix to the *Exercises,* "I will believe to be black if the hierarchical Church thus determines it."[46]

Religion is a system of symbols, the work of culture and passed down from generation to generation. But no symbol is static, no matter how widely it is shared within a religious community. To the contrary, religious symbols evolve and their meanings and importance to a community frequently change over time.

While the doctrine of the Trinity originated as a matter of theology when the Church fathers gathered at the Council of Nicea in 325 to wrestle with the question of Jesus's divinity, the Trinity would eventually become a powerful symbol of religious faith for many Christians, and indeed it came to play a shaping role in the lives of both individual believers and religious communities. Most medieval Christians, for example, evoked the Trinity when they made the sign of the cross and recited the short phrase: "In the name of the Father, the Son, and the Holy Ghost." Priests performed this ritual at baptism, making the sign of the cross on the infant, a seal that participants believed would help hold the Devil at bay. The laity made the sign of the cross during mass. But many also made the sign, always evoking the Trinity, in moments of perceived danger, and in this sense the cross and the Trinity served as talismans against Satan and his works. But against this backdrop of popular practices, many also discovered within the Trinity a gate to a deeper spirituality. A triune God was, as St. Augustine maintained in his early-fifth-century treatise *On the Trinity*, a mystery and one that resonated with the deepest aspects of human experience and psychology.[47] In the early thirteenth century, Joachim of Fiore had drawn on trinitarian teachings to frame the course of sacred history from the Age of the Father through the Age of the Son down to the final earthly state: the Age of the Holy Spirit.

Then, for reasons we do not know, the power of this symbol began to attenuate, and not only for Servetus. In Naples in the 1530s and 1540s, for example, several religious thinkers had begun to question the Trinity. Like Servetus, the most prominent of these figures may have been of converso background. Yet it was Lelio Sozzini, a humanist from Siena, who would offer the most influential critique of the doctrine of the Trinity, insisting on a unitarian God.[48] Nonetheless to many, certainly, the Trinity continued to be a central symbol of the Christian life. Preachers called upon their congregations to ponder its mysteries; and artists depicted the Trinity in paintings of powerful imagination.

For Ignatius too, the Trinity remained a mystery, a powerful symbol. Indeed, it played a pivotal role, as we learn from his *Autobiography*, in his religious journey. Shortly after his conversion, Ignatius would pray daily to the Trinity. At first he felt his understanding of this triune God was blocked. Then, one day, "while praying the office of Our Lady . . . his understanding began to be raised up, in that he was seeing the Most Holy Trinity in the form of three keys on a keyboard." In short, Ignatius had discovered at last a way to grasp the unity of the Father, Son, and Holy Ghost. The text of the *Autobiography* makes it clear that his devotion to the Trinity would remain central to his spirituality throughout his life. The very next day after his vision, he could only talk about it with his companions "in such a way that the impression has remained with him for the whole of his life, and he feels great devotion when praying to the Most Holy Trinity."[49]

There is no evidence that Ignatius of Loyola, as he came to be known, ever knew either Servetus or Calvin in Paris, even though it is certain that these three men were there at the same time. Ignatius, though older than the other two, was still a student. Calvin had returned there after his legal studies, to deepen his knowledge of Hebrew and Greek, the languages of the Bible. Servetus had come to Paris to study medicine. Yet, despite the ferment of the evangelicals in Calvin's circle and the presence of a few radicals who might have been drawn to Servetus, Paris was large enough for Ignatius to cultivate friendships with young men who shared his commitments to the fundamental tenets of the Catholic tradition. Paris in the early 1530s—though the atmosphere was soon to change—was still open to a rich range of religious ideals. And while these three men held fundamentally different views about the nature of salvation, they each were passionate about their convictions—passions that, eventually, would set them on paths of conflict and even violence.

CHAPTER 8

Battles for God

> There is nothing so much to fear in a republic as a civil war, nor among civil wars, as that which is fought in the name of religion.
>
> —ÉTIENNE PASQUIER (1561)

In the summer of 1559, in celebration of the recent peace with Spain, King Henri II of France was presiding over a tournament near his residence in Paris. Henri was in high spirits. Ever confident, at the end of a long day, he insisted on making one last joust. He charged forward at a gallop when suddenly, after a bright flash, everything went dark. His opponent's lance had pierced his helmet, gouging him just above the left eye. The royal physician Ambroise Paré intervened immediately and Andreas Vesalius, the most renowned surgeon in Europe, was rushed from Brussels to the king's bedside. Even though it had seemed at first that he might recover, there was little the doctors could do. Ten days later, on July 10, Henri was dead.[1]

The death of a king in early modern Europe was always a precarious moment, but Henri's was especially perilous. With three minor sons and a widow of foreign birth (the Tuscan Catherine de' Medici), his loss created a vacuum at the center of the kingdom, unleashing powerful centrifugal forces. To contemporaries, it became almost immediately clear that many of France's wealthiest noble houses would try to exploit this moment of weakness to their advantage. These fears quickly materialized. The young new king fell almost immediately under the influence of the Guise, the richest and the most powerful family of the realm.[2] The rapid ascent of the Guise immediately provoked resentment and opposition from other aristocratic families, posing a major threat to the stability of the kingdom.

France was already a tinderbox. Well before Henri's death, growing religious tensions had proven destabilizing to the monarchy. To be sure, the interest in reform in France had at first been largely confined to clerics and humanists.³ Nonetheless, by the 1530s, the circle of those who sought reform had expanded. Throughout the kingdom many merchants, artisans, and aristocrats were reading Luther and, under his influence, had begun to demand a more thoroughgoing reform of the Church and its teachings. Simultaneously criticisms of the mass—many of these inspired by the reformer Ulrich Zwingli of Zurich—also found a following. In 1533 these new ideas were preached publicly with great success. During Lent the learned cleric Gérard Roussel preached a cycle of evangelical sermons in Paris; only a few months later Nicolas Cop, the rector of the university, delivered a powerful evangelical address at the opening of the term. By this point, it had become clear that the evangelicals were no longer merely working within the framework of the Catholic Church. They could no longer hope for reform from within. To the contrary, their "heresies" had—it was obvious—taken a decidedly Lutheran turn. A Reformation movement had put down roots in France, and the growing demands of the reformers would soon make themselves felt in the political sphere.

Then, during the night of October 17–18, 1534, a well-coordinated group of evangelicals put up placards or printed broadsides not only in Paris but also in Blois, Orléans, Rouen, Tours, and even outside the king's bedchamber in the royal château at Amboise. The broadside, entitled *True Articles on the Horrible, Great, and Unbearable Abuses of the Papal Mass, Dreamed Up against the Lord's Supper of Jesus Christ,* denounced, above all, the Catholic eucharist.⁴ For through this ceremony, the placard read, "knowledge of Jesus Christ is completely rubbed out; the preaching of the Gospel rejected and denied; and the time spent in the ringing of bells, howlings, chants, empty ceremonies, candles, censings, disguises, and monkey-business of every kind."⁵ At issue was the Catholic belief that the bread and wine used during the Eucharist were not merely symbols of the Last Supper that, according to scripture, Christ had celebrated with his Apostles shortly before his death. To the contrary, according to Catholic teachings, the bread and wine, through the miracle of "transubstantiation," became the body and blood of Jesus Christ. The Catholic mass, in short, was a moment in which God was immanent in the world and in which Christians came into direct contact with their Savior, bringing Him into their bodies and entering into a profound communion with Him and one another. By contrast, to the evangelicals, this was not merely an absurd teaching but also a superstition that elevated the priests above the laity. Thus in their broadsides, they not only attacked the mass but also lashed out against priests, calling them, among other things, *faulx antichristz*—"false antichrists."[6] In response to the placards, King Francis ordered harsh police measures. French authorities arrested as many of those responsible for the postings as possible. Many were burned at the stake.[7]

It had been this climate, understandably, that Calvin had fled, seeking exile in Switzerland. And eventually, especially from his perch in Geneva, he would come to exercise enormous influence in France itself.[8] There his ideas found a significant following. Not only artisans and merchants but also humanists and monks, nuns and abbesses, nobles and prelates came together to explore Calvin's ideas and to discuss them with one another. In all likelihood, they were attracted less by the finer points of his theology than by his powerful evangelical message that offered a scathing critique of the papacy and the Roman Church. In his writings they found a new understanding of the Christian's relationship to God that made them view most of the traditional trappings of Catholicism as idolatrous. They rejected the Catholic clergy's claim to special privileges, relics, saints, and holy processions. They made fun of the mass, ate meat on Fridays, and mocked the veneration of saints. In a very powerful sense, Calvin's ideas enabled them to try to reclaim the religious life for themselves. Moreover, many embraced Calvin's doctrine of predestination, his insistence that it was impossible to know who was saved and who was condemned. Far more than Luther's ideas, Calvin's were leveling, and not merely in the spiritual sphere. Unlike Lutheranism, which tended to support princes and magistrates, Calvinism would call social hierarchies into question.

Finally, in these gatherings, the followers of Calvin took the first steps, perhaps at first without realizing it, toward forming a church of their own. To be sure, the overwhelming majority of French subjects remained Catholic. But those who embraced Protestantism—concentrated among the nobility in the south of France and among literate craftsmen, merchants, and professionals throughout the kingdom's towns and cities—exercised an outsized influence.[9] Most disconcerting to the French crown, however, was the fact that the Huguenots (as the French Protestants were called) were so well organized, with many of the leading nobles, such as the Admiral de Coligny, offering protection and even military support to the movement.

Protestant forces continued to find strong leadership in the aristocracy. Boldly, in March 1560, Jean du Barry, Lord of La Renaudie, along with several hundred sympathetic nobles, in an attempt to bolster the Huguenot cause, attempted to kidnap the young King Francis II at Amboise, where the royal court had settled in over the winter. But the royal forces led by the Guise family reacted quickly; they intercepted the Protestant conspirators and, in settling scores, executed many of them, in some cases by beheading. To many it seemed that things were spinning out of control. But immediately after the failed attempt to kidnap the young king, it seemed that the Guise party and its policies of repression would continue. Francis died unexpectedly in December. Since his younger brother Charles, who succeeded him, was only ten, French custom required a regent. Henri II's widow, Catherine de' Medici, seized this role. She dismissed the Guise and reinstated the more moderate policies of Francis I and Henri II which, while repressive, sought to accommodate some of the needs of the Huguenots. In this context, moreover, Catherine focused primarily on the means to protect her monarchy. She pushed back against the Guise, at least privately, and worked with those nobles, known as the *politiques* or the *moyenneurs,* who, she believed, might be encouraged to take a moderate path.[10] And it was to that end that Catherine, who was by temperament more interested in practicality than theology, summoned Catholic and Protestant theologians alike for a colloquy at Poissy, a town just outside of Paris, in 1561. Though the colloquy failed to iron out differences between the two parties, Catherine remained optimistic. She was tenacious. Catherine also knew well how to negotiate within court politics; and it was largely her decision in January 1562 to issue an edict of toleration, allowing Protestants to worship freely as long as they did so discreetly, met during the daytime, and were unarmed.

To many devout Catholics, the queen's efforts at a compromise were offensive, for, in their view, the spread of heresy in France served as evidence of an apocalyptic struggle. This is hardly surprising. Over the last decades of the fifteenth century and throughout the sixteenth century, France, much like Germany, was inundated with works—pamphlets, broadsides, and almanacs—that not only located the causes of natural disasters—floods, famines, and storms—in God's wrath but also promised a Beautiful Ending to those who repented of their sins. Preachers, wandering from town to town and from village to village, amplified this message, casting the Huguenots in the role of the Antichrist.[11] In 1525, for example, the Franciscan Thomas Illyricus, who had already attained widespread fame as a prophet in the Guyenne, identified the heretics, the followers of Luther, as signs of the End Times. Deeply influenced by

Joachim of Fiore, Thomas struck an optimistic note. After the tribulations, he argued, an angelic pope would usher in a reformation of the Church and an era of peace.[12] Then, in 1533—putatively the fifteen hundredth anniversary of the Crucifixion and a time in which the French Protestants were preaching publicly—a wave of fierce anti-Protestant fervor swept over the kingdom, as preachers, poets, and astrologers found evidence, in the heavens as well as in the continuing spread of heresy, that the End was near.[13]

Over the next decade the temperatures—the anti-Huguenot passions—continued to rise. In 1556 the priest Artus Desiré published his *The Arguments of Guillot the Swineherd and of the Shepherdess of Saint Denis in France, against John Calvin, Preacher of Geneva, on the Truth of Our Holy Catholic Faith*. Set as a dialogue, the work portrayed two unlettered peasants, both inspired by divine truth, as they parried Calvin's arguments for reform. Desiré set the dialogue within an apocalyptic frame. He associated the Huguenots with the dragon in the Book of Revelation and Calvin with Satan.[14] But this was just the tip of the iceberg. Over the course of his life, Desiré, a tireless poet, composed and published some twenty-five anti-Calvinist works, many of which appeared in multiple editions and which were marked not only by their repetitiveness but also and above all by the violence of their language.[15] As the heresies spread, moreover, Desiré not only continued to dehumanize the Huguenots, whom he saw as a contagion, but also called for their destruction. "Roast them, those preaching false teachings, on hot coals," he implored his fellow Catholics in his *Articles of the Treaty of Peace between God and Man*, adding, "take them from their nocturnal gatherings and bring them to priestly councils and then throw them all to the fire." Only then, he noted, "will we have peace."[16] For Desiré, a Beautiful Ending was only possible once the threat of heresy was entirely eliminated and all French Catholics were once again united under the spiritual guidance of the papacy.

And this was a message that found its way to the public through preaching as well. In 1561 the Dominican Pierre Dyvolé was explicitly calling for violence. Catholics must take up arms to defend their faith against the growing strength of the Huguenots. "Moreover," a contemporary wrote in his account of Dyvolé's sermons, "he predicted the next evil which, in a short period of time, they will carry out in France, when they rise up with arms and seditions against the king, his estate, and the public peace, desolating towns, pillaging churches and holy places, mistreating the priests, trying to abolish true religion, all law—ecclesiastic, political, and civil—and all the sacraments and divine services."[17] Like Desiré and many others, Dyvolé—taking into account what he saw as the threat the Huguenots posed to the kingdom—gave divine permission for the use of arms to protect the Catholic faith.

On the first Sunday in March 1562 a congregation of Huguenots met for worship in a barn just inside the walls of the Norman town of Vassy. Technically, in light of the queen's recent edict, the gathering was legal. But the arch-Catholic duke of Guise, who happened upon this congregation of some five hundred Protestants—men, women, and children—by accident while they were worshiping, saw matters differently. He ordered them to stop their service. When they refused, the duke commanded his retainers to attack. The duke's men burned the barn, killing some fifty and injuring many others. In response, the Protestants, drawing on their already elaborate network of congregations throughout the kingdom, began to organize town militias under the leadership of Louis de Bourbon, the prince of Condé and a leading member of one of the most powerful aristocratic families in France.[18] Throughout the year violence broke out in many areas, as Protestants desecrated Catholic shrines and even sought in some cases to seize control of urban governments. Catholic reactions were fierce. On April 12, militias loyal to the papacy slaughtered hundreds of Huguenots at Toulouse. From late September to October, Catholic forces besieged the city of Rouen, which the Protestants had seized earlier. In December the first major military conflict in what would come to be known as the French Wars of Religion took place at Dreux, just to the west of Paris, with the Catholics easily defeating a ragtag army of Protestants who had risen up to resist them. This battle unleashed the first round of civil war. The violence—much of it subtended by apocalyptic expectations among both Protestants and Catholics—would continue to roil France throughout the rest of the century.

In late summer 1572 hope grew that peace had been achieved. The Bourbon prince Henri of Navarre, a Huguenot, and the king's sister Marguerite of Valois, a Catholic, were to be married in Paris. The warring parties would be reconciled. Yet behind the façade of the marriage celebrations, the city was on edge. Many Parisians—the city was devoutly Catholic—had always viewed the marriage as an unholy alliance; and they grew alarmed when large numbers of Huguenots streamed into the city for the wedding. When on Friday morning, August 22, an assassin's bullet failed to kill the Admiral de Coligny, one of the Huguenot leaders, tensions escalated immediately. The Huguenots suspected that the Catholic duke of Guise was behind the attempt to kill Coligny. The Catholics feared that the Protestants would exploit the attempted assassination to stage an attack against them. On Saturday evening in their apartment in the Louvre Catherine de' Medici and Charles IX met secretly with the city's mayor, whom they ordered to lock the city's gates and not to allow anyone either to leave or enter Paris.

In the early hours of Sunday, August 24—St. Bartholomew's Day—Catholic forces began rounding up Protestant leaders and summarily executing them. Coligny, who had survived the assassination attempt two days earlier, was one of the first victims. Then, as the Guise faction continued its assault on Protestant leaders, slaughtering several of them in the courtyard of the Louvre, larger and larger numbers of Parisians joined in the effort to extirpate the Huguenots. With no one able to escape—the city gates had been locked—Catholics spread throughout the city. They broke down doors and dragged Protestants and even those suspected of being Protestant, women and children included, from their hiding places into the streets, brutally massacring them and casting their bloodied and mutilated bodies into the Seine. In the end—the numbers from this period are imprecise—some two to six thousand perished. Nor was the violence contained to the capital. As news of it spread through the kingdom, it set off similar massacres in some dozen provincial cities: Rouen, Meaux, and Troyes in the north; Bordeaux and Toulouse in the south; and in Angers, Saumur, Orléans, La Charité, Bourges, and Lyon in the more central provinces. These were all Catholic towns with sizable Huguenot minorities. Cities such as La Rochelle, under Huguenot control, and Dijon, under Catholic governance, escaped this secondary wave of violence that claimed the lives of another three thousand or so victims, most of them Protestants.[19]

What has long been clear to historians is that the religious violence that traumatized so many Frenchmen and women in these years was not merely a matter of political calculation, though this came into play, nor merely a response to social and economic tensions, though these too were factors. Clearly religious tensions also proved central to the violence. But—as in the case of the religious wars in Germany in the first half of the sixteenth century—the tensions were not purely a matter of theological differences. To the contrary, as in the empire, what appear to have been decisive rather were deep-seated apocalyptic fears and hopes that had led so many Catholics—not merely elites but also merchants, artisans, and workers—to view the Huguenots as antichrists. Especially in the 1560s, as the Protestant movement had gained strength in France, many Huguenots, who similarly viewed the Catholics as antichrists, had not only continued to denounce the papacy and the mass, but had also carried out attacks on Catholic churches. They lopped off the heads

A Parisian Apocalypse. A near-contemporary painting captures the terror of the St. Bartholomew's Day Massacre.

of statues of saints and desecrated altars. In this context, many Catholic preachers and polemicists stepped up their attacks on the Huguenots, urging their congregations and readers—and even the king—to take violent actions against them. In his *The Origin and Source of All the Evils of this World,* which had appeared just a year before the Parisian massacre, the indefatigable Artus Desiré portrayed the age as one of tribulations and widespread corruption as the growth of the Huguenot movement made clear. He called not only on his readers to repent but also on the king to repress "the great blasphemies and heresies of his realm."[20] Failing to eliminate the heresies, the French Catholics would face eternal damnation. Desiré was far from alone in stirring up attacks on the Huguenots. As Denis Crouzet has argued in his brilliant analysis of the sources of religious violence in France in these years, widespread apocalyptic hopes and fears did much not only to legitimate but even to provoke religious violence in this period. Certainly, such fears served as fuel for the brutal, deadly orgy that consumed Paris in late August 1572.[21]

As soon as 1574 there was again hope for peace when Henri III, the last of Henri II's sons, succeeded his brother Charles to the throne. Like his mother and older brother, the new king attempted to pursue a moderate policy. But many Catholics were not pleased; and in 1576 Henri, the duke of Guise, who had succeeded his father after the latter's assassination in 1563, now formed the Catholic League, aimed, for both political and religious reasons, at extirpating the Calvinists from France. In part, the league drew its strength from the political backing it received from both the papacy and King Philip II of Spain. Yet its members were animated above all by many of the same dreams of Christian unity that had triggered much of the anti-Huguenot violence in previous decades. These were men and women, raised in the Catholic faith, who saw their world as out of joint. Plague, famine, warfare, and confessional strife were all evidence of a seemingly implacable divine justice. Clearly, God was punishing them for their sins. Henri III's assassination of Henri, duke of Guise, in December 1588 was a profound blow. For the next several months, Catholics throughout France sought to placate what they viewed as an irate God through collective penances. They prayed, fasted, and organized processions. In Paris especially, these acts of penitence fostered a deep identification of the city with Jerusalem and the Heavenly Kingdom. Not unlike the Florentines under Savonarola at the end of the previous century, many Parisians believed that these displays of intense piety would accelerate the return of Christ and a restoration of the spiritual unity and holiness that the wars of religion had destroyed.[22]

Yet once again expectations for the End of History set the stage for further violence. On August 1, 1589, the league assassinated Henri III, the last of the Valois. By law, the inheritance of the throne now officially passed to Henri of Navarre. Navarre was a deeply pragmatic man. But he still had to fight off the Guise who had put up a pretender to the throne and who were allied with the Spanish. Eventually he prevailed, but he still had not won over the Catholics in Paris, and in order to enter the capital, in 1593 he renounced his Calvinism and returned, yet again, to the Catholic faith. *Paris vaut bien une messe,* he is said to have proclaimed in announcing his conversion—"Paris is worth a Mass."

The following year Henri of Navarre was crowned king at Chartres. As monarch, Henri would work tirelessly to try to restore order to the French realm. Nonetheless, it took him several years to pacify the most intransigent members of the Catholic League, many of whom had at first been skeptical of this new monarch. Yet, even as Henri recognized Catholicism as the official religion of France, in his *Edict of Nantes* (1598) he granted the Huguenots the right to maintain their own militias and also to worship privately at home throughout the kingdom and publicly in a number of towns in the south-western regions. The Edict of Nantes was not an effort to recognize freedom of conscience. Rather it took a characteristically early modern approach to the question of religious minorities—much as the Ottomans did toward Jews and Christians within their empire and as many Christian states did toward Jews in Europe—by granting clearly defined privileges. In this case, the Huguenots had the right to worship where they already had a church. The king by no means embraced a doctrine of religious toleration, as it would come to be understood two hundred years later in the Enlightenment. Above all, Henri tried to assure partisans from both faiths that he had their interests at heart.

It was not only in France that apocalyptic hopes and fears fueled a politics of violence. They did so elsewhere in Europe and the Mediterranean in these years as well. In western Europe this was most evident in the Low Countries; and the causes were much the same. For there too rulers faced not only immense resistance to their efforts to centralize authority but also growing tensions between Protestant and Catholic factions. On the political level, King Philip II of Spain, who was also the ruler of the Netherlands, was seeking to ensure the loyalty of the people of the Low Countries to his rule. He also insisted that they remain faithful to the Catholic Church.

Shortly after signing the Treaty at Cateau-Cambrésis in 1559—the treaty that Henri II had been celebrating in Paris at the time of his death—Philip had decided to leave the Low Countries for Spain. It was, however, a delicate moment. Heresy was spreading in Holland, and Philip had reason to be concerned about the stability of this territory. It was in this context that he appointed Margaret of Parma, his half-sister, as governor-general of the Netherlands and made Antoine Perrenot, lord of Granvelle and one of his most loyal advisors, her prime minister. Since loyalties to Lutheran, Calvinist, and other Protestant ideas had been growing slowly among the Dutch since the late 1510s and had even picked up in the 1550s and especially in the 1560s, Philip also encouraged a policy of the repression of heresy. Upon his departure for Spain Philip reminded Margaret that it was her sacred duty to be sure that heretics were put to death.[23] And the tribunal against heresy in the Low Countries was the bloodiest in Europe. Between 1523 and 1566, the Inquisition executed some thirteen hundred individuals accused of heresy out of a total population of some two million.[24]

While Granvelle fully supported Philip's measures, many members of the elite in the Low Countries were dubious about the Habsburgs' increasingly repressive policies. Leading merchants feared that a crackdown on nonconformity would threaten their interests; and the magnates also began to give their support more and more to the Protestants. To placate the Protestant leadership—which now included the Count of Egmont whom Philip had earlier appointed *stadtholder* (governor) of Flanders and Artois, and William of Orange whom he had appointed *stadtholder* of Holland, Zeeland, and Utrecht—Philip dismissed Granvelle in 1564. But this action only encouraged more dissent. In 1566 Calvinist leaders—the famous hedge-preachers—began to attract large crowds at their sermons. Unexpectedly, in August of that year, as religious tensions grew, Protestant mobs in Antwerp, Ghent, and other cities smashed sacred images in the *Beeldenstorm,* the iconoclastic fury that swept through the Low Countries at this time. Margaret both sought to channel and to repress this expanding Protestant movement. While granting Calvinists rights to preach and attend services within city walls, she simultaneously bolstered the Inquisition. On this latter measure she almost immediately faced the opposition of many of the nobles who, on April 5, 1566, presented her with the Petition of Compromise, denouncing the Inquisition. Margaret yielded; and, in the wake of her softening, there was a notable upsurge in Protestant activity.

Philip's reaction was swift and decisive. In 1566 he dispatched Don Fernando Álvarez de Toledo, the duke of Alba, to restore order.[25] Alba had already proven himself a strong partisan of the faith in the Schmalkaldic War. He was a ruthless commander. He arrived in the Low Countries with ten thousand troops and immediately established the Council of Troubles—a court with a prosecuting staff of over 170—to bring heretics and others who threatened the stability of Spanish rule to swift justice. In 1568 Egmont was among his first victims. But the Dutch resistance proved powerful. Protestants in the Low Countries may have been fighting for a wide variety of liberties, political and religious, but they did so in the conviction that they were battling the Antichrist. A series of prints and broadsides made this clear. One, an anonymous German print from 1569 portrayed the duke of Alba as the pope's lieutenant seated on a throne. To his right stands the duke of Granvelle, who receives the papal tiara from the Devil, against a backdrop of repression and torture. In 1572 the leaders of the opposition, who continued to draw upon images of their opponents as antichrists, began in earnest. William of Orange had led a militia against the Spanish into the eastern Netherlands. In order to protect his troops, Alba withdrew to the coast only to be surprised by an attack from the sea when, on April 1, the so-called Sea Beggars—a force of slightly over a thousand Dutchmen, who had sailed from England—seized Brill, a small town on the coast of Holland. For the next several years the

Allegory of the Tyranny of the Duke of Alba (1569). On the platform in the background the artist has depicted the executions of Egmont and Hoorne and other scenes of torture and death that had resulted from Alba's repressive policies.

fighting would remain fierce, and the Spanish enjoyed considerable success, especially in the southern provinces. The Spanish troops were also brutal, especially in the south. In 1576, in their attack on Antwerp, they slaughtered more than seven thousand residents.

But the situation was different to the north. There, Orange and the Sea Beggars made significant gains, appealing not only to the religious convictions, but also to the patriotism of the towns. In 1579, with the Union of Utrecht, the Northern Provinces formed a new confederation of states; and in July 1581 these provinces declared independence. "The king of Spain has forfeited his sovereignty," the Estates-General declared in the Act of Abjuration. Drawing on new theories of resistance that had been developed by Huguenots in France, above all Philippe du Plessis-Mornay's *Vindication against Tyrants,* the rebels were convinced of their right to act on their own. They also based their actions on the legal theories of the Dutch humanist Aggaeus van Albada, namely that "all forms of government, kingdoms, empires and legitimate authorities are founded for the common utility of the citizens, and not of its rulers."[26] To ensure that everyone would understand their actions, the new government sought to annihilate any mention or representation of the king within the republic. The king's image was not to appear on coins or seals; his coat of arms was to be removed from buildings. As the authors of the act stated, "we order likewise and command the president and other lords of the privy council, and all other chancellors, presidents, accountants-general . . . and likewise to all other judges and officers, as we hold them discharged from henceforth of their oaths made to the king of Spain, pursuant to the tenor of their commission, that they shall take a new oath to the states of that country on whose jurisdiction they depend . . . to be true to us against the King of Spain and all his adherents."[27]

Over the next decade the northern and southern provinces drifted further apart, as the Spanish continued to make gains in Flanders and Brabant. A new and effective Spanish commander, the duke of Parma, took Antwerp and Brussels in 1585. These victories led Queen Elizabeth of England, worried about the growing power of Spain, to send some six thousand troops into the Netherlands to support the Dutch rebels. It was a risky engagement, especially since it prompted Spain to raise an armada against the queen. In 1588 the Spanish Armada—a fleet of some 130 ships—attempted an invasion of England. But, as they approached London's harbor, the winds worked against them. Not only did the English hold off the invasion but the Spanish lost an enormous number of ships as they retreated. This constituted a major blow to Spain, but ensured the sovereignty of Holland, which effectively became an independent republic at this time. Facing many financial difficulties at home, Spain was unable to reimpose its rule and, in 1609, agreed to a truce with the Dutch. The result was that the Low Countries, once under the rule of Burgundy and then the Habsburgs, had been torn in half. The southern provinces (Brabant and Flanders) were French-speaking and Catholic; the northern provinces (Holland, Zeeland) were Dutch Flemish–speaking and Protestant. What had been essentially one civilization was now two. And the formerly dominant cities of Ghent, Bruges, Brussels, Leuven, and Antwerp gave way to Utrecht and especially to Amsterdam. Peace came in 1609–10, though the revolt would resume, under radically altered circumstances, in 1621.

Moreover, throughout the late sixteenth century, the struggle between the Ottomans and the Christian princes in the Mediterranean continued to play out within an apocalyptic frame that had shaped the relations between Muslims and Christians in the Mediterranean with urgent intensity since the fall of the Byzantine Empire in 1453. In 1556 in Istanbul Murad b. Abdullah, a Hungarian convert to Islam, turned Christian prophecies of the End of the World on their head, predicting that Islam and not Christianity would rule the world. "Then 'Isa will descend to the Earth, and his prayer to God that he be made one of Muhammad's community will be answered," he wrote, adding: "After becoming a Muslim he will kill the *Deccāl*. And then Gog and Magog will come forth, and 'Isa, having said a prayer, will kill them too. After that the twelfth caliph, who is the Messiah of the Age [the Mahdi], will appear, and his reign will be the time of security and piety, abundance and low costs, and the wolf will walk together with the sheep."[28] And similar millenarian expectations would continue to pulsate throughout the century, shaping hopes among the Ottomans for a Beautiful Ending that could be realized in part through their military victories against the Christians in Europe and the Mediterranean.[29]

At the same time apocalyptic visions circulated among Christians. These tensions, which often resulted in violence, were most evident in the nearly constant warfare between the Ottomans and the Christian princes who, from the mid-sixteenth century on, both focused their attention on the Mediterranean. In 1559 the Genoese admiral Andrea Doria led a force of some ten thousand men to retake the North African city of Tripoli from the Ottomans. When the invasion failed, Doria and his men seized the strategically important island of Djerba, some two hundred kilometers to the west.[30] For the next several years, the Spanish and the Ottomans would fight for territory in the central Mediterranean. In 1564, the Spanish managed to take Peñón de Vélez, a coastal stronghold on the Algerian coast. In the following year, when the Turks tried to seize Malta, Spain succeeded in protecting the island, but Ottoman power would continue to grow despite occasional Christian victories.

In 1570 the Turkish navy began its invasion of Cyprus, a bold move that signaled the growing strength of the Ottomans in the Mediterranean. The Venetians, who had controlled Cyprus since the late fifteenth century, persuaded Pope Pius V and King Philip II of Spain, along with several other Catholic powers—Genoa, Savoy, and Malta among them—to join with them in mounting a fleet to take on the Turks. At the height of the Counter-Reformation the response became a religious crusade; and the armada of these states, who had entered into a Holy League, set sail for the eastern Mediterranean prepared to take on the Turkish navy, which had begun a series of operations against Venetian colonies in Cyprus, Albania, and Morea.

On the morning of October 7, 1571, two implausibly large fleets confronted each other in the Gulf of Patras, near the Greek coastal town of Lepanto. The Christian fleet—under the leadership of Don Juan of Austria (the son of Charles V)—numbered more than 200 vessels and nearly seventy-one thousand men. The Turkish fleet, under the command of Müezzinzade Ali Pasha, was even larger. It counted more than 250 vessels and a total of some eighty-one thousand men.[31] But, despite their somewhat smaller numbers, the Christians had an advantage. With cannons mounted on four Venetian galleasses, the Ottomans were outgunned. Within five hours the sea turned red. Nearly thirty thousand Ottomans and some five thousand Christians died. The Christians interpreted their victory with apocalyptic joy. The Holy League attributed their victory to divine protection. Don Juan kept the image of Our Lady of Guadalupe in his cabin—and in the wake of this victory this virgin was transformed into Our Lady of Victory and portrayed standing on a crescent as a sign of the Christian victory over the Turks. News of the victory spread quickly. Reports of the Christian triumph reached Venice on October 19, only twelve days after the battle, and the French court just ten days later. Christians celebrated their success throughout Catholic Europe with the ringing of bells and *Te Deum*s. It is said that the pope was so joyful he couldn't sleep. And the victory unleashed widespread hopes for a Beautiful Ending throughout Christendom, fueling Christian hopes for a Crusade that would restore the entire Mediterranean to Rome.[32] The pope even tried, though in vain, to convince Don Juan of Austria to take Istanbul itself.

Apocalyptic tensions were equally powerful in the western Mediterranean, both in Iberia and in North Africa. In Spain, Christian and Muslim apocalyptic hopes inspired much of the violence that broke out in the late sixteenth century between these two faiths. Christians feared that the Ottomans would exploit the local Morisco population as a fifth column within Spain to help in their drive to conquer the Mediterranean. And they placed their hopes in Philip II, who himself was deeply committed to doing all he could to overcome his religious adversaries—whether, as we have seen, Protestants in the Netherlands or Muslims in the Mediterranean—in order to prepare the world for the Last Days.[33] The Moriscos, who had remained loyal to their Islamic faith, also nourished hopes for the End Times, and, like the Christians, they not only drew on but shaped traditions that assured them of a Beautiful Ending. One of the most powerful Muslim prophecies, first written down in the second half of the fifteenth century, was in fact an Islamic version of a work by the medieval Franciscan prophet John of Rupescissa. As we saw earlier, John had played a particularly important role in shaping hopes for the Millennium among Christians. But in Spain, at least, Muslims would adopt John as one of their own prophets. For the Jews, John fueled messianic hopes, while the Muslim who wrote out an Islamic version of John's prophecy in *aljamiado* (a script that used Arabic letters to spell out words in Castilian), transformed the Turkish Antichrist into an Islamic hero, since his destruction of the Christian churches in Spain would make it possible for Islam to triumph.[34] But many other prophecies also assured the Moriscos who had remained loyal to Islam that their faith would prevail. Popular legends such as those of al-Khidr and of El moro Alfatimí, a figure who, at the End of Time, would lead the Moriscos against the Christians in their last apocalyptic battle, circulated widely among peasants and poor townspeople.[35] In 1568 such stories undoubtedly helped inspire the Moriscos of the Alpujarras who rose up in rebellion against their Spanish lords, convinced also that the Ottomans would lend them assistance. But Philip II was able—with his brother Don Juan of Austria in command—to quash the rebellion within two years.

Then, within a few years after the defeat of the Moriscos in the Alpujarras, a figure who called himself Abrahim Fatimí showed up in Cofrentes, a town in the mountains of Valencia, claiming to be a descendant of the Prophet and the Mahdī.[36] While not everyone in the village accepted his claims, many did. They believed Abrahim, whose Christian name was Joan Crespo, to be "a man of the other world" and "a God," capable of speaking with angels, making himself invisible, and levitating. But above all he captured the villagers' attention through his promise that he would lead them in rebellion against the Catholic monarchs, drive the Catholics from Spain, and transform Iberia into a Muslim peninsula. But Abrahim's mission—once the Spanish Inquisition became familiar with his plot—failed dramatically.

Finally, Christians and Muslims—again each group animated by messianic hopes—clashed at the Battle of Alcazarquivir, a town on Morocco's Atlantic coast, in 1578. At first, the conflict within Morocco had appeared to be a predominantly local matter, sparked by a succession crisis when Abd al-Malik, a brother of the deceased sultan, sought to defeat and kill his nephew Muhammad al-Mutawakkil, the sultan's eldest son who had succeeded his father Abdallah to the throne in 1574. To help counter this challenge Muhammad al-Mutawakkil invited the young and idealistic Portuguese king Sebastian to assist him. Yet Sebastian's goals were not to help Muhammad but rather to defeat the Moroccan ruling families and to bring Morocco to Christianity. This was a fantastic dream. Sebastian's advisors, the Portuguese nobility, and King Philip of Spain all had counseled against the expedition. But Sebastian was headstrong and determined to fulfil what he viewed as a Crusade, one aimed to help bring about the End of History. The battle proved a disaster. With their suits of armor and their expensive carriages, Sebastian's men were an absurd anomaly in the North African summer's sun, and nearly all of them died in the battle. Sebastian himself was hacked to death by the Moroccan troops.[37] But the other "kings" died as well. Muhammad al-Mutawakkil drowned in a nearby river when trying to escape the Portuguese assault. And Abd al-Malik, who had fallen ill after a meal the day before the battle, also died, though of natural causes. When the battle ended, it was Ahmad al-Mansur, Abd al-Malik's brother, who emerged as the sultan of Morocco, basing his power in no small part on his claim to be a descendant of the Prophet Muhammad. And in this same period, rumors began to circulate that Sebastian had not died but had become "the hidden one" who, many believed, would return as a Savior and fulfil the dream of creating a Millennium on earth.

In this period of violence, it was still possible to dream of unity. In the early 1570s Domenego di Lorenzo, an elderly Venetian cobbler who wore a long white beard, talked incessantly about his dream of religious unity in Europe and the Mediterranean. Recent events—the defeat of the Ottomans by the Christian Armada at Lepanto; the massacre of the Huguenots in Paris; the brutal repression of the Calvinists in the Low Countries—had convinced him and several other Venetian artisans that the End of History was at hand. Not only was the Catholic faith triumphant at this moment, Domenego and several of his fellow craftsmen had even fallen under the sway of a mysterious Venetian nobleman, or at least someone who claimed to be a Venetian nobleman, who persuaded them that he was the one for whom they had been waiting.

That such expectations and hopes for a Beautiful Ending found confirmation in the great Catholic victories of the early 1570s was perhaps inevitable. For a long time Domenego and others had gathered from time to time in the dark, cavernous Basilica San Marco where they, like many others before them, studied the story of the Apocalypse as it was depicted in the glittering mosaics. But they also drew inspiration from a mysterious manuscript. A handwritten text of a book had circulated among Domenego and his friends, containing, as a witness testified before the Venetian Tribunal of the Roman Inquisition, "prophecies that the Turks were to be defeated, that the Austrian Empire was to meet its ruin, that Constantinople was to be taken, and that the Turks were to convert to Christianity and that there was to be one sheepfold and one shepherd."[38]

Shortly after Lepanto in 1571 the carder Lunardo, a close follower of Domenego's, wrote to his father-in-law Zuanbattista in the Friuli that a particular man who had identified himself as a member of the noble household of the Priuli, one of the city's most prominent lineages, was "he who is to come." "He had told me all those things we are awaiting through Scripture. He has made me see the truth of it all," Lunardo wrote, adding:

> He showed me his portrait in the Basilica San Marco . . . and he has shown to me and to *messer* Benedetto it is he, and he has shown us his lineage through Holy Scripture. And look yourself in the *Song of Songs* where it says, "Buon cipro diletto mio," and the Gospel where it says, "This is my son who is my beloved (*diletto*)," and in the *Song* again where it says "Viene, diletto mio, viene capriolo mio" ["Come, my beloved; come, my gazelle"] "Viene" signifies Venice, and "capriolo" signifies Ca' Priuli, and he was born miraculously in the house of Priuli, the branch at Santo Stefano, as was his father, and his mother is from Cyprus, of the royal house of Cyprus . . . and the *Song of Songs* also tells us that he will be nourished among the lilies until the new day, and the lilies are the king of France, and he is a knight of the king, and has his rents in France, and thus he is being nourished in France until the new day, that is until dawn . . . which is Alba.[39]

The document is tantalizing, but one of countless such prophecies of this age that connected earlier prophetic texts and images with contemporary events. It was an ancient form of reading signs and deciphering the meaning of history—a practice that the mosaics in San Marco made especially accessible to artisans and craftsmen. But the group's expectations, especially following the St. Bartholomew's Day Massacre, would only intensify and they became more and more convinced that the Venetian "nobleman" was the one they expected. As Domenego di Lorenzo explained to the Inquisition in 1573, this leader would

> bring the whole world under one faith and accomplish the words of Christ who said: There will be one sheepfold and one shepherd. And this man is not Christ, but he is the sun which Malachi speaks of, for Christ is the light and gives light to the sun, that is, to the man who is the son of God and of the Church, it is he to whom the Psalm refers

where it is written, "Ask of me and I will give you your inheritance." This is not said of Christ, but of that captain general who will come to give fulfillment of the law of God, redeeming humanity and bring all to God's law. And this will be the Son of God and of the Church, which will give birth to him in great pain. It is like the death of a pope. The cardinals are pregnant and give birth to another. And similarly with the doge of Venice; when he dies, the *Signoria* gives birth to another. He is given birth to not as a child but as a grown man. And I believe that he is born as all of us are. And the heresies are the birth pains.[40]

What precisely the artisans believed about the "nobleman" is not entirely clear. Did they think he was a great political leader who would usher in the Last Days? Or did they see him as the Messiah? We don't know. What is clear is that they believed he would play a role in uniting all the peoples—Christians, Muslims, and Jews—into one sheepfold. A Messiah or an Elijah or an angelic pope could appear anywhere in sixteenth-century Europe, barely register on the historical record, and then disappear. We have no idea how many such figures there were. What is certain is that the great conflicts of the period encouraged a variety of apocalyptic and messianic dreams. And the dreams of peace, as we have seen, often fueled violence.

CHAPTER 9

The Spiritual Globe

> And this good news of the kingdom will be proclaimed throughout the world, as a testimony to all the nations; and then the end will come.
>
> —MATTHEW 24:14

As the fresh waters of the Orinoco approach the Atlantic along the northern coast of Venezuela they fan out into several tributaries that flow across a vast delta into the Atlantic and the Gulf of Paria, just to the west of Trinidad. Here, on August 1, 1498, on his third voyage to the New World, Columbus was mesmerized by these powerful waters and their turbulence as they crashed into the sea. "This is great evidence of the Earthly Paradise," Columbus wrote, adding, "for I have never read or heard of such a quantity of fresh water so mixed in with the salted sea." "I say," he continued, "that, if this river does not flow from the Earthly Paradise, it comes from a vast land to the south, of which, until now, there have been no reports. But I hold firmly in my soul that the Earthly Paradise, as I have said, lies there."[1]

Again the maps and cosmographies Columbus had studied mattered. Most of them had placed the earthly paradise in Asia. Pierre d'Ailly's *Imago mundi*, which Columbus had examined with special care, described the four rivers running from the earthly paradise—an account that matched perfectly with the Orinoco that Columbus encountered on his third voyage.[2]

It was far from surprising that Columbus at times interpreted the Caribbean in a spiritual key. After all, he was convinced that, in bringing Christianity to the Indies, he was accelerating the End of History, even calculating that the End would come relatively soon, in 1656.[3] And, on this front as well, the maps he had seen and studied reinforced this view. To be sure, the medieval *mappae mundi* had been largely schematic, merely portraying in the celebrated T-O design the disposition of the three continents framed by a circular Ocean. But more elaborate maps took it as their task to connect the earth's geography—God's creation—with the drama of biblical history. At some point in the late thirteenth century, for example, artists and scholars produced just such a map for the cathedral in Hereford, England. This *mappa mundi* focused in part on the history of humanity since Creation. It portrays the expulsion of Adam and Eve from the Garden of Eden, Exodus, the Tower of Babel, Noah's Ark, and the peopling of each of the three continents by the sons of Noah: Shem Asia; Japheth Europe; Ham Africa. But it also focused on the future, displaying the tribes of Gog and Magog, the peoples whom Satan would summon into battle at the End of Time. And at its summit the map reaches outside of time itself, with the promise of salvation.[4]

Similarly, in just the period that Columbus was first sailing, a scholar in the northern German city of Lübeck—we don't know who but possibly a physician who had once cared for pilgrims in Jerusalem—compiled a geography that not only offered detailed illustrations of many features of the earth but also a compelling overview of the earth's apocalyptic fate. Like his contemporary Columbus, this scholar was preoccupied by the growing power of Islam. Over the next century, as he demonstrated in a sequence of symbolic maps, Europe and indeed the world as a whole would fall prey to the growing, militant expansion of Islam, and the Antichrist would take possession of Jerusalem. But beginning in the year 1601, there would be a great reversal—this had been promised in scripture. "The Kingdom of Jesus Christ," the author writes, "will stand here in the world for 45 years and there will be in this time a beautiful ending with lasting peace and every good." And on the map displaying this, he wrote, "there will be one flock and one shepherd."[5] Columbus too held a similar dream. The gold he would gather and the alliance he would forge with the Great Khan would pave the way for a final crusade against the Infidel, and Christ would be restored to his Kingdom in Jerusalem.

Nor were such expectations confined to Christians. In the Islamic world, too, traditional cosmologies such as Ibn al-Wardi's late medieval *Pearls of Wonders and Singularity of Marvels* had also expressed geography in sacred terms. And the Ottoman admiral Piri Reis, even after incorporating knowledge of Columbus's "discoveries" into his celebrated map of the world in 1513, did not abandon this traditional view of cartography entirely. In the second edition of his *Book of Sea Lore*, completed in 1526, he labeled the Mediterranean the "Roman Sea" at a very moment in which Süleyman the Magnificent

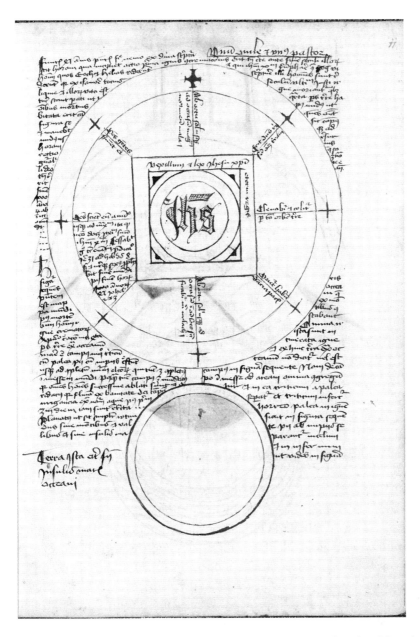

A Christian globe. This map, one of a series illustrating a manuscript from late fifteenth-century Germany, is a cartographic prophecy. After earlier maps that portrayed Christendom under siege first from Islam and then from the Antichrist, this one portrays a future when Christianity would prevail, preparing the way "under one flock and one shepherd" for the End of Time.

was intent on conquering Rome and establishing himself as world emperor, accelerating the End of History.⁶ Similarly, in his *Iggeret orhot 'olam* or *Epistle of the Paths of the World*, completed in 1525, Abraham Farissol, a Jewish scholar in Mantua, found reassurance in the Christian reports of their discoveries—especially when framed within Ptolemaic theory—that David ha-Reuveni's claims about the ten lost tribes were credible and that, therefore, the age of redemption was at hand.⁷

At the beginning of the fifteenth century, an ancient Greek text was translated into Latin and began to circulate in western Europe, at first in manuscript and then later in printed and even illustrated editions.⁸ To be sure, medieval scholars had always had an indirect knowledge of Ptolemy's *Geography*. But it was through the recovery and translation of this text into Latin that scholars in the late Middle Ages and the Renaissance, among them Pierre d'Ailly, first grasped the power of Ptolemy's method for calculating latitude and longitude as well as the value of his method for projecting the three-dimension spherical shape of the earth onto the two-dimensional surface of a map that he and his followers called a planisphere. The second printed edition of his text, published in Bologna in 1477, included such a map, one that made the power of the Ptolemaic projection evident. The reader was now gazing down, as if a god, on the surface of a world mapped with mathematical precision. Even if there were errors in Ptolemy's map, cartographers understood quickly that what mattered were not Ptolemy's particular claims but rather a model of the world that enabled them to continue to correct and refine their maps.

By its very nature, that is, Ptolemy's *Geography* made it possible both to display what was known and to invite continuous updatings, as reports of new discoveries poured into the ateliers of silversmiths, engravers, and printers who worked alongside the cartographers and cosmographers of the age. Moreover, in these displays, early modern men and women could trace the extraordinary speed with which their world picture was changing. Ptolemy's planisphere of 1477 portrayed the then three known continents and famously depicted the Indian Ocean as landlocked, unreachable from the Atlantic. Then, on the basis of Bartolomeu Dias's discovery of the Cape of Good Hope in 1488, Martin Behaim's *Erdapfel*—the first surviving terrestrial globe, produced in Nuremberg in 1492, just as Columbus was embarking on his first voyage—showed that it would be possible to reach the Indian Ocean by sailing south along the African coast. In 1507, inspired by Vespucci's *Letters*,

Martin Waldseemüller explicitly included a fourth continent, which he baptized "America" in his *Universalis cosmographia*. To be sure, the understanding of America and its relation to Asia was still not entirely clear. Franciscus Monachus's double-hemisphere map of the world, published in about 1527, conjoined North America to Asia.[9]

Not surprisingly, in such an environment maps and globes became objects of fascination, introducing a new but always evolving "world picture" to a society undergoing a period of rapid change and astonishing discoveries. More than any other objects of the early modern period, globes and planispheres became proxies of the idea that knowledge itself was always incomplete. Cartography was an emblematic science. "The geographers of this time do not fail to assert that now all has been found and seen," Montaigne wrote in the late sixteenth century, but then wondered, "if Ptolemy was formerly mistaken in the foundations of his reasoning, would it not be foolish for me now to trust what they say? Is it not likely that this great body which we call the earth is something other than we take it to be?"[10] Globes also became proxies of empire. The possession of these new artifacts seemed—at least to some—to promise the possibility of possessing much more. In his *Lusiads*, an enchanting epic celebrating Portuguese exploration, published in 1572, Luís de Camões has a sea goddess display a globe, "uniform, perfect," to the explorer and then tell him, "here I give you / the World for you to see / where you are going and will go / and what you desire"—a clear linkage of the contemplation of the globe to the dream of empire.[11]

But the keen interest in globes and planispheres also in no way disrupted the idea of a map as an apocalyptic agent, even if the gradual and complex recovery of Ptolemy played a role in transforming the rhetoric of apocalypticism. For unlike medieval *mappae mundi*, the maps inspired by Ptolemy not only displayed what was known but also provided an orientation to the unknown, opening up a prophetic space beyond what had been identified in scripture alone, while at the same time making it possible for readers to contemplate humanity—even in all its diversity—from the eye of God. This new, prophetic space would engage the religious imagination of many, as salvation moved to a planetary scale and fostered a long-standing desire to bring the entire world and all its peoples under one faith.

In the age of religious wars, apocalyptic hopes for a better world were especially palpable in the Spanish Netherlands, where mapmakers as diverse as Gerardus Mercator and Abraham Ortelius sought to find in their representations of the world ways of overcoming the confessional violence of the time. To be sure, both cartographers were, in many respects, patently imperialistic. The frontispiece to Ortelius's *Theatrum orbis terrarum* (Theater of the Orb of the World), first published in 1570, portrayed Europe as a majestic female, enthroned between two globes, one terrestrial and the other celestial, and seated atop an arcade. To the left stood the figure of Asia wearing silk and bearing in her hands an urn of myrrh; to the right the figure of Africa, holding a few sprigs of balsam. By contrast, America was portrayed as reclining and eroticized, holding a spear in one hand and the severed head of a European male in the other—a clear allusion to cannibalism—while to her right, not yet fully discovered, was the emerging figure of Australia, at this moment still barely known to European explorers.[12] Moreover, Ortelius dedicated the work to Philip II, "monarch of the greatest empire of all times and of the whole earth." And in his *Atlas,* modeled in many respects on Ortelius's and published immediately after his own death, Mercator wrote: "Here [in Europe] wee have the right of Lawes, the dignity of the Christian religion, the forces of Armes. . . . Moreover, Europe manageth all Arts and sciences with such dexterity, that for the invention of manie things she may be truly called a Mother . . . she hath all manner of learning, whereas other Countries are all of them, overspread with Barbarism."[13]

Yet both Ortelius and Mercator were equally if not more insistent on a global imaginary that, ultimately, would overcome the political and religious divisions of their time. Ortelius made this plain first through the ways in which he framed the planisphere, the map with which the *Theatrum* opens. Ortelius drew on Pliny's *Natural History* to undercut the very desire for empire. The earth, grasped from the vantage point of the cosmos and eternity, is ultimately insignificant, a mere "pin-prick," where men "fill our positions of power and covet wealth, and throw mankind into an uproar, and launch even civil wars and slaughter one another to make the land more spacious."[14] And in his later edition—perhaps rendered even more pessimistic by the horrific religious violence unfolding in the Netherlands—Ortelius added another cartouche, this one citing Seneca, that makes much the same point: "Is this that pin-prick that is divided by sword and fire among so many nations? How ridiculous are the boundaries of mortals."[15] But Ortelius also offered clues to the careful reader that the world he was describing, with its varied peoples scattered around the globe, was also deeply united. On his map of Tartary, for example, he presented the tribes of Israel, who had stopped there prior to spreading out to the rest of the world. In short, all peoples in his view were

Allegory of the five continents. The frontispiece to Abraham Ortelius's *Theatrum orbis terrarum* (1570).

united—a vision that the structure of the atlas reinforced. Ortelius had taken great pains not only to collect the best maps he could find, but also to make them uniform, ultimately rendering the great political and religious divisions insignificant.

Nor was Ortelius alone in this view. In his home city of Antwerp in just these years, the Family of Love, a loose network of radical religious thinkers active in Flanders and Brabant, stressed the interior piety of the believer over the external acts of faith.[16] In a letter he wrote to Ortelius in 1579, the polymath Guillaume Postel, himself a cosmographer and prophet, praised Ortelius's *Theatrum orbis terrarum* as the most important book since the Bible. Divine providence, Postel noted, had established a Beautiful Ending for all humanity. The entire world would be inhabited, as though in the house of Adam, with one sole father and mother for all and with the poor placing their hopes in the name of God and Our Lord Jesus Christ.[17] And Ortelius's work, Postel continued, was the work of Christ. To envision the entirety of humanity was to see a potential harmony of all men and women, of all the faiths, gathered finally into the fullness of a Christian Millennium. The *Theatrum* opened a window onto a world through which the reader could move easily, a harmonious world in which the New Jerusalem would not be found in a particular place but rather in the hearts of the pious everywhere.[18]

For Mercator, it was a darker vision of the End of History that animated his efforts to offer a new cartography. In his correspondence, he wrote frequently of "the End" and "the old age of the world."[19] Convinced that the world would soon face its tribulations, Mercator set out in the 1560s to create a cosmology that would not only map the earth with stunning accuracy, as in his celebrated world map of 1569, but also place the earth within a vast cosmological frame that took both the heavens and time into account. His researches culminated in his *Atlas* of 1594, published a year after his death.[20] While the work was not as complete as he had hoped, Mercator offered a vision of harmony in much the same way that Ortelius had done. As Mercator wrote in his introduction: "For, while we treat of cosmography, we construct it in such a way that from the marvelous harmony of all things towards God's sole end, and in the unfathomable providence in their composition, God's wisdom will be seen to be infinite, and his goodness inexhaustible."[21]

For missionaries in this age, the new cartography opened up a vision of bringing more and more peoples into the faith. In his *Spiritual Exercises,* Ignatius of Loyola placed himself—through an imaginative leap inspired by the

Ptolemaic turn—outside this world, "gazing down upon the whole expanse or circuit of all the earth," where, he continued, "some are white, some black; some at peace, and some at war, some weeping, some laughing; some well, some sick; some coming into the world, and some dying."[22] Jesuits would continue to find the new Apollonian perspective on the earth inspirational in their evangelizing. In 1580, when the first Jesuit mission arrived in the Mughal court, they made a gift to Emperor Akbar of a geography, in all likelihood either Ortelius's *Theatrum orbis terrarum* or Mercator's edition of Ptolemy's *Geography*.[23] And in 1584, in China, the Jesuit Matteo Ricci and his fellow missionaries had displayed a copy of Ortelius's map of the world on the walls of their residence. When Ricci noticed how fascinated his Chinese visitors were with the map, he not only took pains to explain its various cartological features to them, but he produced a Chinese version. He translated Ortelius's Latin into Mandarin, added brief descriptions of the traditions and customs of the world's regions, and even placed China at the center of the world. Local visitors to the residence were impressed, and some even produced printed versions.[24]

The globe had become not only a transmitter of geographical knowledge but also of faith. And, overall, the new cartography, by definition always open-ended, provided a visual representation of a newly revealed world whose peoples Christians were called upon to convert. Deeply impressed by what he saw as the success of the Jesuit missions, Joannes Fredericus Lumnius, a theologian and the pastor of the Begijnhof in Antwerp, warned his parishioners and readers that the End was near.[25] But not everyone believed the End was imminent. Pushing back against the confidence of those who maintained that the Christian missions had reached the majority of the world's peoples, the Jesuit José de Acosta cautioned in his *De temporibus novissimis*, or *On the Most Recent Times*, published in Rome in 1590, that it was impossible to know when the End was coming. His own knowledge of the globe convinced him of this, since, as he observed in this treatise, it was by no means clear that the missionaries had reached everyone in the places that had been discovered, much less that the discoveries were complete.[26]

The new cartography also bolstered dreams of religious unity in the Islamic world. In the Ottoman court, the most striking example of interest in global knowledge was Mehmed el-Su'udi's *The New Hadith*, a work later known as *The History of India in the West*, composed during the reign of Sultan Murad III in the late sixteenth century.[27] Drawing on astronomical accounts of the heavens and the earth, including Ptolemy's *Geography*, the work rejected the narrowly circumscribed world of Ibn al-Wardi and called upon the sultan to seize the territories taken by the Portuguese in the east and by the Spanish in the New World.[28] Were the commander of the Ottoman forces to have a soul "full of the zeal of Islam," the compiler wrote, "in a short period he would seize strongholds and conquer lands [seized by the Portuguese], and he would expel and eliminate the base unbelievers."[29] The manuscript continued: "By the Lord God, we always hope that that advantageous land will, in time, fall [in] conquest to the brave of Islam of exalted lineage, and that it will be filled with the rites of Islam and be joined to the other Ottoman lands."[30] But it is not only the conquests that drew Su'udi's attention. He wrote also of the grandeur and beauty of the cities the conquistadors had discovered, of the rich resources of the new continent: not only of gold and silver but also of its exotic animals and plants. Clearly this was a world worth possessing. "It is requested of His Glorious Majestic Excellency that in the future the bloodthirsty sword of the people of Islam reach that advantageous land and that its regions and districts be filled with the light of the religious ceremonies of Islam and that the possessions and good that have been mentioned and the other treasures of the unbelievers marked by disgrace be divided, with the permission of the Lord God, among the masters of the Holy War and the nations full of driving force."[31] Moreover, in one copy of the manuscript, the compiler gave expression to his apocalyptic hope. As he wrote in his dedication to the sultan, "May God the Sublime make the days of his powerful glory endure; and may He make eternal the years of his prominent grandeur to the arrival of the Hour, the Hour of Resurrection."[32]

In a similar fashion, the new cartography encouraged dreams of the End within the Sa'dian dynasty in Morocco. In the early seventeenth-century Ahmad ibn Qāsim al-Hajarī, a Morisco who had immigrated to this North African kingdom in the early seventeenth century, traveled to both France and the Netherlands, visiting not only Paris and Amsterdam but also Bordeaux and Toulouse with the goal of recovering property that had been stolen from Moriscos during their recent expulsion from Spain. But his activities were not only aimed at securing the release of the stolen property. A man of insatiable curiosity, he entered into several conversations about his religion with both Christians and Jews. Al-Hajarī also offered a deeply Islamic interpretation of the End of History. In his study of the Book of Daniel, he identified the world's final empire with Islam. Drawing on the teachings of the Sufi Ahmad Zarruq, al-Hajarī noted that Muhammad "was sent in the last period of time and he is the one whose prophethood and the sovereignty of whose nation will remain forever, until the Hour of Resurrection. ... He intermingled all the races and made them, notwithstanding their differences in religions and the differences of their languages, into one nation, with one language and one religion."[33] But al-Hajarī did not only draw on religious teachings to prove the superiority of Islam. He also drew on the new geographical knowledge to demonstrate that Islam, as he writes, "has spread in most of the countries of the world." Al-Hajarī notes that in his time, "people are better informed about things of the world than the ancients were."[34] And he makes it plain that he has studied the maps and globes of the Christians and read many of their accounts through which it became evident to him that "most of the inhabited world belongs to the Muslims and they are the ones whose sovereignty will not be changed until the end of the world," once again reinforcing his interpretation that the religious order that the Prophet Muhammad had established would be the last until the final hour.

Jews, too, brought the new cartography to bear on their dreams of redemption. This was particularly the case in Italy, where the Jewish communities were especially aware of the new discoveries. In the 1550s the Genoese physician Joseph ha-Kohen, whose parents had been expelled from Spain at the end of the fifteenth century, wrote about the discoveries first in his *Divre ha-yamin le-malke Tsarefet u-malke bet Otoman ha-Togar* (History of the Kings of France and the Ottoman Turks), though rather briefly and without much accuracy.[35] But ha-Kohen's interest in the New World was most evident in a manuscript that he produced in 1557 and that he copied out in his own hand at least nine times.[36] Less an original work than a collection of his translations of both Johannes Böhm's early sixteenth-century *Omnium gentium mores,* or *The Customs of All Peoples,* and Francisco López de Gómara's writings on the Spanish conquests, ha-Kohen's work simultaneously criticized Spain for its excessive brutality in the New World and sought to establish the view that, by all rights, the new lands should be in Jewish, not Iberian hands. The discoveries of Columbus as well as the new cartography, which he studied in López de Gómara, had not so much disclosed a New World to Europeans, he suggested, as fulfilled the divine knowledge of the world as it was already known in the Bible. And, on this basis, ha-Kohen argued, these lands belonged more properly to the Jews. Throughout, ha-Kohen's central purpose in this section of his manuscript was to underscore divine providence. Like his Christian and Muslim counterparts, he equated the discoveries with a conviction that the messianic age was dawning.[37]

In an age awash in thousands and thousands of maps, Ortelius's *Theatrum* proved especially successful, even electrifying. In it readers could not only gaze, from a bird's-eye view, down upon the world as a whole but also examine each of the world's continents—Africa, Asia, Europe, and America—along with many, delightfully-detailed regional maps. Ortelius's *Theatrum* was translated into Dutch, English, French, German, Italian, and Spanish.[38] Mercator's *Atlas,* by contrast, never found a similarly broad readership, even if it would be his projection, breaking with Ptolemy's, that would do the most to shape cartography in the following centuries. But hopes for a more peaceful world had inspired them both. And Ortelius, in particular, inspired hopes that it would be possible to bring the world under one faith. Perhaps it was inevitable then that a missionary in China in the early seventeenth century would propose a Jesuit atlas. Modeled on Ortelius, it would be called the *Theatrum orbis ecclesiastici,* or *The Theater of the Ecclesiastical World.* Its purpose would be to show the progress of the missionaries in bringing Christianity to the various peoples of the world.[39] The work was never published, but the conception behind it made it clear that the very nature of religious harmony was far from settled. The Family of Love and Ortelius may have dreamt of a cosmopolitan peace, but Christians as well as Muslims and Jews remained convinced that redemption required the triumph of their respective faiths. The spiritual globe not only united, it also divided the world.

Just as the seventeenth century opened, a brilliant, fiery Dominican continued—though now confined in a small cell in one of Naples's most notorious prisons—to dream of a better world. Tommaso Campanella, the friar, had previously faced imprisonment for his heretical writings. But in 1599 it was his actions that posed the greatest concerns for the authorities. In that year Campanella had begun preaching the Apocalypse to peasants and artisans in the towns and villages of Calabria, the region in southern Italy where Joachim of Fiore had first elaborated his vision of sacred history. There, inspired by Joachim and other medieval prophets, as well as by his consultations with astrologers that all the signs of the times pointed to the Second Coming, Campanella even persuaded his fellow friars "that in the year 1600 . . . there had to take place a great revolution and a mutation of the state, especially in the Kingdom of Naples and Calabria; and for that reason they must prepare themselves, and gather armed men."[40] The immediate goal was to overthrow the King and to establish what came to be called "the Republic of the Apocalypse." There they would wait out the turmoil that was engulfing Europe. At the same time they would share in the community of goods until the Second Coming. And in this millenarian vision all of humanity would be converted to Christianity. The testimony of a local monk before the Holy Office in Naples portrayed Campanella as an atheist whose preaching not only cast the pope as Antichrist but also called into question the sacraments, doctrines, and moral teachings of the Church.[41]

On September 6, 1599, royal troops plucked Campanella from his hiding place near the hilltop town of Stilo, where he had been born into a humble family in 1568, his father having been a shoemaker. The Spanish forces rounded up a total of 158 rebels and loaded them, Campanella included, onto a ship bound for Naples, executing several as soon as they boarded. Campanella, who had already been convicted by the Roman Inquisition in 1594 of heresy, would have certainly been sentenced to death had he not persuaded his captors that he was mad. After his trial, which dragged on for two years and during which he was brutally tortured, he was condemned to perpetual prison. There he would spend twenty-seven years. After the authorities released him, now an old man, he moved to France where he spent his last years.[42]

Prison did not break Campanella's spirits nor hopes for a Beautiful Ending, but his ideas did evolve or at least seem to evolve. In prison Campanella appears to have renounced any hint of the political and antipapal radicalism that he had expressed, or been said to have expressed, during the revolt in Calabria. And this shift appears most evident in his growing support for both the papacy and the Spanish monarch. Yet, it is unlikely that the shift was as abrupt as it at first appears. Even before the Calabrian conspiracy, Campanella had closely identified the coming of the Millennium with the Spanish kings. "The monarchy of Spain which embraces all nations and encircles the world," he had already observed in his *Monarchia di Spagna* of 1598, "is that of the Messiah, and thus shows itself to be the heir of the universe."[43] Discerning a deep connection between the political and divine order, Campanella hoped for nothing less than the institution of universal monarchy under Spanish rule. The perfect state could only exist, he argued, "under the most perfect law of Jesus God," adding, "in whom we pray that the kingdom come in which the will of God will be done on earth as it is in heaven . . . and there will be in the world but one flock and one shepherd, and then we shall see the Golden Age the poets sang and the best of the republics as described by the philosophers and the state of innocence of the patriarchs and the happiness of Jerusalem liberated from the hands of heretics and infidels."[44] But it is in his *La monarchia del Messia* or the *Monarchy of the Messiah*, written when he was in prison, that his view of the Millennium is the clearest, the most compelling. Like many millenarian thinkers, Campanella was both distressed by the tribulations of the present and hopeful for delivery. "All the troubles of this world are born either from war, or from plague, or from famine, or from beliefs contrary to the native religion," he observed in his text.[45] But then, he continued, "there shall be no more famine . . . and no more plague . . . and, through the abundance of peace, men will have so much greater knowledge that they will provide for all."[46] At the same time, the growth of Spain's global navigation would contribute significantly to the flourishing of the sciences. "The discovery of the New World," Campanella wrote, "has given birth to marvelous knowledge among us."[47]

Roughly in this same period Campanella writes a utopian text—*The City of the Sun*—first drafted in 1602 but reworked significantly later in his life.[48] The work—a dialogue between a Genoese ship captain, who had sailed with Columbus, and a Knight Hospitaller—shared Thomas More's conceit of a sailor who had happened upon a perfect society on a distant island in this age of discovery, in this case a city on the island of Taprobana (Sri Lanka) in the Indian Ocean. From the very beginning of the work, it is the layout of the city, designed along astronomical principles, that engages the reader's attention. Over the altar located in the middle of the circular temple, which stood in the very center of the City of the Sun—so the Genoese sea captain tells the Grandmaster of the Knights Hospitaller—there are two globes: "one quite large with the heavens painted upon it and the other which represents the earth."[49] Then, reaching out from this center point, the city was composed of seven concentric circles or walls—with each space associated with a particular celestial body: in the center the Sun, then the planets Mercury, Venus, Earth, Mars, Jupiter, and Saturn, beyond which lay the outer circle of the Stars. And on each of the walls artists had painted different areas of knowledge: astronomy, mathematics, geography, mineralogy, fluid dynamics, botany, marine biology, ornithology, herpetology, entomology, and zoology. But the Genoese ship captain was most impressed by the final outer wall. There, he states, "I found Moses, Osiris, Jupiter, Mercury, Muhammad, and many others. In a special place of honor I saw Jesus Christ and the twelve Apostles, whom they hold in great regard. I saw Caesar, Alexander, Pyrrhus, and all the Romans. At this, when I marveled that they knew the histories of these men," the captain continued, "they explained to me that they understood the languages of all the nations and that they dispatched ambassadors throughout the world to learn what was both good and bad in each of them."[50]

The government, like the structure of the city, provided for order and regularity. Ruled by a *principe sacerdote* (a prince-prelate) who was called "Sole" or "Metafisico" and who was assisted by three officials—Pon or Power, who oversaw all the military affairs of the State; Sin or Wisdom, who was responsible for the arts and sciences; and Mor or Love, who brought the inhabitants together to bring forth children—the state Campanella envisioned was, in a political sense, deeply hierarchical, a kind of theocracy, though, as he would frequently make clear in the dialogue, the individuals fulfilling these roles were chosen not on the basis of their birth but rather on the basis of their merit, knowledge, and judgment.

Moreover, like the Utopians, the Solarians—as the inhabitants of the city the Genoese ship captain had encountered were called—had no private property but rather held everything, including wives, in common, with the dispensation of goods, as in a communist system, left in the hands of the officers: "not only food but also the sciences, honors, and entertainments are shared in common in such a way that no one can appropriate anything for himself alone."[51] Property, in Campanella's view, was the outgrowth of a social system in which men created individual households for themselves; and these households, in turn, led to *amor proprio* or pride, a tendency to live in a constant state of comparing oneself to others. And this pride was the root of other sins: rapaciousness and greed. In abolishing private property—again in both material possessions and in women—the Solarians destroyed *amor proprio* and put the common good above the individual person or family. Furthermore, by holding property in common, the Solarians showed great love for their country and to one another, for they forged genuine friendships and helped each other in myriad ways. The City of the Sun was a republic of virtue.

In More's *Utopia* there had been no community of wives. Campanella's *City of the Sun*, by contrast, made this the lynchpin of a world in which sexual relations themselves were regulated in a way that dramatically broke from European customs at that time—radically deprivatizing the family and making reproduction itself a common enterprise. In the City of the Sun, it was the elders, who, through their familiarity with young people, decided who should mate with whom. Love, who oversaw these decisions, aimed to ensure that the offspring would be as healthy as possible and tied the mating to the most propitious moments disclosed by the planets. Moreover, the elders paid careful attention to who was ready for intercourse, which was to be strictly heterosexual, and who was not, pairing the young people off in order to produce the best possible offspring. "And tall and beautiful young women should only sleep with tall and courageous young men; and those who are fat should only sleep with those who are thin; and those who are thin only with those who are fat, in order to strike the best balance."[52] Indeed, in general, the Solarians made a special effort to determine who should mate and with whom, aiming "to improve the natural endowments, not to provide dowries or false titles of nobility."[53]

An elaborate ritual accompanies the mating rituals of the Solarians.

> In the evening the young people come and make up their beds, as the master and mistress command. But they do not have sex until after they have digested their dinner and said their prayers. And they have beautiful statues of illustrious men upon which the women gaze. And then they come to their windows and pray to God in heaven that he grant them fine offspring. They sleep in two cells, separately, until the mistress comes and opens the passages between the cells and they have intercourse. The time is determined by the Astrologer and the Physician, and they force themselves always to take their time until Mars and Venus are to the east of the Sun in a favorable House and until they are gazed upon by Jupiter with a favorable aspect and also by Saturn and Mars.[54]

When children were born, great care was taken to ensure they were well nursed, and then educated from a very young age. The outcome, the Solarians believed, would again benefit the common good. "But, for the most part, those of the same generation, having been born under the same constellation, are similar to one another in virtue, habits, and custom. And this makes for a stable harmony in the republic, and they greatly love and help one another."[55]

Finally, just as the Solarians followed the law of nature in procreating, they did the same in religion. Sole served as their high priest. In this role, he served to expiate the sins of the Solarians, but he also consulted with priests who study the stars, noting "all their movements and the effects these produced." Using these observations, "they establish the hour in which conception should take place, the day on which sowing and harvesting should be done, and in general, serve as mediators between God and man."[56] The priests and the people alike prayed and joined together for worship, which was often festive in nature. Most strikingly, as the captain makes clear, their beliefs, while based on nature, coincided almost perfectly with those of Christianity. They believe in one God, in angels, in a kind of heaven and hell, and even view the three officials—Pon, Sin, and Mor—as a kind of Trinity.[57] And even their study of astronomy leads them toward religious truth. "These people," the captain reports, "are very interested" in astronomy and study it "very closely, for it is important to know how the world is constructed, whether it will end and, if so, when; what the stars are made of, and who inhabits them." Then, remarkably, the captain adds, "they believe that what Christ said about the signs from the stars, the sun, and the moon is true, though fools deny it; but the end of things will come upon them like a thief in the night. Therefore, these people await the renewal of the world or perhaps its end."[58]

As the dialogue draws to a close, the Hospitaller emphasizes how close the religion of the Solarians is to Christianity which, he says, "adds nothing but the sacraments to the law of nature. I draw from this report," he concludes, "that the true faith is the Christian one, and that once it has been ridden of its abuses, it will be the mistress of the world. And even though your fellow Genoese Columbus was the first to discover the rest of the world, it was effectively the Spanish who used this discovery in order to unite the world under one faith."[59] And the sea captain, in agreement, notes that the Solarians themselves draw a similar hope from both astrology and recent developments across the globe. They viewed the inventions of the compass, the printing press, and the arquebus "as signs of the imminent union of the world." They add that the astrological signs point to "a great new monarchy, reformation of the laws and of arts, new prophets, and a general renewal."[60] And other signs as well—the proliferation of women monarchs and rulers, the spread of Christianity and Islam throughout the world, and the growth of heresies, along with the movements of the planets and the stars—point to transformation and renewal of the world.[61]

In his damp, dark, and cramped prison cell in Naples, Campanella had moved his hopes for a great renewal and for the End of History from his overt revolutionary activities in the hills of Calabria of only a few years earlier to a work of his imagination: a utopia that drew explicitly on apocalyptic language and ideas. He drew also on astrology, a science that many believed supplemented scripture, as men and women sought to make sense of the future. But Campanella's insistence that natural law itself made it clear that a Beautiful Ending was on the horizon was deeply original and, along the way, it underscored his hope for the imminent success of Christianity in unifying the globe. As a subject of the Spanish crown and in a world in which the new cartography made it increasingly possible to visualize in its entirety, there was much to suggest that his dream was close to being realized. It had been only two years after the Battle of Alcazarquivir, after all, that the Spanish monarchy had absorbed Portugal. Philip's empire now encircled the globe, with territories reaching out from Europe, where the Habsburgs ruled not only Aragón, Castile, the Balearic Islands, Malta, and Portugal but also Naples, Milan, and the Low Countries, to the New World, with dependencies in the Caribbean, Mexico, Peru, and Brazil, to parts of Africa, Angola, Ethiopia, a significant stretch of the Swahili coast, and the lower Zambezi Valley, and even to the Indian Ocean, where Spain had inherited a foothold in Sri Lanka as well as the territory around Sumatra.[62] Undoubtedly this global empire contributed to Campanella's vision of the end. In his *Monarchia di Spagna*, he had expressed hope to hispanize the world ("spagnolare il mondo").[63] But above all he looked forward to its Christianization. All Jews, Muslims, and others—and this too was a promise of *The City of the Sun*—would be united in the Christian faith, human nature would be perfected, and a golden age would dawn.

CHAPTER 10

Cannibals

Our very image of the world glides away whilst we live upon it.
—MICHEL DE MONTAIGNE, "On Coaches" (1587)

"Those of the Kingdom of Mexico were somewhat more civilized and skilled in the arts than the other nations over there," Montaigne wrote in his essay "On Coaches." "Thus, they judged as we do, that the universe was near its end, and they took as a sign of this the desolation that we brought there," he continued, before offering a concise picture of Aztec sacred history:

> They believed that the existence of the world was divided into five ages and into the life of five consecutive suns, of which four had already run their time, and the one that was then illuminating them was the fifth. The first perished with all other creatures by a universal flood. The second, by the falling of the heavens on us, which suffocated every living thing: to which age they assign the giants, and they showed the Spaniards some of their bones, judging by the size of which these men must have stood twenty-feet high. The third, by fire, which burned and consumed everything. The fourth, by a disturbance of air and wind which beat down several mountains; the men did not die, but they were changed into baboons (to what notions will the laxness of human credulity not submit!). After the death of this fourth sun, the world was twenty-five years in perpetual darkness, in the fifteenth of which a man and a woman were created who remade the human race; ten years, later, on a certain day of their calendar, the sun appeared

newly created, and since then they reckon their years from that day. The third day after its creation the old gods died; the new ones have been born from day to day. What they think about the manner in which this last sun will perish, my author [here Montaigne is referring to Francisco López de Gómara, the historian of Cortéz's conquest of Mexico] did not learn. But their calculations of this fourth change coincided with that great conjunction of stars which produced, some eight hundred years ago, according to the reckoning of the astrologers, many great alterations and innovations in the world.[1]

By the time Montaigne composed this essay in the 1580s, it had become clear to many that European Christians were far from alone in believing that the End Times were at hand. López de Gómara, upon whose *General History of the Indies* Montaigne was drawing, was not only a bestseller but only one of several chroniclers of the New World to offer a description of the Aztec myth of the suns.[2] But Montaigne's treatment of the Aztec cosmology was far more than an ethnographic account. In recounting this story of this New World myth, Montaigne offered a reframing of apocalypticism. The widespread hopes for a great alteration or for a great innovation that were so central to hopes for a Second Coming, if not the Millennium, in Europe in just this period were—Montaigne made this clear—far from exclusively Christian beliefs. To the contrary, they appeared to be universal—an insight that did not validate them but rather called the sacred history of Christianity itself into question.

To be sure, Montaigne did acknowledge that divine providence played a role in human affairs, but he viewed Providence as a quiet force and rejected the efforts of prophets and prognosticators to predict the future.[3] In his essay "On Prognostications," for instance, he expressed his approval of the fact that Christianity had abolished ancient forms of divination based variously on the study of the entrails of sacrificed animals, the patterns of bird flight, or the course of rivers, while noting his discomfort that "there remains among us some means of divination by the stars, by spirits, by bodily traits, by dreams, and the like—a notable example of the fervent curiosity of our nature, which takes delight in worrying about future things."[4] Montaigne conceded that occasionally a prognostication or prophecy appeared to come true, but this was, in his view, pure chance. "For who is there," he asked, quoting Cicero, "who, shooting all day, will not hit the mark?"[5] But most often such prognostications were wrong. This was certainly the case with many of the almanacs that circulated so widely in France at this time. And he even suggested that the prophecies of no less a figure than Joachim of Fiore, still venerated by many of his contemporaries, also frequently missed the mark.[6] Even the discovery of the New World itself had undermined early prophecies. "Our world has just discovered another world," Montaigne wrote, but then immediately asked, "and who will guarantee us that it is the last of its brothers, since the daemons, the Sibyls, and we ourselves have been ignorant of this one up until now?"[7] The New World undermined certainties. A central thread in Montaigne's *Essays* was his insistence that our knowledge and understanding are always at best provisional and partial. In the end, he would reject the views of his contemporaries who believed that they had a monopoly on the truth—a recurrent feature of many of the millennial hopes of this period. Ultimately, writing from within the context of the Wars of Religion and the carnage they had unleashed in France, Montaigne would reject the view that Europeans were in any sense superior to the peoples of the New World.

It is rare that historians can see a new word come into being as clearly as we can in the case of the term *cannibal*. But in this case, the survival of Columbus's journal of his first voyage—again we do not possess the original but rather Las Casas's copy—enables us to watch the word *cannibal* take shape in the midst of Columbus's fears, thoughts, and vacillations as he tried to make sense of a world he was encountering for the first time.

Columbus's first recorded allusion to what he came to believe to be man-eating natives came within a month of his arrival in the Caribbean.[8] In an exchange with several of the Taínos on November 4, 1492, he came to understand from what he believed they were trying to tell him—though here he was drawing primarily on ancient and medieval fantasies of monsters at the edge of the world—"that, far from there, there were one-eyed men, and others, with snouts of dogs, who eat men, and that, when they seized someone, they would slit his throat and drink his blood and cut off his genitals."[9] A few weeks later, on November 23, he gave these alleged man-eaters the name "cannibals." As Columbus reports, the Taíno captives that he was bringing with him as he sailed south told him that one of the islands was inhabited by the Caribs, warning him "that it was very large and that there were people on it who had one eye in their foreheads, and others whom they called cannibals, of whom they showed great fear."[10] How precisely Columbus moved from *Caribs* to *cannibals* rather than to *caribals* is a not clear. Perhaps this verbal slippage occurred because he believed he was close to China and, therefore, assumed that the Caribs were subjects of the great Khan of China and heard "Canibs" rather than "Caribs"—a hypothesis his entry of December 11 seems to support. "The Caniba," Columbus writes, "is nothing else but the people of the Great Khan, who must be here very close to this place. And they have ships and come to capture the people here, and since they do not return they think that those who have been captured have been eaten."[11] At first Columbus was not convinced. "They showed [us] two men who were missing some bits of flesh from their bodies, and they gave the Spaniards to understand that the cannibals had taken bites out of them," he wrote in his logbook, though Las Casas added, "the Admiral did not believe it."[12] Yet, within a month, Columbus seemed prepared to accept the reality of cannibalism; his entry of January 13, 1493, reads: "they must be a daring people since they travel through all these islands and eat the people they can capture."[13]

Ultimately it is this final view of Columbus—namely that the Caribs are man-eaters—that will reach the public. In his famous letter to Luis de Santángel of February 1493, Columbus reported: "Thus I have found no monsters, nor had a report of any, except in an island 'Carib,' which is the second upon entering the Indies, and which is inhabited by people who are regarded in all the islands as very fierce and who eat human flesh."[14] This letter would circulate immediately throughout much of Europe, appearing first in Spanish in Barcelona and then in Latin in Rome, with further Latin editions in Antwerp, Basel, and Paris. But it is Columbus's letter of 1494, about his second voyage—a text that would find even wider circulation—that proves responsible for the introduction of the word *cannibal* into the western languages. Columbus wrote, "and since among the other islands those of the cannibals are numerous, large, and extremely well populated, it would seem that to seize some of the men and women here and send them to Castile would not be without benefit, since they would come to abandon this inhuman custom of eating men."[15] Here Columbus was not simply reporting on cannibals. Rather he had already begun to use this newfound category as a justification for the enslavement of the indigenous peoples of the New World.

It is remarkable how quickly this word takes root and begins to circulate in Europe. We find inhabitants or certain inhabitants of the New World characterized as cannibals in some of the very first texts to report on the Americas. Amerigo Vespucci—the Florentine after whom the New World received its name—wrote of cannibals in his *Mundus novus* of 1504. Describing the inhabitants of northeast Brazil whom he had encountered on his voyage, Vespucci reported, "they eat little meat, except for human flesh: for Your

Magnificence must know that in this day they are so inhuman that they surpass all bestial ways, since they eat all the enemies that they kill or capture, female as well as male, with such ferocity that merely to speak of it seems a brute thing—how much more to see it, as befell me countless times, in many places. And they marveled to hear us say that we do not eat our enemies, and this Your Magnificence should believe for certain: their other barbarous customs are so many that speech fails to describe such facts."[16] A later text attributed to Vespucci also underscored the prevalence of cannibalism. "This you may be sure of, because one father was known to have eaten his children and wife, and I myself knew a man (to whom I spoke), about whom it was spread about that he had eaten more than three hundred human bodies, and I also stayed twenty-seven days in a certain city in which I saw salted human flesh hanging from house-beams, much as we hang up bacon and pork. I will say more: they marvel that we do not eat our enemies and use their flesh, which they say is very tasty, as food."[17] And the Italian scholar Peter Martyr of Anghiera, who drew on the reports of sailors in his *De orbe novo* of 1510, also wrote of cannibals, contributing to the increasingly widespread view among Europeans that the inhabitants of the New World were man-eaters, the lowest form of barbarians, savages—in short, to the fashioning of the myth of primitivism and barbarity in the Americas that would play an important role in legitimating the conquest and exploitation of the New World by the Old.[18] In 1503, Queen Isabella made an exception to an earlier decree in which she forbade the enslavement of the natives in the New World. That exception was the cannibal. For, "if such Cannibals," her new edict pronounced, "continued to resist and do not wish to admit and receive my Captains and men who may be on such voyages by my orders nor to hear them in order to be taught our Sacred Catholic Faith and to be in my service and obedience, they may be captured and are to be taken to these my Kingdoms and Domain and to other parts and places and be sold."[19]

When Montaigne gave his essay the title "Des cannibales" ("Of Cannibals"), he was first and foremost making use of a modern term. The classical term—one Montaigne would have known from such ancient authors as Herodotus and Pliny—was *anthropophagi,* those who feed upon human flesh. By contrast the term *cannibal,* coined by Christopher Columbus, was new. Montaigne's title, therefore, signaled that he was not writing about the ancient world, but rather his own. Although most of his contemporaries viewed the indigenous peoples of the Americas as innocent and as peoples whose conversion could lead to the Millennium, while others viewed them as savages and as satanic, Montaigne found in them a mirror of Europeans. "I am not sorry that we notice the barbarous horror in such acts," Montaigne wrote, writing of the ritual cannibalistic practices of the Tupinambá, but he immediately adds: "but I am heartily sorry that . . . judging their faults rightly, we should be so blind to our own. I think there is more barbarity in eating a man alive than in eating him dead; and in tearing by tortures and the rack a body still full of feeling, in roasting a man bit by bit, in having him bitten and mangled by dogs and swine . . . than in roasting and eating him after he has been found guilty."[20] In this passage, as elsewhere in his *Essays*, Montaigne invited his readers not to be too quick to judge what might appear to be the barbarity of another culture. This is a familiar theme in his work in which he frequently underscores how varied are the customs and beliefs of different peoples and, accordingly, how hard it is to judge others. Indeed, he would rather, in our learning of others, have us hold up a mirror to ourselves. This fascination with the variety of customs had led Montaigne to read the ancient authors avidly; he cited them frequently. But he was also curious about his own world; he read with great interest reports from Asia and the Americas; and he thought hard about what these discoveries meant for his understanding of his own world and, in particular, for his understanding of France, which at the time of his writing, was roiled by civil war and brutal killings and tortures.

Montaigne also developed a genuinely anthropological argument: he was concerned, not simply with ethnography (that is, with the description of a particular culture), but also and above all with anthropology (with a general understanding of the human condition, and in teasing out what diverse cultures, despite their diversity, have in common). And, it is in this context that he developed a distinction between nature and culture that enabled him to overturn implicit hierarchies in which the civilized world of Europeans stood in a superior position to the savage world of the Tupinambá. As Montaigne observes, the Tupinambá "are wild in the same way that we call wild the fruits that Nature produces on her own and in her normal course; whereas really it is those that we have changed artificially and led astray from the common order, that we should rather call wild." And he then adds, "these nations, then, seem to me barbarous in this sense, that they have been fashioned very little by the human mind, and are still very close to their original simplicity."[21]

To Montaigne, the New World offered an example of a culture in which "there is no sort of traffic, no knowledge of letters, no science of numbers, no name for a magistrate or for political superiority; no custom of servitude, no riches or poverty, no contracts, no successions, no partitions, no occupations but those of leisure, no respect of kindred, no respect for any but common kinship, no clothes, no agriculture, no metal, no use of wine or wheat. The very words that signify lying, betrayal, dissimulation, avarice, envy, bad-mouthing, pardon—all things unheard of."[22] And this argument provides Montaigne with his essential point, his deconstruction of the notion of savagery or barbarism: "Now I find that there is nothing barbarous and savage in that nation, from what has been reported to me, except that each man calls barbarism whatever is not his own custom."[23] And then, as if to underscore the reality of a common humanity, Montaigne even points to the existence of cannibalism in the ancient world and in modern Europe. He provides examples of survivor cannibalism and even of the medicinal use of carcasses and mummies, noting that "physicians do not fear to use human flesh in all sorts of ways for our health, whether applying to the inside or the outside of the body."[24]

Montaigne's capacity to view the indigenous peoples of the Americas in this way was, undoubtedly, shaped in part through his long immersion in ancient literature.[25] But, above all, Montaigne's analysis of the Tupinambá benefitted from his reading of two of his contemporaries whose books had been published in France in just these years: *The New Found World, or Antarctike,* of the Catholic cosmographer André Thevet, and the *History of a Journey in the Land of Brazil,* by the Calvinist Jean de Léry. And both these writers, in turn, drew on their direct experience with the Americas. In 1555 the Franciscan Thevet had sailed to Brazil with the Admiral Villegaignon, dispatched there by King Henri II to establish a French colony intended to serve as a check on the Portuguese. Thevet spent slightly over a year there, as chaplain to the fledgling French settlement before coming back to France in early 1557. The following year he published his famous account of his experiences in the New World.[26]

In March 1557, only a few months after Thevet's departure, Jean de Léry had arrived in Brazil as part of the first Protestant mission to the Americas. Then, not long after his arrival, he and several fellow Calvinists were forced to flee the island in Guanabara Bay where Villegaignon had established the colony. Once on the mainland, Léry actually ended up living among the Tupinambá. It was on the basis of this experience that he would offer his rich account of their customs in his *History of a Journey in the Land of Brazil,* though it is important to recall that this work was not published until more than twenty years later. Among the most striking passages in the *History* were those on cannibalism. And it was this theme that Montaigne picked up in his *Essays* and, in so doing, offered a radically different understanding of the

European encounters from his contemporaries who, as we have seen, had viewed them in a largely apocalyptic key.

As Léry wrote in his account of the Tupinambá, "I could add some other examples touching upon the cruelty of the savages towards their enemies," Léry observed, "but it seems to me that what I have said is enough to horrify you, even to make your hair stand on end," and then he invited his readers to consider the savagery in Europe.

> Furthermore, if one wishes to come to the brutal action of really (as one says) chewing and eating human flesh, have we not found people in these regions over here, even among those who bear the name of Christian, both in Italy and elsewhere, who, not content with having cruelly put to death their enemies, have been unable to regain their courage only by eating their livers and their hearts? . . . And, without going further, what of France? . . . During the bloody tragedy that began in Paris on the twenty-fourth of August 1572— . . . was not the fat of human bodies—that had been massacred in ways far more barbarous and cruel than those of the savages—retrieved from the Saône and sold publicly to the highest bidder? The livers, hearts, and other parts of these bodies—were they not eaten by the furious murderers, of whom Hell itself stands in horror? . . . So let us henceforth no longer abhor so very greatly the cruelty of the *anthropophagous*—that is, man-eating—savages. For since there are some here in our midst even worse and more detestable than those who, as we have seen, attack only enemy nations, while the ones over here have plunged into the blood of their kinsmen, neighbors, and compatriots, one need not go beyond one's own country, nor as far as America, to see such monstrous and prodigious things.[27]

Moreover, Léry, like Thevet, offered what we might describe as a social explanation of cannibalistic practices—an explanation that Montaigne himself will largely adopt in his essay on this theme. Essentially both Thevet and Léry presented cannibalism as the ritual core of the vendetta or the feud—Thevet actually specifically calls the conflict a vendetta, underlining the thirst of the "savage" Tupinambá for "vengeance." But it is a passage from Léry that is most compelling. "These barbarians do not wage war," he writes:

> to win countries and lands from each other, for each has more than he needs; even less do the conquerors aim to get rich from the spoils,

ransoms, and arms of the vanquished. That is not what drives them. For, as they themselves confess, they are impelled by no other passion than that of avenging, each for his side, his own kinsmen and friends who in the past have been seized and eaten, in the manner that I will describe in the next chapter; and they pursue each other so relentlessly that whoever falls into the hands of his enemy must expect to be treated, without any compromise, in the same manner: that is, to be slain and eaten. Furthermore, from the time that war has been declared among any of these nations, everyone claims that since an enemy who has received an injury will resent it forever, one would be remiss to let him escape when he is at one's mercy. Their hatred is so inveterate that they can never be reconciled. On this point one can say that Machiavelli and his disciples, against Christian doctrine, teach and practice that new services must never cause old injuries to be forgotten. These atheists, I say, seemly have the courage of a tiger in that they are the true imitators of the barbarians.[28]

The particular feud that Thevet and Léry described was one that had long pitted the Tupinambá and the Tupinikin—two of the major tribes of the Tupi-Guarani peoples—against one another. At times the relationship between these two peoples must have been relatively peaceful, but the arrival of the Portuguese and the French along the coast of Brazil in the sixteenth century and the rivalry of these two European powers to gain greater and greater control over the resources of this territory intensified the hostility between these two groups. The Portuguese allied themselves with the Tupinikin, the French with the Tupinambá. And it was the warfare between these two groups that both Thevet and Léry witnessed, the latter during his exile from Fort Coligny.

And it was in the context of these disputes and feuds that cannibalism took place.[29] The goal of the conflict was not so much to kill as many of one's enemies as possible, as to take a significant number of them captive. And it was these captives who, it appears, would become ritually sacrificed and eaten by the victors. Thevet and Léry each devoted several pages to describing the various stages of the ritual. The prisoners are treated surprisingly well; they are given wives; they are fed well; and, only once they are fattened, are they executed in a ceremony in which both they and their captors declare their valor. But it is by no means clear that Thevet and Léry viewed this behavior as worse than the behaviors they had witnessed in France during the sixteenth century. This was a period in which, largely because of the breakdown of public order during the Wars of Religion, traditional feuds intensified. Moreover, at the siege of Sancerre in 1573 Léry himself had been witness to the practice of survivor cannibalism.[30] And even within feuds in Europe, the violence could be extreme and lead if not to cannibalism, to the ritual dismemberment of the victim's body and, in some cases, the feeding of parts of the body to animals—a practice to which indeed Montaigne himself appears to allude in his essay.

In the end, in his efforts to make sense of cannibalism, Montaigne, like many of his contemporaries, appears to have drawn less on his philological skills than on his familiarity with the practice of feud or vendetta in France. Montaigne, that is, like Thevet and Léry, portrayed cannibalism as a ritual act at the core of New World conflicts which they read in light of their own experience with feuds or conflicts among powerful families in Europe at the time they were writing. This should not be surprising. While the power of government was in general expanding in this period, private feuds continued to provide for public order.[31] And, indeed, during the wars of religion, feuding intensified among France's nobility. In 1565 the *parlement* of Bordeaux, on which Montaigne had served from 1554 to 1562, itself undertook an inquest in the Périgord of an outbreak of "armed assaults, murders, robberies" that were the consequence "more of feuds and private hatreds than the diversity of religion."[32] The *Essays* themselves are filled with references to vengeance and honor. Montaigne clearly deplored the ethos that fueled the outbreak of feuding in French society. Here Montaigne, like Léry, was in part following Innocent Gentillet's celebrated *Anti-Machiavel*, published in 1576, in which Gentillet attributed the growth of feuding to the malicious influence of Machiavelli, citing in particular his famous chapter 7 of the *Prince* as a model for favoring vengeance over reconciliation.[33] And the literary historian David Quint has seen Montaigne's *Essays* themselves as motivated in large part by a desire to replace the ethic of valor, which encouraged feud, with an ethic of forgiveness and reconciliation that would help end the cycles of violence that so deeply scarred France during Montaigne's lifetime.[34]

It would not, however, be Montaigne's account of cannibalism that would prevail. Rather, another narrative, one that stressed the barbarism of the peoples in the New World, would play a central role in shaping European perceptions of the cannibal. By a remarkable coincidence this narrative was another first-hand account of the Tupinambá, written almost contemporaneously

Constructing the image of the savage. European representations of American natives, such as this illustration showing men and women engaged in cannibalism in Brazil, from Theodor de Bry's *Americae tertia pars,* published in 1597, played a key role in legitimating European conquests and colonialism in the New World.

with those of Thevet and Léry. This was the best-selling account by the German adventurer Hans Staden: *The True History and Description of Savage, Naked, and Man-Eating Peoples Situated in a Country of the New World of America,* published in Frankfurt in 1557, and almost immediately translated into Latin, Dutch, and German.[35]

Staden, a native of Hesse who likely served as *arquebusier* in the Schmalkaldic League, had set out from Germany with the original intent of traveling to South Asia. But, in 1547, after making his way to Lisbon, he learned that the fleet to India had already sailed, and he settled for passage to Brazil. He returned to Portugal the next year, and then in 1549 traveled back to Brazil. There he became a gunner in a Portuguese fort where he must also have had the opportunity to pick up the local language, for it was largely his knowledge of Tupi that would save him when he was captured in 1552. At first, Staden, stripped naked and told that he would be eaten, was desperate. But he managed to convince his captors that he was not Portuguese but German and that he could be beneficial to them as a kind of shaman. He ended up living with them for nearly two and a half years, and this provided him with an opportunity to witness their rites firsthand. These he describes in great detail in his book, which he wrote shortly after returning to Europe in 1555.

Staden's *True History* did much to sensationalize the news of New World cannibals among readers in Europe. Its illustrations, above all, would have an important and interesting afterlife. The Flemish engraver and printer Theodor de Bry used modified versions of Staden's illustrations in his anthology of accounts of journeys to the East and West Indies.[36] In volume 3 of this work de Bry published the accounts of both Jean de Léry and Hans Staden. But what was most striking was de Bry's use of Staden's images, which he transformed—at once classicizing their style, and rendering them fiercer. In the end de Bry's images, more than his text, would help fix the European notion of the cannibal as a savage barbarian, an image that would play a role in legitimating the conquest and colonization of the Americas.

For José de Acosta, a Spanish Jesuit who had traveled to Peru in 1572 as a missionary, cannibalism was a sign of a perverted nature. Here was the flip side of natural law theory which over the course of the late Middle Ages had provided a bridge to understanding non-Christian cultures. It did so not only on the assumption that such institutions as marriage and the family were universal but also on the belief that ethical norms—which, in the eyes of the theologians, were rooted in nature itself—were the same everywhere. For example, since sexual intercourse had the end—so Aquinas and others had argued—of producing children, then only a man and a woman, who were in a state of marriage, should engage in sexual intercourse, an act, that outside of marriage, was *contra naturam,* against the very laws of nature. Thus, when Acosta and his fellow Jesuits encountered societies and cultures in the New World in which the ethical order appeared entirely overturned—a world of sodomy and onanism and even of cannibalism—they viewed these practices as forms of self-pollution, and evidence that Satan himself had done much to pervert the natural order of things.

This sense of a gulf between Christian ideals and native practices was a double-edged sword. On the one hand, it constituted one of the factors that underlay much of the violence of this era, with the conquistadors often justifying their crimes on the basis of their presumed superiority to the natives. On the other hand, this gulf opened up a dream of conversion. Through the work of evangelism, Franciscans, Dominicans, and Jesuit missionaries would be able to Christianize the natives and forge a global Christianity. It was this latter dream that inspired millenarian hopes. In the late 1570s, for example, the Dominican Francisco de La Cruz, who had fathered a child with a native woman he called Maria, became convinced that his son, whom he dressed in blue and white cloth—would become a new Elijah and that this Elijah would announce Francisco, his own father, as the new pope. Then La Cruz—as Christianity met its final destruction in Europe—would restore true Christianity in the New World and usher in the Millennium.[37]

In part La Cruz was drawing on the writings of Las Casas, whose ideas he had studied shortly after his entry into the Dominican order in Valladolid in the mid-sixteenth century. But he was also inspired by an angel, who, after appearing to a local creole girl, had "shown him and his accomplices what they would do to cure all the evil things that can be observed in cities and monasteries, stationing him in Lima, another in Cuzco, another in Potosí and another in Quito, to reveal to the men and women the great mysteries of said angel of his; and that in one year's time they were all to come together to preach this in public, with all the soldiers and others who had offered their lives, and defend the angel."[38] While aspects of La Cruz's prophecies struck many of his contemporaries as extreme, he nonetheless inscribed his hopes in a familiar narrative. The Church of Rome was in decline. The New World offered a new hope for the Beautiful Ending. The native peoples, who he believed were descended from the lost tribes of Israel, would welcome the Gospel, and the Millennium would dawn. Above all, he believed that he and his son had been sent by God to rule over the Spanish settlers and the Indians; and that—in the final phase of life on earth before the Last Judgment—all would live in peace and harmony.

The prophecies of Francisco de La Cruz raised enormous alarm. In Lima the Inquisitors feared that his teachings risked taking hold among the Indians and that this could lead to a political rebellion. In 1578 they imposed a sentence of death on him, remanding him to the colonial authorities to be burned at the stake. In the meantime Acosta drew on this trial as an example of the dangers implicit in millenarianism.[39] Acosta did not dispute that the

Millennium would come, and he valued the way in which preaching the Apocalypse would help move Christians toward repentance, but he rejected the idea that it was possible to know when the end was coming—a point he underscored in his *De temporibus novissimis*. In this same work, moreover, he made it clear that there was still much work to do to bring Christianity to many of the peoples of the globe in the age's expanding ecumene.[40]

Acosta's focus—and here he showed certain similarities to Léry and Montaigne—was on making sense of a world most of his contemporaries dismissed as barbarian, though it is clear that his motivations were primarily aimed at the conversion of the native peoples. In fact, many of his texts—above all his *On How to Bring about the Salvation of the Indians* of 1588 and his *Natural and Moral History of the Indies* of 1590—made it plain that he saw his work as essential above all to break down the prejudices of many of his fellow Europeans who assumed the native peoples of the New World were ignorant and more like beasts than humans. While he himself believed that there were certain tribes in the Americas that lacked the qualities of a more "civilized" society and he even devoted considerable attention to describing cannibalistic practices about which he was far less forgiving than Montaigne, he nonetheless wrote his *History* above all "to refute the false opinion that is commonly held about them [the native peoples], that they are brutes and bestial folk and lacking in understanding or with so little that it scarcely merits the name." "Many and very many abuses," he immediately added, "have been committed upon them as a consequence of this false belief, treating them as little better than animals and considering them unworthy of any sort of respect."[41] Acosta sought to accomplish this goal by providing an overview of the political society of the Incas and the Mexica peoples, along with a thoughtful account of their cultures and their languages.

It was, in fact, through his examination of indigenous languages that Acosta, like many other Jesuits of his generation, grappled with the question of how it would be possible to convey the truths of Christianity to the natives. The challenge was daunting. To evangelize in the New World required not merely a knowledge of the languages of the natives—and already this was daunting since, as Acosta noted, there was a "forest of languages" in the Americas—but also an understanding of their culture. To be sure, the priests who arrived in the New World had an advantage. Well trained in Latin and Greek, they were already keenly aware that translation demanded more than mastering vocabulary and rules of grammar. It also required a grasp of Greek and Roman culture. Similarly, to teach Christianity to the Indians, it would be necessary to learn as much about their cultures and institutions as possible. The obstacles were legion. Some of the groups that the Christian missionaries encountered, for example, had no single word for God. How then could one introduce such fundamental concepts as the Son of God or of the Trinity, of the Eucharist and of baptism?

One strategy lay in developing a deep grasp of native culture, and indeed this was one of the primary goals of Acosta's *History*. In Mexico, for example, Acosta had been struck by the prophecies and prodigies that, according to their "histories and annals" foretold "the end of their kingdom." For example, in the final years of Moctezuma's reign, as Acosta reported, "the idol of the Cholulans, who is called Quetzalcoatl, announced that strange people were coming to possess those realms." Similarly, "the king of Texcoco, who was a great sorcerer and had made a pact with the devil, paid Moctezuma an unexpected visit and swore to him that his gods had told him that great losses and travails were being prepared for him and all his realm."[42] And in this same period other prodigies had appeared that presaged the end of Moctezuma's kingdom. But, unlike Montaigne, who had found in the myths of the Aztecs grounds for casting doubt on prophecies and prognostications among his fellow Christians, Acosta not only accepted the reality of such prognostications in general but even believed that God had allowed these prodigies to emerge among the Mexicans as a punishment and a warning to Moctezuma. Above all, the prevalence of such beliefs offered a bridge to Christian teachings. Acosta would draw on the parallels between Christian and native teachings to evangelize the natives.

A second strategy lay in the mastery of the native languages. Some missionaries believed that the languages of the natives were so primitive—so devoid of abstract concepts—that it would be impossible to teach them Christian doctrine in their own languages and that they would have to be taught Spanish first. But other missionaries, especially among the Jesuits who made an effort to master the native languages of the New World, had a greater respect for beauty and richness of the languages they encountered and believed it would be possible to teach the indigenous peoples in their own tongue.[43] And Acosta belonged very much to this second group and played an important role in deepening an understanding of Indian cultures.

Early on, the Spanish conquistadors, missionaries, and settlers had realized that the Incas had a system, radically different from the alphabetic and numeric systems used in Europe, both for the purpose of keeping accounts and records of economic transactions but also for preserving memories, for the history of the community, and for ritual use in their religious ceremonies. Making use of a variety of colored cords or threads that could be arranged and knotted in a variety of ways, *quipus*—the system used by the Inca—stored extensive amounts of information. Acosta's account is helpful: "There were different quipus, or strands, for different subjects, such as war, government, taxes, ceremonies, and lands. And in each bunch of these were many knots and smaller knots and little strings tied to them, some red, others green, others blue, others white: in short, just as we extract an infinite number of differences out of twenty-four letters by arranging them in different ways and making innumerable words, they were able to elicit any number of meanings from their knots and colors."[44] While scholars today have a fairly clear understanding of how these objects were used for accounting purposes, their rhetorical content remains elusive.[45] Yet it is clear that quipus also conveyed historical, religious, and personal information. Within the public sphere, there were specialists, known as *quipucamayocs,* who could interpret the various combinations of colors and knots, but Acosta also notes that they could be used in a more private and personal way: "I saw a bundle of these strings on which an Indian woman had brought a written general confession of her whole life, and used it to confess, just as I would have done, with words written on paper," Acosta reported, "and I even asked about some little threads that looked different to me, and they were certain circumstances under which the sin was required to be fully confessed."[46] But the Incas, as Acosta notes, also made use of other memory systems, some making use of pebbles and others of grains of corn. Having observed Andean peoples make use of these to learn Christian prayers as well as to settle accounts, he was deeply impressed by their abilities. "If this is not intelligence, and if these men are beasts, let whoever wishes to do so make a judgment," he wrote, "but what I truly believe is that, in whatever they apply themselves, they outdo us."[47]

But Acosta did not limit his curiosity to the writing system of the Incas, he was also curious about the systems used among the Mexica peoples. Unlike the Incan quipus, the Mexican system made use of images and figures in objects that the missionaries there readily called books. In these books—as Acosta noted on the basis of an investigation carried out by a fellow Jesuit in Mexico—"for things that had shapes they painted them in their own image, and for things that did not have actual shapes they had characters signifying this, and, in this way, were able to express what they wanted."[48] Acosta was particularly impressed, however, by the way in which the Mexicans used their calendars to record events. "For instance," he notes, "by placing a picture of a man with a red hat and jacket in the Sign of the Cane, which was then dominant, they marked the year when the Spaniards came into their land; and they did the same with other events."[49] And here Acosta was drawing on information he received from his fellow Jesuit Juan de Tovar, who had sent Acosta information about the way that the Mexica used pictographic calendar wheels to record historical events.[50] Yet, as with the quipus in Peru, the Mexican pictures required explication by specialists, and to this end, as Acosta explained, the Mexicans developed schools "where the old men taught the youth these and many other things that are preserved by tradition as fully as if they had been set down in writing. The more famous nations in particular obliged their youths to be rhetoricians and to exercise the office of orators, committing their traditions to memory word for word."[51] Again, a community of specialists elaborated on and explained the books that the Indians kept—skills that Acosta viewed as clear evidence of the intelligence of the natives of the New World.

Finally, like many of his contemporaries, Acosta was intrigued as well by the writing systems of the Chinese and the Japanese that he saw as in many ways quite similar to the system of figures and images used in Mexico. It is likely that Acosta's curiosity about this system came from his interest in New World forms of writing, but he also had the good fortune when he was in Mexico of meeting, not only Alonso Sánchez, a fellow Jesuit who had spent time in the Philippines and China, but also a number of Chinese. From these, he learned that the Chinese writing system was not alphabetic but relied on a highly sophisticated combination of characters that many of his contemporaries viewed as pictograms, though Acosta himself recognized that the written form of Chinese involved more than pictograms, since various combinations of "dots and flourishes and positions" supplied the syntax that made the meaning of the combination of the characters clear.[52]

Ultimately what Acosta presented was a typology of different kinds of non-European civilizations. For Acosta, while all non-Christian societies were, by definition, barbarian, their levels of "civilization" varied considerably, and he placed them in three categories. At the top were China and Japan, both with extremely well developed political and religious orders. Next were the Inca and the Mexica, again societies of considerable religious and political sophistication. But beyond these were the nomadic societies of the New World, groups such as the Chichimecas, whom Acosta characterized as being "very savage forest dwellers," who had "no writing at all."[53] A Renaissance taxonomy of writing and its absence then formed an early hierarchization of human societies—a keystone in the later legitimation of European imperialism.[54]

At the end of his essay "On Cannibals" Montaigne recalled his encounter with three Tupi. Montaigne placed the site of this meeting in Rouen in 1562, noting that he had traveled there to join the young king Charles IX in this port city where several Brazilian natives were to meet the monarch. But Montaigne either dissembled or, more likely, misremembered. The probable venue for the meeting, as one of his biographers has shown, was in Montaigne's own city of Bordeaux and the most likely date was in April 1565, when King Charles made his entry into the city accompanied by "twelve captive foreign Nations, such as Greeks, Turks, Arabs, Egyptians, Taprobanians, Indians, Canarians, Moors, Ethiopians, savages, Americans and Brazilians."[55] We don't know what words were exchanged between the king and the Tupi. We know only that, according to Montaigne, "the king talked to them for a long time; they were shown our ways, our splendor, the form of a beautiful city."[56]

According to Montaigne, after the conversation with Charles ended, there was a general conversation between the Tupi and the courtiers in attendance. Someone in the party (Montaigne does not say who) asked the Brazilians "what of all the things they had seen they most admired." And concerning their response, Montaigne records the following:

> They said that in the first place they thought it very strange that so many grown men, bearded, strong, and armed . . . should submit to obey a child [King Charles was a boy-king], and that one of them was not chosen to command instead. Second (they have a way in their language of speaking of men as halves of one another) they had noticed that there were among us men full and gorged with all sorts of good things, and that their other halves were beggars at their doors, emaciated with hunger and poverty; and they thought it strange that these needy halves could endure such an injustice, and that they did not take the others by the throat, or set their houses on fire.[57]

This is not the only passage in which Montaigne draws on an expanding ethnography to critique his fellow Europeans. To the contrary—through not only this essay "On Cannibals" but also in his essays "Apology for Raymond Sebond" and "On Coaches"—Montaigne had often made similar points.

Montaigne was, to say the least, deeply impressed by what he learned not only about Asia, but also about the Americas. Rather than viewing these other parts of the world as inferior, he saw them as different. He was impressed by what he called the awesome magnificence of the cities of Cuzco and Mexico.[58] But he was not merely a relativist. He was willing to make judgments about other cultures, at times viewing them as superior to his own. Certainly his report of the shock the Tupi expressed at the ravages of inequality in Europe is one example of this. And his fascination with China had been piqued in part by the opportunity he had to view a Chinese book, its characters printed on rice paper, in the Vatican Library during his visit to Rome in 1581.[59] Afterward he would praise China both for its system of government and its technology. "In China—a kingdom whose government and arts, without dealings with and knowledge of ours," he writes, "surpasses our examples in many branches of excellence, and whose history teaches me how much ampler and more diverse the world is than either the ancients or we ourselves have fathomed—the officers appointed by the prince not only punish but also reward."[60] And, in a fascinating passage on technologies, Montaigne reminded his readers that their pride might be misplaced. "We exclaim at the miracle of the invention of our artillery, of our printing," Montaigne wrote, adding, "other men in another corner of the world, in China, enjoyed these a thousand years earlier."[61]

In the end Montaigne, unlike Acosta, would have found it absurd to forge a hierarchy of cultures. Barbarism is ubiquitous. It is found not only among the French but also among the Tupinambá. When the courtiers withdrew from their colloquy with the three Brazilian chiefs during the king's visit to Bordeaux in 1565, Montaigne was able—or so he claims—to converse briefly one-on-one with a Tupi chief. "I talked to one of them for quite some time," he writes, and "asking him what advantage he received from the superiority he had amongst his own people (for he was a captain, and our mariners called him king), he told me, to march at the head of his men in war. Asking him how many men followed him, he showed me a stretch of ground, to signify as many as could march in such a space, which might be four or five thousand men; and putting the question to him whether or not his authority expired with the war, he told me this remained: that when he went to visit the villages of his dependence, they cut him a path through the thick of their woods, by which he might pass at his ease." And, then, in the closing sentence, Montaigne quips: "That's not so bad. But what else can they do? They don't even wear breeches."[62]

CHAPTER 11

The Restitution of All Things

Elijah is indeed coming and will restore all things.
—MATTHEW 17:11

Safed, a scramble of stone houses on a hill in Galilee, overlooked a holy landscape. With its thriving textile industry, this Ottoman city—in an age of exile—became a magnet for Jews from Italy, North Africa, the Middle East, and even Poland.[1] But by the 1560s—with nearly two thousand Jewish households in the city—it was Safed's spiritual reputation that proved attractive. In the nearby hilltop town of Meron stood the presumed tomb of Shim'on bar Yochai, a second-century rabbi who, many believed, had been the author of the great kabbalistic text, the *Zohar*. Accordingly, some of the most brilliant teachers of Kabbalah made this city home. Moses ben Jacob Cordovero, a charismatic mystic of Spanish origins who arrived there in the 1540s, stressed the ethical teachings of the Kabbalah and drew many followers. Then, in 1570, just before Moses died and after years of ascetic practice in Cairo, Isaac Luria arrived in Safed. Known as "the Ari" ("the Lion"), Rabbi Isaac would transform the teachings of the Kabbalah into a powerful myth of exile and redemption—one with messianic overtones—that would come to shape Jewish spirituality throughout Europe and the Mediterranean over the next several centuries.[2]

Kabbalah was esoteric, but, for many Jews, its appeal was palpable. On the most abstract level, it was a reservoir of myths, and it did much to restore the symbolic dimensions of Judaism in the wake of what many felt was an overly rationalistic interpretation of the faith in the writings of the great Jewish scholar Maimonides in the thirteenth century. For, as a tradition that had allegedly begun with Moses, Kabbalah offered its hearers a path to a deeper grasp of the Torah as a living symbol of God's power, with each of the words in the Hebrew Scripture resonating with multiple meanings and the divine presence. In Kabbalah, that is, Jews encountered the mythic power of their faith and could dream of a mystical union with God.

As a principal work of Kabbalah, the *Zohar*—an extensive text that we now know was not the work of Shim'on bar Yochai but rather the product of currents in Judaism in the thirteenth century—also operated not as a work of logic but rather of myth. As we have seen, the *Zohar* offered its adepts nothing less than a cosmology through which they were inspired to grasp the divine reality not as the changeless and perfect Being the philosophers had imagined but rather as a dynamic force that, through the *sephirot* (the emanations of the varied aspects of God), sustained all of Creation. The *Zohar* did not present this cosmology abstractly but rather through a series of conversations in which Shim'on bar Yochai, his son, and several of their companions probed the hidden meanings of the Torah, the first five books of the Hebrew Bible. Finally, as was often the case with the mystical imagination, the ultimate goal was union with God—an aspiration that often took on erotic overtones. God was hidden, yet God would reveal himself to his lover. At one point in a charming parable told by an anthropomorphized Wisdom who had taken on the form of a donkey-driver, Torah becomes a beautiful maiden who "opens a little window in that secret palace where she is, reveals her face to her lover, and quickly withdraws, concealing herself."

> Come and see! This is the way of Torah: At first, when she begins to reveal herself to a person, she beckons him momentarily with a hint. If he perceives, good; if not, she sends for him. . . . As he approaches, she begins to speak with him from behind a curtain she has drawn, words suitable for him, until he reflects little by little. This is *derasha* [the search for meaning]. Then, she converses with him from behind a delicate sheet, words of riddle, and this is *haggadah* [the telling of parables]. Once he has grown accustomed to her, she reveals herself to him face-to-face, and tells him all her hidden secrets and all the hidden ways, concealed in her heart, since primordial days. Then he is a complete man, husband of Torah, master of the house, for all her secrets she has revealed to him, concealing nothing.[3]

Luria had absorbed this tradition but would transform it through his visions. Like the earlier kabbalists, he continued to teach that God, whose reality overflowed into and filled every part of the cosmos, first had to make room for the created world, which He did in an act of *zimzum* (withdrawal or contraction) into himself.[4] This generous act of self-exile created a great emptiness into which Creation itself could flow, as the light of the Godhead radiated outward in the luminous emanations of his divine attributes, filling the ten spheres or vessels of the sephirot. But Luria gave new emphasis to the chaos that was embedded in this pulsating process of continuous divine withdrawal and emanation. As the sephirot radiated outward into the world, cascading into the cosmos—so he taught—they shattered the vessels God had fashioned to make his attributes manifest. Then, pieces of these vessels—fragments that Luria had called "shells" or "peels" and identified with evil—crashed into the created world, even as rays of divine light remained trapped within them. The Shekhinah—the divine presence, the tenth of the ten sephirot—was now in exile in the world, longing to return to the Godhead.

In this highly visual and even eroticized myth of the creation, Jews not only glimpsed the imperfections in the cosmic order but also came to understand that exile (*galut*) was the underlying condition of all reality. At a time in which the expulsions from Spain and much of western Europe continued to cast long shadows over the experience of Jews throughout the Mediterranean while at the same time fueling messianic hopes, this myth resonated deeply. But it was not one that, according to Luria and the other kabbalists in Safed— above all Chayyim Vital, Luria's most articulate disciple— they were expected to accept passively. To the contrary, Jews were called upon to do all that they could to repair this broken world. Luria's followers were encouraged to engage in a variety of spiritual practices—often visiting and prostrating themselves on the tombs of holy figures such as Shim'on bar Yochai—with the goal of purifying themselves, rituals that were aimed at releasing the divine sparks trapped in the fragments of the vessels that had brought evil into the world. "By his works," Gershom Scholem, the twentieth-century historian of the Kabbalah, has written, "the Jew healed the sickness of the world and reunited the scattered fragments."[5] This *tiqqun* (or repair) was the path to redemption. Following the divine law took on a deeper meaning, since in carrying out the rituals of their faith, believers engaged in acts of making whole a fragmented universe, while putting into order a broken world.

For centuries Hebrew prophets had envisioned the coming of the Messiah as an event entirely outside of history, as a breaking into history of a priest-king who would not only deliver the Jews from political repression and injustice but also shape a new world. Eight centuries before the Christian era, for example, the prophet Isaiah had promised a great transformation in which "the wolf

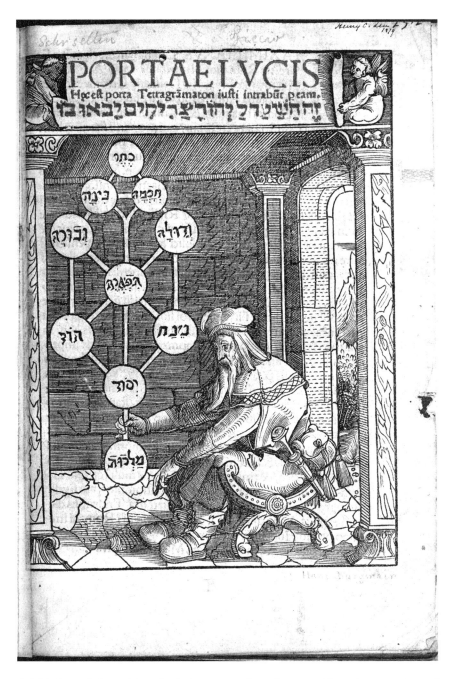

A kabbalist and the sephirot. A scholar contemplates the ten sephirot, the emanations of the divine reality. Title page to the *Portae lucis,* Paolo Riccio's early sixteenth-century translation of the medieval kabbalist Joseph ben Abraham Gikatilla's *Sha'are Orah*. Riccio, a Jewish convert to Christianity, published this work in Augsburg in 1516.

shall live with the lamb, the leopard shall lie down with the kid . . . for the earth will be full of the knowledge of the Lord as the waters cover the sea" (Isaiah: 11:6–9). In this and other biblical stories and in the medieval commentaries, Jews had awaited the arrival of a Messiah as a divine or miraculous event in which God intervened in history to rescue humanity from its sins and sufferings. By contrast, the Lurianic kabbalists offered a radically new conception of messianism and, therefore, of the End of History. In the cluster of Luria's new ideas about Kabbalah that developed in Safed over the course of the late sixteenth century, redemption was no longer seen, as Scholem has written, as coming "suddenly" but rather "as the logical and necessary fruition of Jewish history," adding, "the messianic king, far from bringing about the *tiqqun,* is himself brought about by it: he appears after the *tiqqun* has been achieved."[6] An earlier conception of history that had conceived of redemption as God's rescue of humanity in crisis gave way to a conception of history that called upon humans themselves to participate actively in reestablishing the divine order through their own actions, redeeming themselves. Essentially, within the vast artifice of his teachings, Luria had offered his fellow kabbalists a new vision of the messianic, shifting it from the expectation of an individual savior to the collective acts of *tiqqun* or repair of the world by the Jewish people.[7]

Luria did not intend for his ideas to be widely shared. Moreover, he died a young man in 1572, shortly after arriving in Safed, and only a few texts composed by him have survived.[8] But the publications of his disciple Chayyim Vital played a central role in bringing his teachings to a broader public. In his writings, which circulated primarily in manuscript, Vital made it clear that Luria—as he strove to deepen his understanding of redemption and of God's relation to humanity—offered a cosmic vision that would resonate profoundly with the Jewish experience of exile in the early modern world. Indeed, what thrilled Vital about Luria's teaching was the latter's insight that the Kabbalah offered Jews a way of making sense of their experience of exile. Jews had long struggled with the traumas created by their expulsion from Spain and their forced conversions in Portugal at the end of the previous century. At first, they were convinced that the catastrophes they had suffered signaled that the End of History was at hand. But the Jewish prophets who had predicted the coming of the Messiah in the aftermath of 1492 had been proven wrong. Exile had become a permanent state. And the Jewish people had become the Messiah.

Nearly a century earlier, Christians too had turned to Kabbalah in order to deepen their understanding of their own faith. In late fifteenth-century

Florence the brilliant young thinker Pico della Mirandola—a phoenix—studied Kabbalah with the goal of converting Jews to Christianity. Somewhat later the German theologian Johannes Reuchlin also became a leading student of Kabbalah, presenting his own views of its significance not only to Jews but also to Christians and Muslims first in his *On the Wonder-Working Word* of 1494 and then in his *On the Art of Kabbalah* of 1517. Reuchlin's interest in Kabbalah lay primarily in its exegetical power, as it offered Christian students of the Bible a new method, through attention to individual Hebrew letters and their numerical value, to tease out the hidden meanings in scripture. But other Renaissance figures began to bring the messianic strains within Kabbalah to the surface. Giles of Viterbo did so not only in his *Historia viginti saeculorum*, his sacred history of time before and after Christ, but also in his explicitly prophetic text, the *Scechina* (Shekhinah), which he completed in 1530.[9]

It would be a later Christian kabbalist, however, who expanded these ideas most decidedly in a messianic direction. Across the mid-sixteenth century the French humanist Guillaume Postel drew on Kabbalah in a quest not only to refute the Jews but also to help lay the foundation for "the restitution of all things" and, ultimately, a Beautiful Ending. Postel had first turned his attention to Kabbalah in Rome in the mid-1540s, after being expelled from the Jesuit order, in which he briefly served. But it would be in Venice, where he moved in 1547, that he devoted himself to the study of the *Zohar*. That year he purchased a manuscript copy of the text from the Christian printer of Hebrew texts Daniel Bomberg, with the goal of translating it into Latin.[10] Like Pico and Reuchlin, Postel too had grown convinced that Kabbalah contained the secrets that would disclose the ultimate religious truths behind the diverse faiths. But Postel sought to dive even more deeply into the kabbalistic texts. His remarkable linguistic skills—he read both Hebrew and Aramaic—would enable him to bring out in great relief the value of Kabbalah for a deeper understanding of religious truth. And he would benefit as well from contact and conversations with the famed rabbis living in the city.

Postel's immersion in the *Zohar* and his experience in Venice transformed him. After a trip to Palestine in 1549, Postel, after a brief sojourn in Venice, settled in Paris. There, just after Christmas 1551, he slipped into a debilitating depression. He couldn't even get out of bed. With his body wasting away, he feared he was at death's door. Then on the Feast of the Epiphany, January 6, he awoke to a blinding light; his body suddenly quickened—as he put it, each of his interior parts, through "thousands of Angelic virtues and operations," came back to life.¹¹ Postel wasn't merely revived, he was also reborn, as Madre Zuana—a woman he had encountered several years earlier in Venice and whose spiritual teachings had mesmerized him—came to dwell in his body. In Postel's view, Zuana was not a mere prophet; she was the *anima mundi,* the Soul of the World, the New Eve, the spiritual mother to what many hoped, in this age of religious divisions and conflict, would be nothing less than a new age of religious unity in Europe and Mediterranean.¹²

As Postel had quickly learned through his years in Venice, making sense of Kabbalah inevitably required more than linguistic skill, especially since the kabbalists themselves rejected the view that the believer could understand either God or reality on the basis of language alone. Since Kabbalah was a mystical practice, it required not merely the reading of the kabbalistic texts but also initiation into its secrets. What he hadn't anticipated was that his guide into Kabbalah would be *vne petiote vielle femmelette de l'eage de cinquante ans ou enuiron*—"a little old woman about fifty years of age."¹³ But Madre Zuana's spiritual gifts captivated him. She had operated a kind of soup kitchen on the *campo* Santi Giovanni e Paolo and eventually had convinced several Venetian noblemen to establish a permanent hospital there for the poor and dying. Postel was impressed by her charitable works; and she, in turn, sought him out to be her confessor. It was then in this relationship of confessor and penitent that Postel quickly learned that Zuana possessed gifts that far exceeded his own, noting that in his conversations with her he seemed, as he saw things, "dead and outside myself, considering that I, who passed as having read so many and such excellent theologians, and who through the benefit of contemplation had (thanks to the infinite goodness) tasted something of divine grace, had never reached such rational or exalted conceptions."¹⁴ Even her learning was great. As he noted, "And as for her feminine knowledge [*sçavoir feminine*], it was very great and eminent in her. . . . Although she had never learned any Latin or Greek or Hebrew or any other language, she nonetheless knew how to open up and explain to me the *Zohar*—a very difficult book containing the ancient evangelical doctrine—when I was rendering it into Latin."¹⁵

Though Postel never published his translation of the *Zohar,* his life was a continuous quest for a deeper understanding of religious truth. Indeed, in virtually all his writings, Postel directed his energies toward a vision of a world united under Christianity. A brilliant linguist, he saw the growing interest among sixteenth-century humanists in Hebrew—in his view, the original language under which all had been unified—as evidence of reemerging unity for all humanity. His foray into cosmology, especially his *De orbis terrae concordia* or *On the Harmony of the Earth,* also laid out his hope for a world unified under the Christian faith.[16] But it was in his studies of Kabbalah that Postel gave his most powerful expression of his messianic hopes. Like Pico and Reuchlin he viewed Christianity as the true, revealed religion. At the same time he was convinced that the careful study of languages would enable believers to overcome their differences. And he was buoyed by the changes taking place around him. In his view, the revival of Latin, Greek, and Hebrew, the knowledge of which had expanded more in the past fifty years than in the previous thousand, pointed to the eventual triumph of Christianity. So too did the recent inventions of the printing press and of artillery—evidence that divine providence was again favoring the Christians.

But Postel was especially inspired by Europe's expansion. "We see a greater mutation and marvel in these last fifty years," he wrote, "when we consider how through the power of sailors and merchants the New World, which is larger than our own, was not only discovered and conquered but also converted to the Christian religion under the power of Spain, after the Portuguese had first begun their navigations under a Venetian gentlemen from Ca' da Mosto and the Spanish had begun theirs under Columbus, a Genoese citizen."[17] The great changes taking place in his world contributed decisively to his optimism and his vision of an emerging *harmonia mundi,* harmony of the world. Yet he did not approach this task from an exclusively scholarly perspective. He was himself also a mystic, and he came to see himself as chosen by God for this divinely-inspired role. This was, ultimately, the meaning of his epiphany. Zuana, whom he had identified as *Eliana* or *Elianus secundus,* the reincarnation of the Old Testament prophet Elias (the Latin version of Elijah), had come to dwell in his body and thereby passed this gift on to him. Postel was to be the third Elias, the prophet of the restitution of all things.

Yet, ultimately, Postel cast himself not only into a prophetic role but also into a quasi-messianic one. He believed that he would emerge as the angelic pope and that, under him, the entire world would be united. Drawing on various medieval prophetic traditions, from the pseudo-Methodius to Joachim, he had come to see himself as the agent who would bring into existence a world empire, in which, after so many years of religious warfare and strife, there would be one shepherd and one flock. Jews and Muslims would be converted to a deeply spiritualized and universal Christianity in which what mattered was not dogma or external practices but rather that God dwell in the hearts of the faithful, uniting them in peace and charity.[18] However, despite Postel's emphasis on the internal transformation of the faithful, he prophesied the radical reordering of society. It would be the French king who, in cooperation with a Church council, would forge a new universal order under the leadership of three sovereigns: a king, a pope, and a judge, while missionaries would bring the teachings of a true Christianity to all the world. At the same time, private property would be abolished and humanity would be restored to a universal language. Thus the golden age, "an original perfection," would return—a restitution which, Postel believed, was imminent.[19]

After Postel had been tried by the Venetian Inquisition in 1555 and imprisoned in Rome until 1559, he was tried in Paris and confined to the monastery of Saint-Martin-des-Champs. While Postel shared many of the impulses of his fellow Catholics, his deep interest in both Islam and in Judaism raised suspicion. And his examination of the Kabbalah had raised the greatest suspicion of all, for in it he had discovered the secrets of a transcendent faith. Yet, even from his prison cell, Postel continued to foster his connections with others he believed shared his aspirations. In particular, while in protective custody in the 1560s and 1570s, he remained in contact not only with several of the leading members of the Family of Love—among them Abraham Ortelius and Christophe Plantin in Antwerp—but also with several of the leading messianic thinkers in Palestine. His letters, in short, provide a glimpse of an epistolary network, perhaps one of many, that enabled some of the most creative thinkers about religion to interact and share ideas, often across religious boundaries, in the early modern period.

If Kabbalah provided an opening to those who, experiencing a broken world, sought to make it whole, alchemy provided a similar path. To be sure, we are likely to think of alchemy as an occult and superstitious practice, its adepts focused on the quest to transmute lesser metals, copper or lead, into gold. But such an image is misleading and captures only one facet of the broader import of this science. For, like Kabbalah, alchemy too was a wisdom of repair.[20] Starting from the perspective that nature is incomplete, alchemists held that it was up to humanity to complete or perfect what God had created. This was evident at the most basic level. God had provided grain and trees, but it is men and women who must harvest the grain to make bread or fell trees to build houses, while miners extracted the ores God had created from the ground which they then transformed into useful metals. Ultimately, one could even use this knowledge to transform the world, creating a more just society. Alchemists, in short, could improve, even perfect, the broken and incomplete world that God had bequeathed humanity. As the sixteenth-century physician and alchemist Philippus Aureolus Theophrastus Bombastus von Hohenheim, better known as Paracelsus, wrote in his *Paragranum,* "God has indeed created nothing that cannot be perfected.... For God does not want to be accused of having placed incomplete things before the human being, except for those things the human being can bring to perfection for himself out of that which is incomplete."[21]

Alchemy certainly provided Paracelsus, an innovative physician, with insights into the ways to treat disease. To be sure, Paracelsus's approach to disease was expansive. Working within a broad naturalistic frame, one that connected chemical processes to the heavens and to the individual, he identified five essential causes of disease. The first he located in the "force and efficiency" that the stars exercise over our bodies; the second he detected in "the influence of poisons"; the third in the "natural constitution" of the body; the fourth in the imagination; and the fifth in the divine will.[22] To some degree his emphasis on the influence of the stars over the body was traditional, combining astrology with a rich tradition, dating back to antiquity, that maintained that the entirety of the universe—what contemporaries called the "macrocosm"—was a great unity, while each individual, each person constituted a microcosm that was believed to reflect or mirror the universe as a whole. Thus, like the kabbalists and many others in the Renaissance, Paracelsus drew on and developed the ancient and medieval belief, not only that "man" is a microcosm in a vast universe, but that every aspect of the human person—from physical to spiritual—is connected to the macrocosm or the entirety of creation. Just as the earth itself was folded within the cosmos as a "world-egg," so the heavens themselves were also reflected or mirrored within each individual.[23] And within this frame, as the fifteenth-century Renaissance philosopher Marsilio Ficino had taught, the physician's calling required that he be able to interpret both spheres.[24]

Yet Paracelsus's approach to disease, while in many ways traditional, was radically innovative. Since antiquity European medicine had been based primarily on what scholars call the "humoral" theory of disease. Most fully developed by the second-century Greek physician Galen of Pergamon, humoral theory, which remained dominant throughout the Middle Ages and much of the Renaissance, assumed that an individual's health depended above all on the proper balance of the four humors: blood, phlegm, yellow bile, and black bile. In this model an illness was an expression of an imbalance.[25] By contrast, Paracelsus located the origins of disease outside the human body in various "poisons" that, invading the body, weakened the individual's system. Salts, for example, could cause ulcers, while mercuries could result in apoplexy and tumors. The physician's role was, therefore, to identify the causes of specific diseases and then to treat these with the proper antidotes, often locating the correct drug in various chemicals—mercuries, arsenics, and antinomies—that could be administered in proper doses. For example, in his recommendations on the treatment of syphilis—the sexually transmitted and disfiguring disease that had recently been introduced to Europe from the Americas—Paracelsus rejected the use of guaiacum for its cure and proposed instead an ointment distilled from mercury as the best antidote.[26] To be sure, Paracelsus's own chemical medicine was always embedded in a larger metaphysics, but his ideas would play a major role in reshaping pharmacy and the treatment of diseases over the next century.[27] Nonetheless, it was Paracelsus's location of the causes of illness outside the human body in various poisons that weakened the individual's system that was his most radical innovation. Thus for Paracelsus and the Paracelsians, the physician's role was to identify these poisons and then, working with a new pharmacology—one based on animal, plant, and mineral remedies—to counterbalance the poison, all along with close attention to the patient's own particular internal anatomy, which Paracelsus studied with attention to digestion and evacuation.

Given the cosmic reach of Paracelsus's studies, it is hardly surprising that he also gave attention to humanity's future. And, like many in his generation, he was convinced that a *güldene Welt* ("a golden age") was on the horizon. It would be, as he wrote in his popular *Prognosticatio* of 1536, a time in which there will be "such a total renewal and change of all things that the golden age will soon seem to have returned; the candor, simplicity, and integrity of children will reign; and all cunning, deceit, and intrigues of men will be destroyed."[28] And in many of his other works, he would continue to develop his ideas about a messianic age that—following upon the age of the Father and of the Son—would be the age of the Holy Spirit. It would be an earthly kingdom of the blessed poor, woven of justice and brotherly concord, and it would precede the Last Judgment.[29] Furthermore, the history of humanity itself made this clear: "From Adam on," he wrote in his *Volumen medicinae paramirum*, "there has been such a long time with so many people without anyone of them having been like another," that "when the Final Day comes, all the colors and customs of people will be fulfilled; for it awaits the point in time when all colors, forms and figures, and customs of people are exhausted and none more can be born without having to look like another; then the hour of the course of the first world will have struck."[30]

There was much in Paracelsus's view of the End of Time that reflected the teachings of such figures as Joachim of Fiore and many of his German followers in the early sixteenth century. A contemporary of Luther, he had developed his ideas in the midst of a period in which many had believed that the End was near and in which some had come to develop radical millenarian ideas. And while Paracelsus never fully embraced the teachings of either Luther or the Anabaptists and indeed never left the Catholic Church, the radical social thought of the Anabaptists and the mystical teachings of the Spiritualists inspired him. Paracelsus's views of the End of History also reflected his belief that humans would play a major role in bringing about the Kingdom. He was convinced that human beings, within the larger framework of their role in perfecting nature, offered their greatest service in mastering the secrets of nature. This was the ultimate act of repair and of the restoration of the wisdom of Adam. On this front, Paracelsus believed that this repair would be achieved above all by *Elias artista*.

Elijah or Elias, the ancient prophet who had played a key role not only in Judaism but also in Christianity and Islam, figured prominently in the eschatologies of the sixteenth century. Many had called Luther "Elias," and many prophets from Hubmaier and Hoffman to Postel and even the kabbalists of Safed had envisioned themselves in this role. What was novel about Paracelsus was his association of this prophet not only with a savior but also with the one who would restore humanity's full understanding of the natural world. For the messianic age would not only be a time of justice and concord but also one in which men and women would have full understanding of themselves and the cosmos as a whole.[31] As Paracelsus wrote in *On Natural Things*, "many arts are withheld from us because we have not ingratiated ourselves to God so that he would make them manifest. . . . God has allowed only that which is lesser to emerge; what is sublime is still hidden and shall remain so up to the time of the arts of Elias."[32] Ultimately, Paracelsus believed that, with a clearer understanding of nature, man had the capacity to recover his Adamic state and, in so doing, he would hasten the perfection of society.[33] As Paracelsus brilliantly demonstrated, not only scripture but also the Book of Nature was teeming with signs that pointed the way toward a *güldene Welt*—one that would bring peace and harmony to the world for a thousand years prior to the Second Coming, the Last Judgment, and the Final End of History.

While many had appreciated his teachings in his lifetime—Paracelsus died at the age of forty-seven in 1541—few of his writings were published while he was still living, most of them circulating only in manuscript. An acerbic personality, he publicly burned the books of Galen at Basel and frequently alienated his fellow scholars to whom he was often condescending, bluntly attacking them and dismissing their theories as useless. Constantly on the move, he sought medical knowledge not in books but from practitioners. Wherever he went he sought wisdom from anyone who had had genuine experience in treating disease. He sought out "not only doctors, but also barbers, bath attendants, learned physicians, old wives, magicians (or *schwarzkünstlers,* as they call themselves), among the alchemists, in the cloisters, among the nobles, the common people, among the clever and the simple."[34] But within a generation after his death, attitudes toward Paracelsus shifted dramatically; by the 1560s and 1570s an intense interest in his work had developed, and many began to publish not only his works but works attributed to him.[35] Many now found his medical ideas compelling. But what was especially exciting to his readers in the late sixteenth century was the fact that Paracelsus had put alchemy at the center, not only of medicine and natural philosophy, but of religion as well—in a body of work that constituted nothing less than a Theory of Everything.[36] While providing a coherent picture of the world, Paracelsus's work was also profoundly optimistic.

This interest in a more positive outlook derived in large part from a major shift that was taking place in German culture in the mid-sixteenth century. In the first decades of the Reformation, most Lutherans had resigned themselves to a sense that history was in its Last Days, that there would be no reform on earth.[37] But in the mid-sixteenth century these views had begun to shift, and a more optimistic eschatology emerged. More and more writers began to share prophecies and to find signs in nature that humanity was on the verge of a Beautiful Ending. In 1545, for example, Melchior Ambach, a pastor in Nuremberg, published *Vom Ende der Welt* (Concerning the End of the World). In this anthology of prophetic poems, which he claimed to have found in an old book, he presented the reader with a veritable chorus not only of biblical prophets but also of medieval visionaries, casting their teachings in verse and presenting such major figures as Methodius, Hildegard, Joachim of Fiore, and John of Rupescissa to a broad public.[38]

The growing prestige of Paracelsus's teachings undoubtedly gave an edge in the highly competitive world of early modern alchemy to those who could claim a special knowledge of his ideas and his methods. Certainly the unlikely band of alchemists who presented themselves in the early 1570s to the court of Julius, duke of Braunschweig-Lüneburg, exploited such claims. In addition to promising the duke gold, beneficent medicines, assistance with his mining enterprises in the Harz Mountains, and even theological advice, the alchemist and former Lutheran pastor Philipp Sömmering—along with his assistants Heinrich Schombach, a former court jester, and Schombach's diminutive twenty-five-year-old wife, Anna Zieglerin—claimed a close relationship with a certain and undoubtedly fabricated nobleman, Count Carl von Oettingen, whom, they convincingly maintained, Paracelsus had fathered. But, most crucially, Paracelsus had prophesied that Count Carl, who had inherited his writings, would provide his seed to Anna; and, together, using the alchemical arts, they would produce a large number of children who would help usher in the End of Time.[39] Anna and Count Carl von Oettingen would use a secret substance that they called the lion's blood to conceive children and help prepare for the End Times.[40]

Late sixteenth-century Europe teemed with prophets and visionaries—many of them laborers and artisans, male and female. But learned commentaries on the End of Time also played a significant role in shaping hopes for a better future. Jacopo Brocardo, who had earned his living as a tutor to the sons of nobles in Venice, was tried for heresy in the 1560s. After a brief period in prison, from which he escaped in 1565, he fled Italy and established friendships and contacts in France, Switzerland, England, and the empire. A Calvinist, deeply inspired by Joachim and convinced that Savonarola had been the Elias or Elijah of the Age of the Spirit, he elaborated his hopes for an evangelical Christian republic in his commentary on the Apocalypse, first published in 1580 and translated into English in 1582. Buoyed by the growing harmony among reformers throughout much of Europe, he dreamed of an ecumenical council to be held in Venice and of the Christianization of the world. All signs pointed to this. "The preaching of the Gosple shalbe in the whole worlde," he wrote, adding, "No other religion, no other lawe, and rule to heare then that of the Gosple shall be heard." The final stage of history in Brocardo and in so many others would be a Beautiful Ending. "Then shall be the kyngdom of God in the state of the Holy Ghost untyll that when the Saboth is fynished in this worlde, hee bryngeth us in his thyrde comming to Heaven."[41]

In the early seventeenth century a powerful story of cultural crossings circulated in Europe. At the end of the fourteenth century, a German monk by the name of Christian Rosenkreuz—so the story went—had set out on a journey in the company of an older brother to the Holy Land. When the older monk died in Cyprus, Christian continued his journey. He never reached Jerusalem.

Rather, he first spent three years among scholars in Damascus, studying physics and mathematics and mastering Arabic. He then traveled across the Sinai to Egypt, where he studied indigenous plants and animals, before sailing west to Fez, where he spent two years studying Kabbalah and alchemy before returning to Europe. He came first to Spain where he had hoped to share his knowledge, but he was greeted by scorn. He then returned to Germany where he formed a secret society—the so-called Rosicrucian Brotherhood. This brotherhood sought both a furtherance of religious reform and the advancement of the knowledge of nature. It looked forward to the Apocalypse and the Heavenly Jerusalem. Rosenkreuz died in 1484.

According to a short text entitled the *Fama Fraternitatis, or A Discovery of the Most Noble Order of the Rosy Cross,* first published in Kassel in 1614, the story of Christian Rosenkreuz, which had been neglected and then forgotten over the course of the sixteenth century, had come to light when, in 1604, his tomb had been unearthed while alterations were being made on a building where his vault had been located. There, along with his body, various kabbalistic and alchemical writings, all pointing to the promise of a spiritual and intellectual revival, had been discovered. Then, in 1615, a second Rosicrucian text, the *Confessio Fraternitatis*, elaborated the deeper hopes of this imagined fraternity for a great renewal.[42] The destruction of the world was at hand, but before the End came, God will bless the world with "such truth, light, life, and splendor as the first man, namely Adam," had enjoyed before the Fall.[43]

The story of Christian Rosenkreuz was a fiction, but—by drawing on the hopes for restitution and renewal in the writings of such figures as Postel, Paracelsus, and Brocardo along with many of the more optimistic Lutheran prophecies of the late sixteenth century—it gave expression to many of the longings for renewal that intensified not only in the empire but in much of Europe in these years. Political circumstances also fueled hopes for a Beautiful Ending. Tobias Hess, a visionary and—along with Johann Valentin Andreae—likely one of the authors of the *Fama* and the *Confessio,* had looked to Duke Frederick of Württemberg as the political leader who would give decisive support to the great renewal.[44] In the meantime, other writers looked to Frederick IV, the "Winter King" of the Palatinate who had recently married Princess Elizabeth, the daughter of James I of England, as the ruler who would usher in a great renewal.[45] Their court at Heidelberg became a beehive of cultural activity, much focused on Paracelsian, kabbalist, and Rosicrucian longings for a new Reformation and for a great transformation into a happier world. Indeed, many contemporaries—among them the astrologer Paul Nagel of Leipzig and the mathematician Johann Faulhaber of Ulm, some of them unsure whether the story of Christian Rosenkreuz and the announcement of the Secret Brotherhood were true or not—picked up on these themes and elaborated them in profoundly optimistic works of their own. In 1619, for example, Nagel predicted in his *Stellae prodigiosae* a new golden age that would dawn under the conjunction of Saturn and Jupiter in 1623 and last for forty-two years.[46] But such ideas were widespread. From 1600 to 1630, according to the recent calculations of Leigh Penman, more than two hundred such works poured off the presses in the Empire, with the numbers spiking from 1618 to 1625.[47]

It was also in this context, for example, that Philip Ziegler, a Paracelsian who fell under the sway of the *Fama Fraternitatis* in 1614, proclaimed himself a new Elias. Anticipating a major reform of society and at one point even contemplating killing all the nobles in Europe—a group he saw as an obstacle to justice—he prophesied the coming of a New Jerusalem. The outbreak of the Thirty Years War in 1618, a conflict that was accompanied by a bright comet, generated both new fears and new hopes about the End of Time. In 1622, the Lutheran pastor Paul Felgenhauer articulated his dream of Frederick V's restoration to Prague, which he envisioned as a New Jerusalem. And in 1626 Wilhelm Eo Neuheuser, a former metallurgist who had long viewed himself as a prophet, laid out his hopes for a radically reformed Holy Roman Empire in his *Sacrosanctum et unitum imperium*.[48]

Not everyone had a political program. Johann Arndt's *Wahres Christenthum*, or *True Christianity*, published in four volumes between 1605 and 1610, stressed not social or political reform but rather the interior piety of individual Christians and their unity with Christ. Inspired by late medieval mystics, Arndt's work gave particular emphasis to the need for the Christians to reflect on their hearts or their interior state. The great struggles between good and evil took place, not in the external world, but rather within each individual, and the world that Arndt looked forward to was one in which individual Christians would focus on their own personal piety and their relationship with Jesus.[49] It was a popular message, offering believers a more intense personal piety that the theological formalism of late sixteenth-century Lutheranism had largely eclipsed. Also in the early seventeenth century the Paracelsian cobbler and visionary Jakob Böhme also foresaw a new golden age, though, in his thought, it would take a more political, though not a perfected form. "Zion shall indeed be discovered," he wrote, "and heaven will give its dew, and the earth its fat, but not to the extent that evil shall cease altogether."[50]

One of Arndt's popularizers—Johann Valentin Andreae, who published a digest of *True Christianity*—also turned away from the millenarian hopes of many of his contemporaries. Earlier Andreae had embraced more radical expectations. Indeed, he likely played a role in the drafting of the two Rosicrucian manifestos. Certainly he had shared many of the ideals of these texts and embraced the aspirations of their authors for a new reformation that would lead not only to a spiritual but also to a scientific reform of the world. Born in 1586, Andreae had come of age in a period of deep disillusionment among many Lutherans with what they perceived to be the rigidity and sterility of the Lutheran church and hierarchy. Like many of his generation, he was intellectually restless and, at least in his youth, open to many of the new philosophical and theological ideas—circulating not only in Germany but throughout all of Europe—that placed an increasing emphasis on the inner self. At first he explored these ideas at the University of Tübingen, where he studied under the hermeticist and millenarian Christoph Besold, and then in his *Wanderjahre,* which carried him not only throughout the empire but also to Rome and even to Geneva, where he was impressed by the city's regulation of morals. While we are not sure whether or not Andreae contributed to the Rosicrucian manifestos, he was undoubtedly the author of the *Chymische Hochzeit Christiani Rosenkrantz,* which made it clear he did share many of their hopes for scholarly, ethical, and religious renewal and was open to a variety of alchemical, even Paracelsan ideas, in his pursuit of a general reformation. He was also open to the ideas of Campanella, several of whose works, including *The City of the Sun,* had found their way in manuscript into Tübingen itself and played some role in inspiring his hopes for a better future.[51]

Yet, by the late 1610s, Andreae had begun to reject his earlier, more expansive enthusiasms, which had often taken him beyond the bounds of Lutheranism, and returned to a stricter Lutheran outlook. We are not sure what led to this change in outlook. Perhaps it was a matter of Andreae's having settled down. He had married and he had been ordained Lutheran pastor at Calw, a town some forty kilometers northwest of Tübingen. Or perhaps he was responding to the growing fears in Germany of the breakdown of the political order in the years just before the outbreak in 1618 of what would come to be known as the Thirty Years' War, one of the deadliest conflicts in modern history. But, while Andreae firmly rejected the millenarianism of both Besold and Campanella, he nonetheless dreamed of a better world, as his utopia—the *Reipublicae Christianopolitane descriptio* (Description of the Republic of Christianopolis), published in 1619—makes clear.[52]

Like More and Campanella before him, and Francis Bacon in the following decade, Andreae's utopia was the relation of a traveler or pilgrim. Shipwrecked in the southern hemisphere, the pilgrim washed ashore upon Capharsalama, the site of a battle in the Old Testament, in a sea south of the equator, and there he is given permission to visit what proves to be the astonishingly harmonious city of Christianopolis. Beautifully laid out—the very image of a Heavenly Jerusalem on earth—the city is surrounded by a rich countryside that supplies its every need. While life is in many respects communal, the family remains the single most important social institution and the individual citizens are for the most part craftsmen-scholars who seek both to learn as much as possible and who express themselves in their exquisite works. They conduct experiments in a vast *laboratorium* and attend lectures on, among other topics, grammar, logic, math, music, astronomy, astrology,

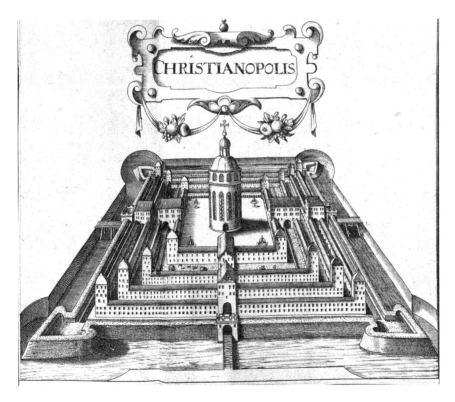

Utopia as the Heavenly Jerusalem. From Johann Valentin Andreae, *Reipublicae Christianopolitane descriptio* (1619).

civil and church history, ethics, theology, and prophecy. The city is ruled gently by three governors: one in charge of religion, another of justice, and a third of learning. Finally, the community achieves its harmony under the sway of its deeply-felt devotion to Christ. In his dedication of the work to Johann Arndt, Andreae described Christianopolis as "a very small colony taken out of that great Jerusalem that you have built up."[53] Comparing his own age to that of Luther in his preface, Andreae made it clear he was living in a period in which, once again, "the light of a purer religion" would shine upon his generation.[54]

Above all, Andreae makes it plain that salvation lay not in the Millennium so many others eagerly awaited but rather in the more traditional Lutheran hope for individual salvation. Final happiness would come in the promise of eternal life "through which we shall possess perfect light, completion, true peace, knowledge, contentment and happiness."[55] Nonetheless, like Arndt and Böhme, Andreae had not abandoned hope for a better future but channeled it away from millenarian dreams of collective salvation to the individual plane. Nonetheless he preserved in his utopia a vision of a harmonious, peaceful, and fulfilling world, even if the world of which he dreamed was not perfect. Andreae may have abandoned his more enthusiastic hopes of his youth, but he offered nonetheless a vision of a better world, a vision that would play a role in shaping at least one current of Christian expectations for a Beautiful Ending in the first part of the seventeenth century not only in the empire and the continent but also—and perhaps especially—in England.

CHAPTER 12

Crossing the Pillars of Hercules

> Accordingly I must open and lay out my conjectures which make hope in this business probable, just as Columbus did before his epic voyage across the Atlantic, when he gave reasons why he believed he could discover new lands and continents beyond those known then, reasons which, through rejected at first, were afterwards vindicated by his experiment, and were the origin and cause of vast consequence.
>
> —FRANCIS BACON, *The Great Instauration* (1620)

The fall was sudden, precipitous. After many years of ingratiating himself to his monarch—first to Queen Elizabeth and then, after her death in 1603, to James I—Francis Bacon reached the very pinnacle of power when, in 1618, the king made him lord chancellor. Courtiers always run the risk of a fall from favor. In 1621 Bacon, hated by many, was impeached in Parliament and then, even after tendering his resignation, tried and sent to the Tower. Bacon's demise stunned London. "Nothing like it has ever been seen in any other Parliament," the Tuscan ambassador to the English crown observed, "and, being so sudden, it is even more fearful."[1] But Bacon was quickly released, and managed to transform what would have been a disaster for others into an advantage. From his unceremonious impeachment down to his death five years later, Bacon—whom the king insisted retire to his estate, a day's carriage ride from London—wrote tirelessly.[2] The texts that Bacon produced or completed in this period would form a critical addition to his earlier philosophical work, bringing his insights to a broader and broader public. Yet even before his fall, Bacon had published what would prove to be his most important work: *The Great Instauration,* published in London in 1620.[3] Ever since, readers have looked back at this text as constituting nothing less than a revolution in the understanding of the natural world.

Throughout the Middle Ages and down through much of the Renaissance, humanists, philosophers, and theologians had based their arguments about nature largely on the authority of ancient texts—whether religious or secular. Theologians looked back to scripture; astronomers to Ptolemy; anatomists to Galen; and almost everyone to Aristotle. Even as a student at Cambridge, Bacon had been deeply critical of this reliance on ancient texts, a practice that, as he would later write, resulted in men "forever working over and talking about the same things," culminating in what he saw as pointless, endless "repetitions."[4] But what Bacon saw as the failure of earlier generations to understand the natural world was not merely a result of their overreliance on the ancient authorities; it was also a question of method. Traditionally, as Bacon made clear, natural philosophers did not investigate nature so much through the study of natural phenomena—though they did indeed do this—as through the exploration of logical relationships, deducing what they viewed as new understandings of the natural world through deductive reasoning. On this front, the logical treatises of Aristotle, who was viewed in the Middle Ages as the greatest of the philosophers, had exercised enormous sway. After all, his *Organon*—a work that anthologized all his treatises on logic from the *Categories* to the *Prior* and *Posterior Analytics*—offered powerful tools of reasoning and laws of inference that, as they began to be recovered and to circulate in the Middle Ages, seemed to bring order into a chaotic world. Bacon, by contrast, rejected the largely deductive approach of Aristotle and the Aristotelians. In its place he put the direct examination of natural phenomena at the very heart of his philosophy, a method that he introduced and explicated in his *Novum Organum*. The title of this work, which constituted part 2 of Bacon's *Great Instauration,* would have immediately captured his readers' attention. With it Bacon made an explicit play to replace the *Organon* of Aristotle with an entirely new set of methodological principles.[5]

Bacon's emphasis on method stemmed also from his view—and here he was doubtless influenced both by his reading of Montaigne and, perhaps more deeply, by Calvinist teachings—of the frailty of the human mind, which he regarded as "fallen," subject to many forms of self-deception, and prone to error. "The root cause of practically all the evils in the sciences," Bacon wrote in the *Novum Organum,* "is but one thing: that . . . we mistakenly admire and magnify the powers of the human mind."[6] To overcome this, Bacon delved into an analysis of what we might call a cognitive psychology, identifying four modalities—what he called "idols"—in which the human mind was prone to misinterpret the world around it. For most of his contemporaries, "idols" referred, of course, to false gods, but Bacon here used this word to mean "illusions" or predispositions that make the clear apprehension of reality difficult.

In part, this was a limitation of human nature itself, "for the mind of Man is farre from the Nature of a cleare and equall glasse, wherein the beames of things should reflect according to their true incidence; Nay, it is rather like an inchanted glasse, full of superstition and Imposture."[7]

In particular Bacon warned his readers against the idols of the *Tribe*, of the *Cave*, of the *Marketplace*, and of the *Theatre*. All human beings, Bacon argued, suffer from the illusions of the Tribe. That is, we are all inclined to assume greater order in nature than is actually found there, to ignore evidence that contradicts our most deeply held beliefs, and to believe true what we would like to be true. But if the idols of the Tribe are shared by all of us, the idols of the Cave are particular to each one of us. Some of us, Bacon suggests, tend to focus on differences; others on similarities; some of us are attracted to antiquity, others to novelty; and some focus on form, others on structure. But our understanding, Bacon continued, is also hampered by ordinary language itself—by its vagaries and even its naming of things that do not exist in reality—and Bacon referred to this illusion as the Idol of the Marketplace. By contrast, the Idol of the Theatre focused not on common language but rather on reigning philosophical constructs, abstract arguments that often—based on limited knowledge—ended up mixing philosophy with theological concerns.[8]

It was both to escape the enchantments of antiquity and these weaknesses of the human mind (the "inchanted glasse") that made the construction of a new method essential, and in his *Novum Organum* Bacon offered what would prove to be a transformative proposal for a new path to the construction of human knowledge. Dismantling traditional and often metaphysical theories of form, such as those found in Plato and Aristotle on the underlying structures of the physical world, Bacon sought the principles of the natural world not in the heavens or in logical inferences but in the identification of the natural causes of various particular phenomena such as motion, gravity, color, and heat. To grasp these forms, the philosopher must follow, Bacon argued, a rigorous method of induction—thus the philosopher must begin with specific observations of natural phenomena and not from received generalizations. Thus, the investigator begins, first, by listing a "Table of Presences," all those factors—or "causes"—that appear to contribute to a certain form; then, secondly, makes a "Table of Divergences," a list designed to demonstrate, through a principle of falsification, whether the possible contributing factors identified in the Table of Presences are always present when the form under investigation occurs and are *always* absent when not; and finally, a "Table of

Degrees"—"a presentation to the understanding of instances in which the nature under inquiry is found in different degrees."[9]

Thus, in his celebrated investigation into the form of heat, for example, Bacon included in his Table of Presences twenty-seven probable causes of heat, among them the sun's rays and the insides of animals—certainly places in which we are likely to find warmth. But, upon examination, as he shows in his Table of Divergences, the sun's rays are not always consistent with heat, since at the polar regions, even when the sun is shining, conditions are often cold. By contrast, the insides of animals are consistently warm. Finally, in exploring the nature of heat in animals, when studying them by degree, he determined that the heat increases when the animal is in movement and decreases when the animal is less active. Bringing these observations together, Bacon then offered his preliminary account of the causes of the form of heat in what he calls his "first vintage"—his phrase for his first findings. "Heat," he concluded, "is a motion, expansive, restrained, and acting in its strife upon the smaller particles of bodies."[10]

Bacon's theory of induction proved thrilling. Indeed, in offering an approach that combined an empirical and a rational approach, the *Novum Organum* constituted the most serious blow yet to Aristotelianism and effected nothing short of an "epistemological" revolution—a fundamental shift in the understanding of how we know what we know—while establishing a "scientific" method that came to replace the most influential methods of both the ancient and the medieval world. It was a science, moreover, explicitly directed toward dominion over nature that Bacon believed would result in a plenitude of benefits for humankind. He saw himself, that is, as an architect of a better world that could be created through human agency. His meticulous attention to "scientific" method, his recognition of the need for collective and collaborative models for the investigation of nature, and his emphasis on the "utility" of knowledge would ultimately do much to shape the modern idea of progress—the belief that the understanding and domination of nature would be a great benefit to humankind.[11] The *Novum Organum*, then, constituted a major milestone in the institution not only of modern science but of modernity itself. Certainly, many of Bacon's followers in late seventeenth-century England and, later, in eighteenth-century France perceived him this way. And, by the early nineteenth century, this view had become historiographical orthodoxy.[12]

The image of ships passing back through the Pillars of Hercules that appeared on the frontispiece to the *Great Instauration* with the motto *multi pertransibunt, et augebitur scientia* ("many will pass to and fro, and knowledge will increase")—served as a powerful metaphor for the explosion of knowledge that was taking place in Bacon's time.[13] The image brilliantly captured Bacon's goal of crossing boundaries, of establishing something new. Moreover, from his own vantage point and that of his contemporaries, ever since Columbus's first crossing, the growth of global commerce had fundamentally transformed the ways in which Europeans understood the world. And for someone like Bacon who long lived in London, the reality of global trade was palpable.

Like Montaigne in Bordeaux, Bacon would have frequently seen ships putting into harbor after their voyages across the ocean. He and his contemporaries also heard stories of the adventures of mariners who had seemingly accomplished the impossible. In the late 1570s, when Bacon was in his teens, the English courtier and adventurer Sir Francis Drake had circumnavigated the globe, and the stories of his heroic travels, of exploration and of piracy, stirred the imaginations of Londoners and other Englishmen eager to hear of his most recent feats.[14] Bacon was twenty-four when Queen Elizabeth, in 1584, chartered Sir Walter Raleigh to explore and settle "any remote, heathen and barbarous lands, countries, and territories, not actually possessed of any Christian prince, or inhabited by Christian people."[15] Bacon had undoubtedly been thrilled by stories of his adventures: his establishment of Roanoke Colony in North Carolina in 1587 and his two journeys up the Orinoco River in Venezuela in search of El Dorado, the legendary city of gold, the first in 1595, and, after thirteen years in prison, the second in 1617. To be sure, Raleigh's ventures failed—the colony at Roanoke had inexplicably vanished upon his return there in 1598—and he never found El Dorado. After his final expedition, he was tried for treason and executed in 1618.[16]

English expansion overseas was by no means limited to such celebrated figures as Drake and Raleigh. From the late sixteenth century on England had begun to establish plantations, first in Ireland, subjugating an Irish population with brutality. Then, from the early seventeenth century the English began to establish colonies in the Americas, eventually creating settlements that reached from the Caribbean to Newfoundland. The first attempt by Raleigh at colonization, at Roanoke, failed. In 1607, however, Captain John Smith succeeded in establishing the first permanent English settlement in North America in Virginia at Jamestown. And there, and along the Chesapeake, English settlers—often through their violent displacements and repression of the native peoples—would rapidly grow in numbers, creating plantations, cultivating tobacco. While many came for economic reasons, others came for religious freedom.[17] According to a famous story, the passengers on the *Arabella*—Calvinists or Puritans who established the first English colony in New England—were inspired by their religious hopes. It was during their crossing in the summer of 1630 that John Winthrop, so it is said, underscored the role of Providence in bringing them to the New World and laid out his hopes for a new society, his "city upon a hill."[18]

Moreover, in this same period England also began to compete directly in the Indian Ocean for the Asian trade. And within only a few years after Queen Elizabeth chartered the East India Company, the English had established a network of trading fortresses in such major ports as Calcutta, Madras, and Mumbai.[19] English expansion in this period was clear evidence of a growing economy and growing contacts with the world beyond its borders. The expansion, as Bacon recognized, led to new forms of knowledge. We have already seen how, over the course of the sixteenth century, Europeans came to embrace the model of an inhabited earth, continually adding more and more detailed knowledge of the seas and landmasses that stretched around the globe. But the new forms of knowledge extended well beyond cartography. Missionaries and travelers brought back reports of new flora and fauna, greatly expanding such fields as botany and zoology. And they also produced new ethnographies, creating an increasingly rich library of the diverse customs of the world.[20] As Bacon observed in his text, "surely it would be a disgrace to mankind if, while the expanses of the material globe, i.e. of the lands, seas, and stars, have in our times been opened up and illuminated, the limits of the intellectual globe were not pushed beyond the narrow confines of the ancients' discoveries."[21]

To be sure, Bacon was not the first to make use of the image of the Crossing of the Pillars of Hercules to underscore the rapid growth of new knowledge in his own age. In the first sections of his *Natural and Moral History of the Indies*, for example, José de Acosta had criticized not only Plato and Aristotle but also St. Augustine and Gregory of Nazianzus for having viewed the Pillars of Hercules as a boundary beyond which no humans could be found.[22] And he had invoked the crossing of the Pillars of Hercules to convey his sense of the great changes not only in empire and commerce but also in culture that the era of oceanic reconnaissance had brought about. But none had gone as far as Bacon in emphasizing the linkages between commerce and culture. Nor was this the first time Bacon had made use of this image. In his first major book, *The Advancement of Learning*, published in 1605, Bacon had made use of the same metaphor. "For why should a fewe received Authors," Bacon wrote, "stand up like *Hercules Columnes,* beyond which, there should be no sayling, or discovering, since wee have so bright and benigne a starre, as your Majesty, to conduct and prosper us."[23] Yet as Bacon's frontispiece also makes clear, his project was not a purely secular one, as tempted as we might be, at least at first, to read the motto *multi pertransibunt, et augebitur scientia* in this way. To the contrary, Bacon's contemporaries would have immediately recognized its religious and even eschatological significance. For the motto was a phrase Bacon derived from a familiar apocalypse in the Book of Daniel in which an angel reveals that "many running to and from" and "knowledge increasing" would be signs of the End Times. For when these two conditions would converge, the angel announced, then "many of those who sleep in the dust of the earth shall awake, some to everlasting life, and some to shame and everlasting contempt. Those who are wise shall shine like the brightness of the sky, and those who lead many to righteousness, like the stars forever and eve" (Daniel 12:2–3).[24]

Nor was this the only moment in which Bacon underscored his view that the growth of commerce and knowledge was a sign of the End of History. In his *Valerius terminus* of 1606, he had attributed this growth of knowledge to "God's own planting," and its spreading and flourishing to divine "providence" that had been designed by "a special prophecy" to occur in "this autumn of the world." "So it is not violent to the letter," Bacon continued, "and safe now after the event, so to interpret that place in the prophecy of Daniel where speaking of the latter times it is said, *Many shall pass to and fro, and science shall be increased;* as if the opening of the world by navigation and commerce and the further discovery of knowledge should meet in one time or age."[25] He made a similar point in the body of the *Novum Organum* itself. "And we must not forget the prophecy of *Daniel* concerning the last ages of the world: that *Many shall go to and fro and knowledge shall be increased,* which manifestly hints and signifies that it was fated (i.e. Providence so arranged it), that thorough exploration of the world (which so many long voyages have apparently achieved or are presently achieving) and the growth of the sciences would meet in the same age."[26] To be sure, Bacon's apocalyptic vision had little in common with traditional forms of millenarianism or with anticipations of a Second Coming—expectations held by Columbus as well as many other individuals and communities in the early modern world—that served as motivating forces through their promises of a Beautiful Ending. Nonetheless, Bacon's faith in divine providence was as foundational to his undertaking as Columbus's had been in his. Columbus, to whom Bacon frequently compared himself, had been inspired by his faith that he had himself been called to fulfill the divine prophecy that, as Jesus had reassured his disciples in Matthew, the "end will come" when the Gospel had been proclaimed throughout the world. And it was in this light that Columbus fashioned himself as *Christopherens,* as the one who would bring Christ to the New World. Bacon's faith was more sober, as his was an apocalypticism of retrieval, recovery, restoration. He believed that in his own day the growth of knowledge would enable—as he made clear in his *Novum Organum*—a restoration of its "God-given authority over nature." Such a vision, in Bacon's case, involved a return to Man's Estate before the Fall and was, therefore, deeply inscribed in Christian History. "For by his fall," Bacon wrote in his conclusion to the *Novum Organum,* "man lost both his state of innocence and his command over created things. However, both of these losses can to some extent be made good even in this life: the former by religion and faith, the latter by the arts and sciences. For the curse did not quite put creation into a state of unremitting rebellion, but by virtue of that injunction *In the sweat of thy face shalt thou eat thy bread,* it is now by various labours (not for sure by disputations and the idle ceremonies of magic) at length and to some degree mitigated to allow many his bread or, in other words, for the use of human life."[27] Man's scientific striving would restore Eden. Man would be again innocent and knowledgeable and thus, by going backward, would be assured a Beautiful Ending. Bacon's understanding of Providence and of Christian history, therefore, provided a powerful framework for the way he understood the changes taking place around him. He saw time not only as providential in the sense of leading to the Apocalypse but also as circular, as restorative, taking humanity back to its pristine beginnings, to Adam.[28] Like many of his contemporaries in Germany, especially the Rosicrucians, Bacon believed that a deepening of the knowledge of nature would serve as a portal to the retrieval of Eden and a great renewal. Thus, with the *Great Instauration* we are not in the world of modern secular science. Rather we are in a far more curious world in which linear, circular, and providential time coexisted and informed one another—in a world, that is, marked by a plurality of temporalities.

For Bacon, moreover, Providence instilled great confidence in the ability to improve humanity's state in the future. The growth of knowledge, in his view, was taking place in nothing less than a quiet unfolding of Providence. As Bacon writes, "we must begin with God; for what we are about is, on account of the excellent nature of the good in it, something which comes from

God, who is the author of good and father of lights. Now in divine workings, beginnings however slight have an inexorable outcome. Indeed what is said concerning spiritual matters that *the Kingdom of God cometh not with observation,* is also found to take place in every great work of divine providence, that everything comes in quietly and without sound and fury, and the thing is done before men suppose or take notice of it."[29] "Now because these things are not ours to command, at the beginning of the work," Bacon wrote in the preface to the *Great Instauration,* "I pour forth most humble and hearty prayers to God the Father, God the Word, and God the Holy Ghost that, having in mind the afflictions of the human race and the pilgrimage of this our life in which we wear out days short and evil, they will think fit through my hands to endow the human family with new mercie."[30] Providence, that is, played a significant role in shaping his own optimism about man's ability to bring "new remedies to the human family." And it was in this frame that Bacon addressed a prayer to God: "Therefore, *Father,* who has given us visible light as the first fruits of Creation, and breathed intellectual light into the face of man as the culmination of Your works, protect and reign over this work which, proceeding from Your goodness, returns to Your glory. . . . Wherefore if we sweat and strain in Your works, You will make us to share in Your vision and sabbath."[31]

Bacon's faith was also the basis of his hope—a virtue Bacon viewed as foundational to the improvement of humanity's lot, for, as Bacon observed, "by far the greatest obstacle to the advancement of the sciences . . . lies in men's despairing belief that the job is impossible." Accordingly, he continued, "I must open and lay out my conjectures which make hope in this business probable, just as Columbus did before his epic voyage across the Atlantic, when he gave reasons why he believed he could discover new lands and continents beyond those known then, reasons which, though rejected at first, were afterwards vindicated by his experiment, and were the origin and cause of events of vast consequence."[32] Indeed, it was hope that, ultimately, became the springboard of Bacon's undertaking. Bacon not only saw his entire enterprise as one of hope—a hope inspired by faith in Providence and by recent accomplishments in natural philosophy and a confidence that "nature's recesses still conceal[ed] secrets of excellent use," but also by his own example.[33] He points to his own experience and activities "not to show off but because it helps to say it," but because if he had been able to accomplish so much while "engrossed in the affairs of the state" and "not in very good health." How much more, then, could those who followed Bacon's teachings and methods accomplish—especially those "with plenty of spare time" and capable of working together. On the matter of hope, then, Bacon was not offering merely one element among many in his new philosophy of nature. Rather he was pointing to hope—shaped above all by faith in God—as a wellspring of action in the world.

In many ways, Bacon's years at St. Alban's were a gift. Here, for the first time in his adult life, he was able to devote himself fully to his studies. He was immensely active and productive, both writing and conducting experiments. It is even possible, if the early accounts can be trusted, that it was his overzealousness in his investigations of nature that led to his death. One cold evening in February 1526, Bacon had stayed out for too long while trying to determine if packing a hen with ice would preserve the meat. He took ill, traipsed through the snow, and sought shelter in a friend's home, where he died. But earlier at St. Albans Bacon had completed what would prove to be his most popular work: the *Sylva sylvarum,* a collection of a thousand "experiments" meant to illustrate for a more popular reading public the power of his ideas.

Toward the end of the volume, which was published slightly over a year after Bacon's death, William Rawley, his literary executor, included another of Bacon's previously unknown works, the *New Atlantis,* as a kind of appendix.[34] In this work—which Rawley called a fable, clearly noting that Bacon had not "perfected" or fully completed it—Bacon offered a vision of an ideal society on an imaginary island. But, unlike More and Campanella, whose utopias—in their largely egalitarian and communitarian social arrangements—offered a clear counterpoint to the brutal inequalities and hierarchies of European society, Bacon paid very little attention to social and political arrangements. Indeed, his description of the island of Bensalem, the utopian society of his *New Atlantis,* makes it clear that Bacon was not intent on breaking with the social and political hierarchies of his own day. For Bacon, the utopian turn lay above all in the fashioning of a society in which natural philosophers, engaged in continuous and sustained investigations into nature, would share their benefits with the residents of the island. Bacon's utopia in short was one of health and abundance, of luxury and pleasure.[35] What is most striking about Bacon's vision is its commitment to "scientific" experimentation in furthering the welfare of humanity. It would be a world in which, rather like Plato's Republic, philosophers would be kings.

To accomplish this goal, the residents of Bensalem enjoyed and supported the work of an institute called Salomon's House, dedicated, as the narrative reports, to the "knowledge of Causes, and secret motions of things; and the enlarging of the bounds of Human Empire, to the effecting of all things possible."[36] The research utopia Bacon then described was breathtaking. Salomon's House had a network of natural laboratories for the conduct of experiments. Some were located in caves deep underground; others in towers that soared high above ground; and still others at natural sites such as lakes, ponds, and wells—all used to explore such matters as refrigeration, the production of "new artificial metals," the study of weather, and the prolongation of life.[37] Still other sites—"great and spacious houses," chambers, baths, orchards, gardens, parks, brewhouses, bakehouses, kitchens, dispensatories, furnaces, perspective houses, sound-houses, perfume houses, and engine-houses—were used for various experiments in the alteration of nature for the benefit of humanity. In addition, Salomon's House sent out six of its fellows every twelve years to other parts of the world to collect "books, and abstracts, and patterns of experiments." It is true, the governor stated, that the king had forbidden contact with the outside world and had decreed that strangers could not land on the island unless they were ill and in need of replenishment. But he made one exception:

> Every twelve years there should be set forth out of this kingdom two ships, appointed to several voyages. That in either of these ships there should be a mission of three of the Fellows or Brethren of Salomon's House, whose errand was to give us knowledge of the affairs and state of those countries to which they were designed, and especially of the sciences, arts, manufactures, and inventions of all the world, and withal to bring us books, instruments, and patterns in every kind. . . . Thus you see, we maintain a trade, not for gold, silver, or jewels; nor for silks, nor for spices, nor any other commodity or matter, but only for God's first creature, which was *Light:* to have *light* (I say) of the growth of all parts of the world.[38]

Crucially—much as in his frontispiece to the *Great Instauration*—Bacon connected the growth of knowledge and the improvement of humanity with crossings. A period of frequent commerce between previously distant lands such as the one that had emerged in Bacon's own time would lead to an augmentation of knowledge. Most striking, however, is the rich way in which Bacon's vision of his utopia derives its energy from the sacred narrative in which it was embedded. For in the *New Atlantis,* as in the *Great Instauration,* Bacon dreamed of a restoration, enfolding his account of the island and its customs within a providential history. In the *Great Instauration,* the emphasis had fallen on the connection of commerce with the growth of knowledge in a passage from the Book of Daniel that evoked the dream of a Beautiful Ending. In the *New Atlantis,* by contrast, Bacon enfolded his account of the island and its customs within a sacred narrative, one that pointed both back to a golden age and forward to a Beautiful Ending in which human beings would bring the natural order under their control for the benefit of humanity.

In most narratives of this nature we would expect the inhabitants of this unknown island, which the sailors learn is called Bensalem, to be pagans. But even before landing the sailors had learned that the Bensalemites were Christians, and, as we might expect, were deeply curious about how these men and women, in such a remote place and unknown to Europeans, had received the Gospel. What they learned astonished them. Some fifteen hundred years earlier, shortly after the death and resurrection of Christ, as the governor explained to them, a mysterious pillar of light had appeared off the island's shore. Several residents boarded boats to investigate, but only one was able to draw near. As the boat approached, the pillar vanished, turning into what Bacon calls an ark, a floating chest made of walnut. Inside, the residents of Bensalem found various writings: a letter from the apostle Bartholomew and the canonical books of the Old and the New Testament—a true hierophany since, as Bacon notes, many of these books, and he names the Book of Revelation in particular, "were not at that time written."

But if the Gospels became an ark for their own age, an earlier ark, that of Noah, had transformed the world some three thousand years earlier, for it had inspired confidence that it was possible to sail upon the waters, thus fostering what the sages of Bensalem described as a golden age of commerce. At this time, the governor explained—far surpassing the comparatively vague knowledge that Plato had conveyed concerning Atlantis in the *Timaeus*—"the navigation of the world (specially for remote voyages) was greater than at this day," noting that "the Phoenicians, and especially the Tyrians, had great fleets. So had the Carthaginians, their colony, which is yet further west. Toward the east, the shipping of Egypt and of Palestine was likewise great. China also, and the great Atlantis (that you call America), which have now but junks and canoes, abounded then in tall ships. This island (as appeareth by faithful registers of those times) had fifteen hundred strong ships, of great content."[39]

Then, in an act of revenge, God visited a great inundation upon Atlantis, which reduced the inhabitants to a "primitive" state so that, as the Bensalemites teach, they are a thousand years younger than the Europeans and the

Asians. Nonetheless overseas contacts continued to weave together many parts of the globe, and it had been in this ancient period that the Bensalemites had first learned of the Hebrew King Solomon, after whom they had dedicated their foundation, Salomon's House, for the "study of the Works and Creatures of God." Indeed the Bensalemites even had "some parts of his works" unknown to the Europeans: namely that Natural History which he wrote.[40] Now, in Bacon's own age, with a further revival of global commerce, it becomes possible to envision a future in which the knowledge of God's works would rapidly increase. And this increase, along with the "passing to and fro" across the oceans, would serve, as we have seen, as signs of the Apocalypse—the Beautiful Ending that Bacon sought to achieve through science. Even in Bacon's view utopia remained an apocalyptic project.

In the *Great Instauration* Francis Bacon famously pointed to three inventions as the source of the modern age. "Again it helps to observe the force, virtue and consequences of what has been discovered," Bacon wrote, "and that is nowhere more apparent than in those three things which were unknown to the ancients and whose origins, though recent, are dark and inglorious: namely the *Art of Printing, Gunpowder,* and the *Mariner's Compass.* For these three have altered the whole face and state of things right across the globe: the first in things literary, the second in things military, and the third in navigation. From these countless other alterations have followed so that no empire, no sect, and no star seems to have exerted a greater effect and influence on human affairs than these mechanical innovations."[41]

Bacon was far from original in pointing to these "innovations" as transformative. Montaigne had underscored the significance of printing and gunpowder in his *Essays;* Campanella had praised all three in *The City of the Sun;* and each of these inventions figured prominently in Stradanus's *Nova reperta* (New Discoveries), a collection of prints published toward the end of the sixteenth century and widely circulated.[42] Yet Bacon gave greater emphasis than his contemporaries to these new technologies. In stating that "no empire, no sect, and no star seems to have had a greater effect and influence on human affairs than these," he was making a strong claim about historical causality. Rather than locating change in the actions of great men, the focus of most historians in his age, Bacon looked beneath the surface events in order to explain the emergence of the modern. In this book I have also sought to step outside predominant narratives of change. In our own time the majority of scholars have located the origin of modernity in the vast economic,

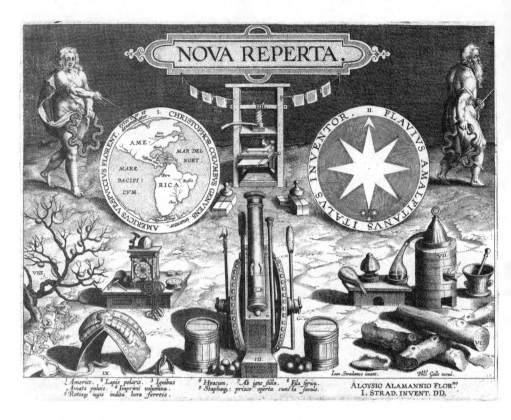

Modern inventions (ca. 1591). The printing press, the compass, and the cannon were only three of the most prominent inventions of the age. Highlighted here are also the discovery of America, the silkworm, the stirrup, the clock, tools for distilling, and guaiacum.

social, intellectual, and economic transitions of the late medieval and early modern period. Without rejecting the view that changes in these spheres were important, my own goal has been to highlight another key factor: namely, the widespread diffusion of apocalyptic hopes and expectations that, first intensifying in the second half of the fifteenth century and then accelerating in the Age of Reform, came to play such important roles in the cultures of Europe and the Mediterranean in the early modern period.

To be sure, the early modern apocalyptic imagination varied in form, content, and intensity. In part, this diversity was shaped by different faith traditions. Jews still awaited the Messiah, while Christians looked forward to the Second Coming, and Muslims hoped either for the return of the Hidden

Imam or the Mahdi and/or the return of the Prophet Jesus. Additionally, within each tradition, a wide range of ideas and a constantly shifting constellation of social and political circumstances shaped different apocalyptic visions. Nonetheless, a common thread of hope ran through this diversity.

For early modern men and women, hope—whether conceived as an emotion, a disposition, or a capacity for optimism—was not a purely individual phenomenon. To the contrary, early modern hope was also always collective and rooted, above all, in the apocalyptic imagination.[43] The consequences of this were powerful. First, the apocalyptic imagination encouraged men and women to understand history not as an academic discipline but rather as a living sacred force that connected them to the past as well as to the future. History in this sense reached from Creation to the End of Time. The world in which men and women lived—the *saeculum*—could perhaps be understood or even analyzed on the basis of archives, documents, and memories of a verifiable past. It is possible to discern an emerging linear notion of secular time from the Renaissance on, yet sacred time remained dominant. Within this framework, the reading of scripture, the interpretation of the stars, and the visions of prophets, living and dead, provided occasions over and over again for men and women to try to make sense of their experience. Second, and perhaps more decisively, the apocalyptic imagination was pregnant with a promise for the future: the promise that, despite the sufferings and traumas of the present tribulations, there would be a Beautiful Ending. In such an environment it was inevitable that some of these men and women would seek to take history into their own hands and do all that they could to shape a future in which their hopes might be realized.

To be sure, apocalyptic expectations were not solely responsible for the transformations that Bacon and his contemporaries had observed. The compass, indeed, had enabled long sea voyages. "When men only had the stars to sail by, they could indeed coast along the shores of the Old World or cross lesser and mediterranean seas," Bacon had observed earlier in his *Great Instauration,* "but before they could cross the oceans and discover the regions of the New World, the use of the mariner's compass, as a more trustworthy and certain guide, had first to be found out."[44] Yet it was not only the compass that gave many the confidence to sail upon open seas, far from land. Equally important was the faith sailors placed in divine providence. Those who first embarked upon open seas were convinced that they were engaged in something much larger than themselves and that their voyages were part of a providential plan in which they themselves played a role in bringing about the

End of History. Columbus, as we have seen, made this aspect of his voyages explicit in his *Book of Prophecies* as well as in many of his letters and other writings, but he was hardly alone. Throughout the early modern period, thousands of missionaries, merchants, and settlers embarked on crossings, in conditions and with risks that are virtually impossible to imagine today, confident that they were fulfilling a divinely sanctioned and providential role. Certainly, many of the friars who journeyed to the New World in the sixteenth and seventeenth centuries viewed their undertakings in a decidedly apocalyptic key. But merchants and settlers also often viewed their undertakings as part of a providential plan. Thus the tendency or habit in the early modern world to see one's actions as part of a sacred drama, and one that often promised the faithful a Beautiful Ending, was undoubtedly no less reassuring than the compass to many who boarded ships in Europe leaving for distant lands across often perilous seas.

In a similar fashion, gunpowder played a significant role in the ability of early modern emperors to expand their authority over larger and larger swaths of territory. First invented in China in the ninth century, this explosive powder took several centuries to reach Europe, and considerable investment and experimentation were necessary before it became possible to use it in ballistics and in military operations. But the increasing centrality of the cannon and the arquebus to the battlefield did much to strengthen imperial authority and underlay the power not only of the Habsburgs in Spain and the Holy Roman Empire but also of the Ottomans in southeastern Europe, the Near East, and North Africa, as well as that of the Dutch and the English as they first began, with cannons now routinely mounted on their galleons, to expand their power across the oceans.

Yet on this front as well apocalyptic hopes also played a role. In both the courts of Süleyman and Charles V astrologers and advisors assured their prince that God had chosen him for a special role. Drawing frequently on the Pseudo-Methodius, other medieval prophets, and the counsels of court astrologers, each portrayed himself as the Last World Emperor. Such a vision undoubtedly motivated many of the efforts from the imperial center to conquer more and more territory, but it was also a vision that justified and legitimated imperial and royal authority. Yet, on the political front, the apocalyptic imagination not only encouraged the growth of empire. As we have seen, from Savonarola's Florence and the Anabaptist uprising in Münster to the rebellion led by Tommaso Campanella in Calabria, dreams of a Beautiful Ending also shaped various utopian dreams of community. What is essential is the recognition that the history of early modern politics, which scholars have tended to interpret in an almost exclusively secular key, was shaped and shaped profoundly by the political theologies of the age. Indeed, it is these theologies and the popular forms that they often assumed that best explain the growth of popular political passions as well as violence not only in the early modern world but also in more recent times.

The printing press, too, as much recent scholarship has made clear, played a key role in enabling the emergence of new forms of knowledge. At the most basic level, print not only increased the number of books and made possible their more rapid circulation but also played a part in raising literacy rates throughout Europe over the course of the early modern period. And print undoubtedly contributed to the broad shifts in the understanding of the natural world, as scholars were able to share their findings with a broader and broader public—a process that played a key role in shaping the great changes that we associate with the scientific revolution.[45] Yet print, which had always been tied to the prophetic, had also done much to enable the diffusion of apocalyptic ideas. And these ideas, in turn, contributed to the drive to discovery within the natural philosophies of such writers as Paracelsus, the Rosicrucians, and Bacon. The conquest of nature was never, that is, a purely secular quest. To the contrary, it too was a path to the recovery of an Edenic state and even, at times, a prelude to millenarian dreams.

In short, from the late fifteenth through the first part of the seventeenth century, the apocalyptic imagination played a central, animating role in many of the more significant transformations of the period. Over and over again, dreams of a Beautiful Ending proved inspirational, as individuals and communities sought ways to fulfill their prophetic traditions. In these efforts, they brought into being many of the major features of our world. Modernity is rooted in the Apocalypse.

Epilogue

Well into the seventeenth century, men and women continued to look to the stars and to scriptures to guide their lives. Astrologers cast horoscopes for princes and generals. Farmers studied the skies to determine when to plant and harvest. And prophets continued to foretell a Beautiful Ending: the Millennium or a comparable era of peace, and the coming (or return) of the Messiah. In Germany such prophecies fueled much of the violence in the deadly and widespread conflict that would become known as the Thirty Years' War, while in England prophets of all social backgrounds stirred up many of the political tensions that resulted in a protracted civil war. But there were also moments of elation, even if fleeting. In 1665 Jews from Cairo to Amsterdam danced in the streets when they learned from credible reports that the Messiah had come. The news created waves of excitement; thousands sold their belongings and traveled by boat and wagon to Istanbul to meet their savior. When Sabbatai Ṣevi, the young kabbalist in whom so many now placed their hopes, suddenly converted to Islam, most Jews were sorely disappointed and deeply disillusioned. For others, however, their faith was not shaken; they believed his conversion was part of a mysterious spiritual journey and they continued to accept Sabbatai as their Messiah, convinced that the End of History was imminent.¹

Clearly, not only for Christians but also for Jews and Muslims at this time both the natural order of things and political life were bound together with faith in a cosmic nexus. The early modern world, that is, pulsated with correspondences between the Word of God and the movement of the stars, on the one hand, and between both these spheres and human activity, on the other. Accordingly, it would have been extremely difficult for most in this period to discern, much less define, boundaries between phenomena that many in later generations would come to distinguish variously as either "political" or "religious" or "scientific." Yet, already in the seventeenth century, a number of shifts were underway in patterns of social, institutional, and intellectual life that would gradually lead to a new ordering not only of experience but also of knowledge. Increasingly, politics, science, and religion each came to be seen as constituting largely separate spheres of human endeavor, breaking the myriad links that had, over the previous centuries, bound men and women to a profound sense that God was an active force in shaping every aspect of human experience from the study of nature to the forging of governments and empires.

A transition of this magnitude was inevitably complex and multifaceted. Nonetheless, the fact that the earlier nexus was unraveling became evident in part when, in 1651, Thomas Hobbes, a former secretary to Francis Bacon and now a noted philosopher in his own right, published *Leviathan*.[2] In this work Hobbes broke with a long tradition of political theory that had almost invariably viewed humanity's political organization as an expression of God's will. Hobbes, rather than grounding the state in political theology, turned to observable experience, basing his theory of the best form of government on what he believed was a realistic assessment of human nature.[3] This required a thought experiment. Imagining a world in which there was little or no political authority, Hobbes argued that—since individuals invariably seek to protect and advance their own interests—not only would a "war of all against all" prevail but also life itself would be "solitary, poore, nasty, brutish, and short."[4] Given this natural state, reasonable people, Hobbes argued, would enter into a covenant—a social contract—and agree to invest a ruler with undivided sovereign power. Thus, the best form of government would be the establishment of a powerful monarch—"of that great LEVIATHAN, or rather (to speak more reverently) of that *Mortall God,* to which we owe under the *Immortall God,* our peace and our defense."[5]

But the most radical, even scandalous aspect of *Leviathan* was Hobbes's argument—and he was one of the first to make this point—that the sources of religious beliefs and practices were not divine but lay entirely in human psychology. In part, such beliefs and more general ideas about divinity, he argued, derived from human curiosity about the origins and causes of things.

But religious faith was, above all, a response to anxieties about the future, offering believers the hope that, no matter what trials they faced in life, the grace of a merciful God would guarantee them peace at the end.[6] Thus prophets, most of whom Hobbes characterized as charlatans and madmen, were influential not because they received a divine revelation but rather because they offered reassurances to those who listened to them. But prophets were also dangerous; they played upon the fears of the people and stirred up animosities both by convincing their followers that they were "chosen" and by vilifying their opponents. Apocalyptic beliefs—Hobbes felt this in his bones—led to violence. The civil war in England, largely fueled by a new wave of apocalypticism that swept through England in the generation after Bacon, made this destructive dynamic clear. Indeed, it was his growing fears for his own safety in this environment that led Hobbes to flee England in 1640 and to seek refuge in Paris. There, as Puritan attacks on both the monarchy and the established church spilled over into violence and warfare across the Channel, he wrote *Leviathan*. In arguing that "Prognostiques" undermine the loyalty of the people to political authority, Hobbes offered a devastating critique of the apocalyptic imagination.[7]

The seventeenth century also witnessed another rupture in the cosmic nexus, as science too became increasingly untethered from both religion and politics.[8] A major step had been taken in this direction when the Italian astronomer Galileo Galilei made one of the first forceful arguments for the independence of natural philosophy from theology. "The intention of the Holy Ghost," he wrote in his celebrated letter to the Archduchess Christina of Lorraine in 1615, "is to teach how one goes to heaven, not how heaven goes."[9] But in this same period social and institutional change also fostered an increasingly independent natural philosophy. Throughout Europe, the investigation of nature, now less and less tied to universities, was increasingly centered in learned academies in which scholarly gentlemen came together to discuss their study of the natural world in an age of proliferating scientific instruments, from the telescope and the microscope to the air pump.

The air pump, a device perfected by Hobbes's younger contemporary Robert Boyle, garnered enormous attention. Boyle used this device (essentially a glass globe out of which it was possible to suck the air) to create a vacuum, enabling him to make several significant discoveries about air pressure and the role of air in respiration, combustion, and even the transmission of sound. Within this experimental globe, natural philosophers could not only measure the movement of air pressure but also observe, as the air was removed, a candle flicker out or a lark suffocate. Since such experiments often had a spectacular quality, they did more than establish new scientific "facts"; they also served as advertisements for the power of the experimental method as well as the growing independence of science. For Hobbes this independence, given his metaphysical commitments to a unified state, constituted a threat to the undivided power of the sovereign. He feared the potential subversion of the monarchy by learned gentlemen coming together in academies to establish the truth on their own.[10] With respect to what he perceived as the proper relation of science to politics, therefore, he feared the unraveling of the connection between the two even as his vision of the ideal state hastened the separation of politics from theology.

Accordingly, by the end of the seventeenth century, and certainly across the eighteenth—even if there were many ways in which these different spheres of endeavor continued to inform and enfold one another—religion, politics, and science were, for various reasons, becoming increasingly disentangled from one another. The links between heaven and earth, if not entirely broken, had frayed. And this shift has long been seen as enabling nothing less than the emergence of the modern world. For once science and politics were liberated from religion, so the argument goes, each of these spheres, no longer encumbered by irrationality or superstition, could advance the interest of humanity. The modern state, now conceived as a secular institution, became an agent of progress, while science too, independent of religion, promised if not a Beautiful Ending at least the betterment of the condition of humankind.[11]

Yet even with the separation of these three spheres, which was especially pronounced in western Europe, hopes for a Beautiful Ending persisted. Religion may have withdrawn from the public square, at least in the West, but faith by no means vanished. On the one hand, many continued to read history as a sacred drama in which the faithful were to play an important role. On the other hand, even those who explicitly rejected religion did not abandon faith so much as transform it, placing it not in religious beliefs but rather in reason, science, and politics. For these, the transition of their hopes from religion to politics and science was at first almost seamless. Within the political realm, Hobbes himself had done much to redirect the hopes for a better life away from the Christian promises of Revelation to the state itself, his *Leviathan* or "Mortall God" perceived as the guarantor of security.[12] Similarly, within the scientific realm, much of the effort to understand, explain, and master the natural world continued to derive its energy from a dream of the restoration of humankind's original knowledge of the world as exemplified by

The Heavenly Jerusalem (1712). This engraving, likely by the Dutch artist Jan Luiken, looked forward to a Beautiful Ending. The city of Jerusalem, its "radiance like a very rare jewel, like jasper, clear as crystal," promised a better future.

Adam, the first man, in the Garden of Eden.[13] Finally, the very idea of progress, which was embedded in the politics and science of the seventeenth, eighteenth, and nineteenth centuries, was itself based on the belief that history is not a series of random, often catastrophic events, but rather a meaningful progression toward a better world.[14]

Such a view of history is providential. In the late medieval and early modern period it had served as the foundation or the underlying mental architecture of what I have called the apocalyptic imagination. Often terrified by the intensification of warfare, famine, and disease, Jews, Christians, and Muslims alike in this period not only placed their hope in the promise of a Beautiful Ending, but also took actions to help accelerate its coming—actions that did much to shape modernity. But this providential vision, though not always in its apocalyptic form, has persisted into the late modern world as well. Whether in the growing sense that the conquest of nature will result in benefits to humankind or in the sense that nations themselves have a manifest destiny, modern men and women have clung to the conviction that history itself has meaning and to the faith that, no matter what obstacles, setbacks, and catastrophes we have faced and continue to face, history, should we engage with the present in a constructive way, will lead to a better future. Clearly the idea that we are not simply made by history but also make history continues to stem from faith, and it matters little whether or not this faith is religious.

These imaginaries have remained powerful even in recent times. Writing in the mid- and late nineteenth century, the staunchly atheist Karl Marx, for example, did nothing less than bring Joachim of Fiore's spiritual model of sacred history down to earth, uncovering the hidden dynamic or providential logic of social and economic change. Rather than the ages of the Father, Son, and Holy Spirit, Marx wrote of the ages of feudalism, capitalism, and communism—this final stage, no less than Joachim's age of the Holy Spirit, a consummation in which the sufferings of humanity would be eliminated.[15] And in the early twentieth century, Marxist ideas would come to play a central role in Russia, where the leadership of the revolution identified the Elect with the proletariat.[16] In Germany, by contrast, the Nazis—and here the resonances of Joachim of Fiore were also palpable—described themselves as avatars of a "new world." To Hitler, the collapse of the German military in the First World War was "in the nature of eternal retribution" and out of its ruins would emerge a new order: the Third Reich—a millennial kingdom.[17] Perverting this providential logic, Nazis invested the Aryan "race" with a sense of near-divine election, and demonized the Jews.

Providence, at its core, may promise hope, but it can also, especially when it takes on apocalyptic tones, be dangerous, as the political theologies of the early twentieth century make clear. As Hobbes would have foreseen, the renewal of apocalyptic passions would prove catastrophic. Such passions contributed to a resumption of hostilities in the late 1930s and the outbreak of the Second World War. The devastation in this conflict even surpassed that of the Thirty Years' War. Above all, the vicious Nazi assault on Jews actually made the Apocalypse a reality for millions. Europe, three hundred years after its Wars of Religion, was still haunted by the powerful myth of a Beautiful Ending. Then, once Nazism and fascism were defeated, the apocalyptic braid shifted to the Cold War, the titanic struggle between the United States and Russia, which squared off against each other with the theological certainties that each alone could serve as the global bulwark of a saving ideology to fulfill the goals of history. Early on, contemporaries understood the threat of nuclear destruction in apocalyptic terms. Nuclear destruction is not only mass destruction, the U.S. diplomat Hans Morgenthau observed in the mid-twentieth century; it destroys all meaning as well. As Morgenthau noted: "It destroys the meaning of death by depriving it of its individuality. It destroys the meaning of immortality by making both society and history impossible. It destroys the meaning of life by throwing life back upon itself."[18]

When the Soviet Union collapsed in 1989, some believed that history had ended, reaching its final consummation in the triumph of liberal democracy.[19] But history is never one-sided. While some celebrated 1989, others were less sanguine about this moment. For precisely in the decade leading up to this year, researchers around the globe had discovered that the modern reliance on fossil fuels had begun to heat the planet at an alarming rate. Indeed, it was in November 1989 in Noordwijk, a town on the Dutch coast, that the first major international conference on climate change took place.[20] To those who had begun to focus on global warming, the apparent triumph of capitalism in the last years of the twentieth century was thus not something to celebrate but rather to fear. How would it be possible, they wondered, under such an economic system, to contain the ongoing destruction of our planet?[21]

That the apocalyptic continues to shape contemporary imaginaries is hardly surprising. The world we have constructed over the last two centuries seems no more permanent to us than it did to those who lived in the late Middle Ages and the early modern era. On the one hand, the threat of a nuclear holocaust has only continued to increase in recent decades as the arsenal of weapons, whose destructive power is unimaginable, stands at the ready not only in underground silos in the Urals and the Rockies but also in submarines and aircraft which can continue to deliver them even once land-based weapons are neutralized.[22] But the fragility of the world is even more apparent in the inexorable warming of the planet. The polar icecaps are melting; sea levels are rising; forests are going up in flames; storms are ever more destructive; and many regions of the earth, that have long supported human communities, are no longer able to do so. Our reliance on fossil fuels, no less than our embrace of nuclear weapons, casts a dark, apocalyptic shadow across all our lives.

Yet Providence is not easily swept aside. People continue to hope for a Beautiful Ending. The "modern" impulse is to continue to place our hope in politics, science, and human agency, the belief that we can act as masters of our own fate.[23] But there are also those who locate the potential solutions to the crises confronting us within religious traditions and the embrace of the transcendent. In the most promising scenarios, such hopes draw on a variety of sacred practices aimed to foster justice and to ensure the sustainability of the earth—practices that are most evident in many of the nontheistic religions of South and East Asia.[24] Within Jewish, Christian, and Muslim communities some leaders have also called for a spiritually-informed stewardship of the planet, one that also seeks to make sustainability of the earth a priority.[25] Yet in the Abrahamic traditions religion can present another face. Not only within Islam but also within Christianity and Judaism, a minority of believers continue to live under the sway not only of Providence but also of the apocalyptic imagination. At times violence or near violence ensues. When al-Qaeda warriors flew planes into the World Trade Center in New York in 2001, they did so in the conviction that the United States was the Antichrist and that their sacrifice would help bring about the End of History. At the same time, within the United States, a strong current of apocalypticism—bolstered by widespread evangelical beliefs that the End Times are upon us—continued in the first part of the twenty-first century to inform American foreign policy. And in Israel messianic dreams have persisted among a significant number of Jews. Hobbes dreamed of "disarming the prophets," but prophecy has a way of returning and haunting us. The enduring attraction of apocalyptic hopes to relatively large numbers of men and women throughout the contemporary world makes it clear that modernity, despite the frequent tendency to believe otherwise, is far from purely secular or rational.

Originating some three millennia ago in the ancient Near East and then finding its first systematic articulations among the Jewish ascetics at Qumran, the early Christian sects at Nag Hammadi, and the first followers of Muhammad at Mecca, the apocalyptic vision of a Beautiful Ending came to permeate the cultures of Europe and the Mediterranean, its symbols and images nourishing hope even in times of despair. But it was in the early modern period—largely as a consequence of the spread of print media but also as a result of the overseas discoveries, the growth of empire, and the passions unleashed by the Reformation—that the apocalyptic became a chorus, diffusing hopes for the End of History throughout all of society, even as some writers like Montaigne pushed back against it. Yet ultimately the Apocalyptic became a stamp of modernity. It was, as we have seen, an animating force in Europe's overseas expansion, in the growing reach of empire, and in the development of new technologies and new understandings of the natural world. At the same time, the early modern apocalyptic imagination contributed significantly to the violence and religious hostilities of this era. The apocalyptic imagination of the early modern world, in short, was woven deeply into the making of the modern.

Perhaps it is inevitable, therefore, that the apocalyptic imagination continues to exercise such a powerful hold on contemporary societies. The apocalyptic braid is woven into the very fabric of our world, our culture, our technologies, our politics. This doesn't mean that we do not have other ways of imagining history, above all within the frame of the tragic, to make sense of the catastrophes that have littered and continue to litter human history without having to have recourse to the apocalyptic. Yet, even when the secular path seems the more certain, the continuing appeal of apocalyptic thought to significant segments of contemporary societies across the globe makes it clear that an understanding of our world requires that we be able to grasp not only the secular but also the religious dimensions of human experience. As *A Beautiful Ending* has sought to demonstrate, the apocalyptic is a fundamental feature of the modern itself. Whether or not we can hold it at bay is uncertain, but the evidence I have gathered in this book suggests that our efforts to do so will be only partially successful at best. Even as I write, many around the globe continue to put their faith in revelation. How can they not? In a world full of suffering, many will hope for a New Jerusalem where "God himself will be with them; he will wipe every tear from their eyes. Death will be no more; mourning and crying and pain will be no more, for the first things have passed away."

Acknowledgments

I carved this book out of the side of a mountain, a mountain of my own making. For nearly a decade I had struggled to write a history of early modern Europe from the age of Columbus to the French Revolution. The text grew far too long, the arguments too diffuse. Yet a particular stratum in my narrative pushed its way to the surface and arrested my attention: namely, the role of apocalyptic in the history of Europe and the Mediterranean from the fifteenth to the seventeenth century. I hadn't self-consciously set out to work on this theme. Nonetheless, once this stratum emerged, I found I couldn't look away. In part, I was fascinated by the hold of the apocalyptic on the imaginations of Jews, Christians, and Muslims in the early modern period. Moreover, this story seemed also to speak to the present with our own renewed anxieties about the End of History. Above all, as I came to see, the apocalyptic imagination played not an insignificant role in the making of the modern world.

Writing a book calls for solitude, stolen moments, time to reflect. But the making of a book also depends on community and is enriched by the give and take with students, friends, and fellow scholars. On all these fronts I have been blessed. My students, undergraduate and graduate, both kept me company and raised challenging questions as I subjected them to some of my earliest ruminations on the history of apocalypticism. I am forever grateful to them for their patience and insights. And throughout I have benefited from the flourishing scholarship—evident in a seemingly endless sea of books and articles—that has transformed the histories of messianism, millenarianism, and Mahdism in recent decades. Indeed, my debt to scholars who have explored these themes before me is immense, as the notes to this volume should make clear. But it is a special pleasure to thank by name the large number of friends and colleagues who have generously contributed to this project in more direct ways. For their reading of recent drafts of the manuscript either in its entirety or nearly so, I am deeply grateful to Robin Barnes, Douglas Biow, Prasenjit Duara, Kenneth Gouwens, Malachi Hacohen, Sina Rauschenbach, and David Harris Sacks, as well as to the readers, whose names I do not know, for the press. Each of these scholars not only saved me from many egregious errors and offered fresh insights but also provided me with new perspectives, pushing me to clarify my arguments. In addition many other friends and colleagues have also generously offered feedback of various kinds throughout the life of this project. Several read early drafts of portions of the book; others provided critical leads and suggestions; and still others—often in unexpected conversations—provided encouragement when encouragement was sorely needed. In light of these many kindnesses, it is pleasure to thank Charles Bartlett, Jodi Bilinkoff, Flora Cassen, Erasmo Castellani, Yossi Chajes, Roberto Dainotto, Jean-Pierre Filiu, Ashley Elrod, Susan Ferber, John Freeman, Sara Galletti, Juan Gil, Michael Gillespie, Mayte Green-Mercado, Mona Hassan, Kristin Huffman, Lyle Humphrey, Claire Judde de Larivière, Vasant Kaiwar, Gary Kates, Timothy Kircher, Lloyd Kramer, Derryl MacLean, Jehangir Malegam, Seymour Mauskopf, Sucheta Mazumdar, Elisabeth Narkin, Kristen Neuschel, Agata Paluch, Mary Pardo, Luca Pes, Sumathi Ramaswamy, William M. Reddy, Andrea Robiglio, Trish Ross, David Ruderman, Phil Stern, John Taormina, Mustafa Tuna, Maartje van Gelder, Consuelo Varela, Brett Whalen, and Giovanni Zanalda. I also thank the late Ronald G. Witt not only for his remarkable intellectual generosity but also for his warm and outgoing welcome when I arrived at Duke now nearly fifteen years ago. Finally, Rochelle Rojas and Matthew Lubin worked as research assistants on this project, and I am grateful to both for their enthusiasm and ability to trace down the most obscure of references. I also thank Carla Ivey. While serving as my assistant when I was department chair, she graciously provided critical tactical support as I struggled to carve out time for research, reading, reflection, and writing.

For providing me with translations of several key passages in Hebrew I am grateful to Malachi Hacohen; and I thank Mona Hassan for her translations of several passages in Arabic. Sara Galletti offered help with translations from French, Bill Lewis from German, Rochelle Rojas from Spanish, and Olivia Merli from Italian and Latin. And throughout this process Duke librarians Elizabeth Dunn, Carson Holloway, Kelly Lawton, Liz Milewicz, and Lee Sorensen have each offered assistance at critical junctures, while Mark Thomas kindly developed the maps for Chapter 5. I also received considerable support from librarians abroad. In particular I would like to thank the staffs of the Biblioteca Marciana in Venice, of the Biblioteca Capitular y Colombina in Seville, and of the Herzog August Bibliothek in Wolfenbüttel for helping me orient myself to their rich, indeed invaluable collections.

Institutional support has also been decisive. The deans of Trinity College at Duke University enabled leaves and provided research funds that made the completion of this work possible. Furthermore, I remain grateful to several colleagues for opportunities to share versions of my work as I sought to make sense of the materials I was studying. I owe a special debt of gratitude to Mario Infelise and Claudio Povolo who arranged for me to give an early workshop on my research at the University of Venice. Andrea Robiglio kindly hosted me at the University of Leiden where I was able to share some of my early ideas on both Montaigne and Bacon. I thank Damien Tricoire who facilitated a presentation of an early version of my emerging argument on the role of the apocalyptic imagination in the making of the modern world at the "Last Days" Conference at the Martin-Luther-Universität in Halle, Germany. There I benefited enormously from the opportunity to discuss my ideas with Moti Benmelech and Evrim Binbaş, from whom I learned much more than they could have realized. But I owe a special debt of gratitude to the Providential Modernity Seminar at Duke University where I was able to give a preview of my book to a lively gathering of colleagues and guests. Indeed, this seminar has been a most welcome intellectual home where colleagues from a wide array of disciplines have explored the role of religious ideas in the shaping of social and political life in a variety of times and places around the globe—intellectual community at its best.

My agent Don Fehr first expressed interest in this work when it was in its earliest stages. I am appreciative to him for seeing its potential and helping me, in the crafting of the story, not to lose the forest for the trees. At Yale University Press, I have benefited immensely from the encouragement and feedback of my editor, Jennifer Banks, as well as from the indispensable logistical support of Abigail Storch and Ann-Marie Imbornoni. And I owe a special debt of gratitude to Andrew Frisardi not only for his invaluable and meticulous copyediting but also for pushing me, as he drew on his own deep knowledge of late medieval spirituality, to rethink my arguments at several key junctures. Finally, I thank Dana Hogan, who—through her patient and attentive readings of the proofs—has made this a decidedly better book.

My brother Thomas and his wife, Virginia, have remained enthusiastic about this book throughout this project. And my children, Margaret and Junius, have, as ever, been immensely supportive in more ways than I can count, not least—through their commitments and their own hopes for a Beautiful Ending—to make this a better world. Finally, Olivia Merli, my companion and best friend, has brought immeasurable joy into my life. It is no accident that I first shared my idea for this book with her on an Italian hillside overlooking the Mediterranean in the summer of 2017. We share so many passions, intellectual and otherwise, and she has been my most constant supporter. My dedication to her can only be a small gesture of my love and appreciation.

Hillsborough, NC
June 12, 2021

Notes

INTRODUCTION

1. Citations from the Bible are from *The New Revised Standard Version: The New Oxford Annotated Bible with the Apocrypha*, eds. Bruce M. Metzger and Roland E. Murphy (New York: Oxford University Press, 1994).
2. For a pivotal reconceptualization of the way early modern men and women understood history, see Stuart Clark, *Thinking with Demons: The Idea of Witchcraft in Early Modern Europe* (Oxford: Oxford University Press, 1997). Rejecting both claims about the typicality of Renaissance humanist and Ciceronian concepts of history in this period, Clark makes the compelling point that, for the early modern period, "we can now see that the Bible and Augustine's universal theodicy continued to offer an entirely satisfying framework for the interpretation of the historical process. The importance of theological time, the 'time of the church,' reflected the attention given to salvation, its channel and its processes. Providence remained fundamental to conceptions and the causation and the purpose of events" (318).
3. Scholars have long recognized the role of apocalyptic ideas—especially those rooted in Christian eschatology—in the making of the modern world. On this theme, Ernest Lee Tuveson, *Millennium and Utopia: A Study in the Background of the Idea of Progress* (Berkeley: University of California Press, 1949) is foundational. For the early modern period, the work of Robin B. Barnes is essential; see especially his "Images of Hope and Despair: Western Apocalypticism ca. 1500–1800," in *The Continuum History of Apocalypticism*, ed. Bernard J. McGinn, John J. Collins, and Stephen J. Stein (New York: Continuum, 2003), 323–53. Finally, both Arthur H. Williamson, *Apocalypse Then: Prophecy and the Making of the Modern World* (Westport, CT: Praeger, 2008), and John R. Hall, *Apocalypse: From Antiquity to the Empire of*

Modernity (Cambridge: Polity, 2009), offer sweeping overviews of the history of the apocalyptic. Hall's work, refreshingly, gives considerable attention to Islam.

CHAPTER ONE. THE APOCALYPTIC BRAID

1. Friar Giovanni Caroli, cited in Armando F. Verde, "Le lezioni o i sermoni sull' 'Apocalisse' di Girolamo Savonarola (1490): Nova dicere et novo modo," in *Immagine e parola: Retorica filologica—retorica predicatoria (Valla e Savonarola), Memorie Domenicane* 19 (1988): 6–7. Unless otherwise noted, translations from Italian, French, German, Latin, and Spanish are my own, though in certain instances (notably in the cases of Acosta, Bacon, More, and Montaigne) I have drawn heavily on earlier translations, while always revisiting and examining the works in the original. By contrast, for texts originally written in Arabic, Hebrew, and Ottoman Turkish—languages I have not studied—I have relied exclusively on translations.
2. Cited in Nancy Bisaha, *Creating East and West: Renaissance Humanists and the Ottoman Turks* (Philadelphia: University of Pennsylvania Press, 2004), 63.
3. Ibid., 158; and Alison Knowles Frazier, *Possible Lives: Authors and Saints in Renaissance Italy* (New York: Columbia University Press, 2005), 91, n. 171. Historians now doubt the story of the eight hundred Christian martyrs, said to have died bravely for their faith.
4. Fernand Braudel, *The Mediterranean and the Mediterranean World in the Age of Philip II*, trans. Siân Reynolds (New York: Harper and Row, 1973), 2:657–81.
5. Pasquale Villari, *La storia di Girolamo Savonarola e de' suoi tempi* (Florence: Le Monnier, 1926), 1: appendix, xv–xvi, provides a summary of Savonarola's sermons at this time. Savonarola preached on the Turks in his sermons on the Hebrew prophet Haggai—see his *Prediche sopra Aggeo, con il Trattato circa il reggimento e governo della città di Firenze*, ed. Luigi Firpo (Rome: A. Berladetti, 1965), 319.
6. Donald Weinstein, *Savonarola and Florence: Prophecy and Patriotism in the Renaissance* (Princeton, NJ: Princeton University Press, 1970), 39.
7. While I have not been able to find direct documentation of the immediate reaction of the court to the news of the fall of Granada, it is clear that there was close communication between the Moriscos of Spain and the Muslim powers of the eastern Mediterranean at this time. See Pieter Sjoerd van Koningsveld and Gerard A. Wiegers, "An Appeal of the Moriscos to the Mamluk Sultan and Its Counterpart to the Ottoman Court: Textual Analysis, Context, and Wider Historical Background," *Al-Qantara: Revista de estudios arabes* 20 (1999): 161–89.
8. Colin Imber, "A Note on 'Christian' Preachers in the Ottoman Empire," *Osmanlı Araştırmaları / The Journal of Ottoman Studies* 10 (1990): 59–67.
9. Gülru Necipoğlu, *Architecture, Ceremonial, and Power: The Topkapi Palace in the Fifteenth and Sixteenth Centuries* (Cambridge, MA: MIT Press, 1991).
10. Year 1 on the Islamic calendar is equivalent to Year 622 according to Christian reckoning—the date of Muhammad's *hajj* or migration to Mecca. The Islamic year is based on lunar cycles and is, therefore, shorter than the Christian one. A Muslim century is roughly equivalent to ninety-seven years in Christian timekeeping.

11. Not everyone believed the tenth century would be the last. When pressed in 1493 on his view of the end of time, the Egyptian jurist and Sufi Jalal al-Din al-Suyūti rejected the view that the End was imminent. Nonetheless, he remained watchful for the Signs of the End and he did believe that the beginning of the tenth century would be a time of renewal. Cornell H. Fleischer, "A Mediterranean Apocalypse: Prophecies of Empire in the Fifteenth and Sixteenth Centuries," *Journal of the Economic and Social History of the Orient* 61 (2018): 50–51.
12. Ibid., 51–52; and Hüseyin Yılmaz, *Caliphate Redefined: The Mystical Turn in Ottoman Political Thought* (Princeton, NJ: Princeton University Press, 2018), 141–43. Both these studies offer superb depictions of the role of eschatology in the shaping of the Ottoman Empire.
13. Ahmet Tunç Şen, "Reading the Stars at the Ottoman Court: Bāyezīd II (r. 886/1481–918/1512) and His Celestial Interests," *Arabica* 64 (2017): 557–608.
14. Jonathan S. Ray, *After Expulsion: 1492 and the Making of Sephardic Jewry* (New York: New York University Press, 2013); and Dean Phillip Bell, *Jews in the Early Modern World* (Lanham, MD: Rowman and Littlefield, 2008).
15. Isaac Abravanel, "Sources du salut" (selections from his *Sefer Ma'ayenei ha-Yeshua*), in *Isaac Abravanel: La mémoire et l'espérance*, ed. and trans. Jean-Christophe Attias (Paris: Cerf, 1992), 80. On Abravanel's messianism, see in addition to Benzion Netanyahu, *Don Isaac Abravanel: Statesman and Philosopher*, 5th ed. (Ithaca, NY: Cornell University Press, 1998), Seymour Feldman, *Philosophy in a Time of Crisis: Don Isaac Abravanel, Defender of the Faith* (London: Routledge, 2003), and Eric Lawee, "The Messianism of Isaac Abarbanel, 'Father of the [Jewish] Messianic Movements of the Sixteenth and Seventeenth Centuries,' " in *Jewish Messianism in the Early Modern World*, vol. 1 of *Millenarianism and Messianism in Early Modern European Culture*, ed. Matt Goldish and Richard H. Popkin (Dordrecht: Kluwer, 2001), 1–39. Netanyahu offers an interesting comparison between Abravanel and Savonarola, see 249 ff.
16. Abravanel, "Sources du salut," 153–56. Abravanel also assured his readers that the Messiah would return no later than 1573, and he ultimately favored 1531 as the most likely date.
17. On Lemlein see especially Elisheva Carlebach, *Between History and Hope: Jewish Messianism in Ashkenaz and Sepharad*, Third Annual Lecture of the Victor J. Selmanowitz Chair of Jewish History (New York: Graduate School of Jewish Studies: Touro College, 1998); and Rebekka Voß, *Umstrittene Erlöser: Politik, Ideologie und jüdische-christlicher Messianismus in Deutschland, 1500–1600* (Göttingen: Vandenhoeck and Ruprecht, 2011), 52–87. Münster is cited in Carlebach, *Between History and Hope*, 8.
18. Norman Cohn, *Cosmos, Chaos, and the World to Come: The Ancient Roots of Apocalyptic Faith*, 2nd ed. (New Haven: Yale University Press, 2001); and Robert Gnuse, "Ancient Near Eastern Millennialism," in *The Oxford Handbook of Millennialism*, ed. Catherine Wessinger (Oxford: Oxford University Press, 2011), 235–51.
19. John J. Collins, *Daniel: With an Introduction to Apocalyptic Literature*, vol. 20 of *The Forms of the Old Testament Literature* (Grand Rapids, MI: Eerdmans, 1984), and especially his *The Apocalyptic Imagination: An Introduction to the Jewish Matrix of Christianity* (New York: Crossroad, 1984).

20. Bart D. Ehrman, *Jesus: Apocalyptic Prophet of the New Millennium* (Oxford: Oxford University Press, 1999).
21. Daniel J. Crowther, "Qumrān and Qur'ān," *Journal for the Study of the Old Testament* 43 (2018): 117 ff.
22. Citations from the Qur'an are from *The Study Quran: A New Translation and Commentary*, ed. Seyyed Hossein Nasr et al. (New York: HarperCollins, 2017).
23. Abdulaziz Abdulhussein Sachedina, *Islamic Messianism: The Idea of the Mahdī in Twelver Shi'ism* (Albany: State University of New York, 1981); Moojan Momen, *An Introduction to Shi'i Islam: The History and Doctrines of Twelver Shi'ism* (New Haven: Yale University Press, 1985), 75, 166–68; and David Cook, *Studies in Muslim Apocalyptic* (Princeton, NJ: Darwin Press, 2002), 137–38.
24. Norman Solomon, ed. and trans., *The Talmud: A Selection* (London: Penguin, 2009), 513–17.
25. Joachim of Fiore, *Expositio in Apocalypsim*, selections translated in *Visions of the End: Apocalyptic Traditions in the Middle Ages*, ed. and trans. Bernard J. McGinn, rev. ed. (New York: Columbia University Press, 1998), 130.
26. Marjorie Reeves, *The Influence of Prophecy in the Later Middle Ages: A Study in Joachimism* (Notre Dame, IN: University of Notre Dame Press, 1993).
27. Cited in Brett E. Whalen, *Dominion of God: Christendom and Apocalypse in the Middle Ages* (Cambridge, MA: Harvard University Press, 2009), 100.
28. John of Rupescissa, *Liber secretorum eventuum*, ed. Robert E. Lerner and Christine Morerod-Fattebert (Fribourg Suisse: Editions Universitaires, 1994), 138.
29. Ibid., 175.
30. Ibid., 179.
31. Cited in Gershom Scholem, *Major Trends in Jewish Mysticism*, 3rd ed. (New York: Schocken, 1971), 128.
32. Moshe Idel, *The Mystical Experience in Abraham Abulafia*, trans. Jonathan Chipman (Albany: State University of New York, 1988), esp. 2–4.
33. Ibn al-'Arabī, *The Bezels of Wisdom*, trans. Ralph W. J. Austin (New York: Paulist Press, 1980), 282–83.
34. Jean-Pierre Filiu, *Apocalypse in Islam*, trans. M. B. DeBevoise (Berkeley: University of California Press, 2011), 31–34.
35. Ibn al-'Arabī, "The Mahdi's Helpers" (selected passages), in *The Meccan Revelations*, ed. and trans. William C. Chittick and James W. Morris (New York: Pir Press, 2002–4), 1:67–100. For an important analysis of the structure of this chapter, see James W. Morris, "Ibn Arabī's 'Esotericism': The Problem of Spiritual Authority," *Studia Islamica* 71 (1990): 37–64. A complete, but not a scholarly, translation of the entire chapter is available in French: Ibn al-'Arabī, *Le Mahdi et ses conseillers: Une sagesse pour la fin des temps*, trans. Tayeb Chouiref (Paris: Mille et Une Lumières, 2006).
36. The classic account of these movements remains Norman Cohn, *The Pursuit of the Millennium: Revolutionary Millenarians and Mystical Anarchists of the Middle Ages*, rev. ed. (Oxford: Oxford University Press, 1970).
37. Ibn Khaldūn, *The Muqaddimah: An Introduction to History*, trans. Franz Rosenthal, 3 vols. (New York: Pantheon, 1958), 1:6.

38. Ibn Khaldūn, *Le livre des exemples,* trans. Abdesselam Cheddadi (Paris: Gallimard, 2002–12), 1:906–7.
39. Ibn Khaldūn, *Muqaddimah,* 2:156: "It has been well-known (and generally accepted) by all Muslims in every epoch, that at the end of time a man from the family (of the Prophet) will without fail make his appearance, one who will strengthen the religion and make justice triumph. The Muslims will follow him, and he will gain domination over the Muslim realm. He will be called the Mahdî. Following him, the Antichrist will appear, together with all the subsequent signs of the Hour (the Day of Judgment), as established in the sound tradition of the Sahîh. After (the Mahdî), 'Isâ (Jesus) will descend and kill the Antichrist. Or, Jesus will descend together with the Mahdî, and help him kill (the Antichrist), and have him as the leader in his prayers."
40. Andrew Cunningham and Ole Peter Grell, *The Four Horsemen of the Apocalypse: Religion, War, Famine, and Death in Reformation Europe* (Cambridge: Cambridge University Press, 2000) offers a compelling analysis of early modern European crises and their relationship to the apocalyptic imagination. Here, drawing on a medieval tradition, they interpret the horseman with the bow not as the Turkish conqueror, as I have here, but rather as Christ the Conqueror (13). While such an interpretation no doubt circulated in learned circles on the basis of a careful exegesis of the Book of Revelation, most early modern men and women would have been more likely to view the bow-carrying horseman, given his headdress, as a Turk. Of course, the power of the Book of Revelation lies in the way its images can be both universal and particular simultaneously.
41. Ibid.: on war, chap. 3; on famine, chap. 4; and on plague, chap. 5. The citation from Paré is from 280. See also Jean Delumeau, *La peur en Occident: XIVe–XVIIIe siècles* (Paris: Fayard, 1978).
42. Alexander Perrig, *Albrecht Dürer, oder, die Heimlichkeit der deutschen Ketzerei: Die Apokalypse Dürers und andere Werke von 1495 bis 1513* (Weinheim: VCH Acta Humaniora, 1987).
43. For an analysis of the dialect of hope and despair within Christian apocalypticism, along with an analysis of their role in shaping modernity, see Robin B. Barnes, "Images of Hope and Despair: Western Apocalypticism ca. 1500–1800," in *The Continuum History of Apocalypticism,* ed. Bernard J. McGinn, John J. Collins, and Stephen J. Stein (New York: Continuum, 2003), 323–53.

CHAPTER TWO. PUBLISHING THE APOCALYPSE

1. Edward Schröder, *Das Mainzer Fragment vom Weltgericht: Ein Ausschnitt aus dem deutschen Sibyllenbuche* (Mainz: Gutenberg-Gesellschaft, 1908).
2. Bernard J. McGinn, ed. and trans., *Visions of the End: Apocalyptic Traditions in the Middle Ages,* rev. ed. (New York: Columbia University Press, 1998), 18–21.
3. Jonathan Green, *Printing and Prophecy: Prognostication and Media Change, 1450–1550* (Ann Arbor: University of Michigan Press, 2012), 31.
4. *Gutenbergs Türkenkalender für das Jahr 1455: Rekonstruierter Typendruck von hölzerner Handpresse mit Handrubrizierung* (Mainz: Gutenberg-Gesellschaft, 1928).
5. Green, *Printing and Prophecy,* 15–38.

6. See Andrew Pettegree, *The Book in the Renaissance* (New Haven: Yale University Press, 2010), and Elizabeth L. Eisenstein, *The Printing Press as an Agent of Change: Communications and Cultural Transformations in Early-Modern Europe* (Cambridge: Cambridge University Press, 1979), for bibliography on Gutenberg.
7. For a fascinating history of printing technology in China, see T. H. Barrett, *The Woman Who Discovered Printing* (New Haven: Yale University Press, 2008).
8. Eisenstein, *Printing Press as an Agent of Change*, 31.
9. For more on the technology involved, see part 1 of Joseph A. Dane, *What Is a Book? The Study of Early Printed Books* (Notre Dame, IN: University of Notre Dame Press, 2012).
10. "Non vidi Biblias integras sed quinterniones aliquot diversorum librorum mundissime ac correctissime littere, nulla in parte mendaces, quos tua dignatio sine labore et absque berillo legeret. . . . Conabor si poterit fieri aliquam huc bibliam venalem afferri eamque tui causa comparabo. Quod timeo ne fieri possit, et propter distantiam itineris et quia antequam perficerentur volumina paratos emptores fuisse tradunt." Martin Davies, "Juan de Carvajal and Early Printing: The 42-Line Bible and the Sweynheym and Pannartz Aquinas," *The Library*, 6th ser., 18 (1996): 196.
11. Johannes Trithemius, *In Praise of Scribes: De laude scriptorum*, ed. Klaus Arnold, trans. Roland Behrendt (Lawrence, KS: Coronado Press, 1974), 65.
12. Pettegree, *Book in the Renaissance*, 43.
13. Cited in Eugene F. Rice, Jr., with Anthony Grafton, *The Foundations of Early Modern Europe, 1460–1559*, 2nd ed. (New York: Norton, 1994), 10.
14. Cited in Anthony Grafton, *Leon Battista Alberti: Master Builder of the Renaissance* (London: Penguin, 2000), 331.
15. Immanuel Wallerstein, with a bias that privileges the western regions of the continent, has called this belt of cities Europe's "dorsal spine," with less urbanized regions stretching out to both the west and the east. Wallerstein, *The Modern World System* (New York: Academic Press, 1974), 1:165. The phrase "urban archipelago" is from Fernand Braudel, *The Perspective of the World*, vol. 3 of *Civilization and Capitalism, 15th–18th Century*, trans. Siân Reynolds (London: William Collins, 1986), 124.
16. Pettegree, *Book in the Renaissance*, 33, 44, 49.
17. Eltjo Buringh and Jan Luiten van Zanden, "Charting the 'Rise of the West': Manuscripts and Printed Books in Europe; A Long-Term Perspective from the Sixth through Eighteenth Centuries," *Journal of Economic History* 69 (2009): 409–45.
18. Eisenstein, *Printing Press as an Agent of Change*, 163–302, 453–682.
19. Miriam Usher Chrisman, *Lay Culture, Learned Culture: Books and Social Change in Strasbourg, 1480–1599* (New Haven: Yale University Press, 1982), esp. fig. 22, but see also her "Printing and the Evolution of Lay Culture in Strasbourg, 1480–1599," in *The German People and the Reformation*, ed. R. Po-Chia Hsia (Ithaca, NY: Cornell University Press, 1994), 74–102.
20. Alexander Perrig, *Albrecht Dürer, oder, die Heimlichkeit der deutschen Ketzerei: Die Apokalypse Dürers und andere Werke von 1495 bis 1513* (Weinheim: VCH Acta Humaniora, 1987).
21. David Hotchkiss Price, *Albrecht Dürer's Renaissance: Humanism, Reformation, and the Art of Faith* (Ann Arbor: University of Michigan Press, 2003), 34.

22. Jane C. Hutchison, *Albrecht Dürer: A Biography* (Princeton, NJ: Princeton University Press, 1990), 57.
23. Joseph Leo Koerner, *The Moment of Self-Portraiture in German Renaissance Art* (Chicago: University of Chicago Press, 1993).
24. Desiderius Erasmus, "Paraclesis [An Exhortation to the Study of the New Testament]," in *The New Testament Scholarship of Erasmus: An Introduction to the Prefaces and Ancillary Writings,* ed. Robert D. Sider (Toronto: University of Toronto Press, 2019), 211.
25. Cited in Roland H. Bainton, *Here I Stand: A Life of Martin Luther* (New York: Abingdon-Cokesbury, 1950), 103.
26. Ulrich Zwingli, "On the Clarity and Certainty of the Word of God," in *Zwingli and Bullinger: Selected Translations,* ed. Geoffrey W. Bromiley (Philadelphia: Westminster Press, 1953), 75.
27. Henry Joseph Schroeder, trans., *Canons and Decrees of the Council of Trent: Original Text with Translation* (St. Louis: Herder, 1941), 17.
28. Richard Tresley, "Annotated Bibliography of Commentaries on the Book of Revelation to 1700," in *The Book of Revelation and Its Interpreters: Short Studies and an Annotated Bibliography,* ed. Ian Boxall and Richard Tresley (Lanham, MD: Rowman and Littlefield, 2016), 125–272.
29. Riccardo Calimani, *Storia del ghetto di Venezia: Gli ebrei e la Serenissima Repubblica* (Milan: Mondadori, 2018); Avner Shamir, *Christian Conceptions of Jewish Books: The Pfefferkorn Affair* (Copenhagen: Museum Tusculanum, 2009).
30. Flora Cassen, *Marking the Jews in Renaissance Italy: Politics, Religion, and the Power of Symbols* (Cambridge: Cambridge University Press, 2017), 38.
31. David Stern, *The Jewish Bible: A Material History* (Seattle: University of Washington Press, 2017).
32. The second edition appeared between 1526 and 1539, and the third between 1543 and 1549; see David C. Kraemer, *A History of the Talmud* (Cambridge: Cambridge University Press, 2019), 216.
33. Reprints appeared in 1546–48, 1568–69, and 1617–18.
34. On the Soncino: Herbert C. Zafrin, "Bible Editions, Bible Study, and the Early History of Hebrew Printing," in *Eretz-Israel: Archaeological, Historical, and Geographic Studies,* vol. 16: Harry M. Orlinsky volume, ed. Baruch A. Levine and Abraham Malamat (Jerusalem: Israel Exploration Society, 1982), 240–51. Abraham ben Hayyim dei Tintori published the Torah (the first five books of the Jewish scripture) with commentaries in Bologna in 1482 and, six years later, Joshua Solomon ben Israel Nathan Soncino published the Tanakh (the entire Hebrew Bible) in Brescia. But it was Gershom Soncino, Joshua's nephew, who took the initiative to bring the Bible to a broad readership, printing only the text, without commentaries, in a portable edition. He made it clear in his preface that he hoped readers would have the books of the Bible with them "day and night, study them and never walk more than four cubits without the Torah." Cited in Pavel Sládek, "The Printed Book in 15th- and 16th-Century Jewish Culture," in *Hebrew Printing in Bohemia and Moravia,* ed. Olga Sixtová (Prague: Jewish Museum—Academia, 2012), 13.

35. Cited in Moshe Halbertal, *People of the Book: Canon, Meaning, and Authority* (Cambridge, MA: Harvard University Press, 1997), 78–79.
36. Kraemer, *History of the Talmud*, 209–36.
37. On the challenge to rabbinical authority, see David B. Ruderman, *Early Modern Jewry: A New Cultural History* (Princeton, NJ: Princeton University Press, 2010), 100–103; and on the possession of Hebrew books in Jewish households, see Shifra Baruchson-Arbib, *La culture livresque des juifs d'Italie à la fin de la Renaissance*, trans. Gabriel Roth (Paris: CNRS, 2001), 77–83.
38. Zafrin, "Bible Editions," 241.
39. Levi ben Gershom, *Perush a'l sefer Daniel* ([Rome]: [Obadiah, Menasseh, and Benjamin of Rome], 1469-1572); Isaac Abravanel, *Perush a'l sefer Daniel* (Ferrara, 1551).
40. Gershom Scholem, *The Messianic Idea in Judaism and Other Essays on Jewish Spirituality* (New York: Schocken, 1971), 41–42.
41. Cited in David B. Ruderman, "Hope against Hope: Jewish and Christian Messianic Expectations in the Late Middle Ages," in *Essential Papers on Jewish Culture in Renaissance and Baroque Italy* (New York: New York University Press, 1992), 299–323.
42. Ahmet Tunç Şen, "Reading the Stars at the Ottoman Court: Bāyezīd II (r. 886/1481–918/1512) and His Celestial Interests," *Arabica* 64 (2017): 557–608.
43. Frances Courtney Kneupper, *The Empire at the End of Time: Identity and Reform in Late Medieval German Prophecy* (Oxford: Oxford University Press, 2016), and Monica Azzolini, *The Duke and the Stars: Astrology and Politics in Renaissance Milan* (Cambridge, MA: Harvard University Press, 2013), are both excellent on the circulation of prophecies among elites in the late Middle Ages and the Renaissance. On popular texts, see Ottavia Niccoli, *Prophecy and People in Renaissance Italy*, trans. Lydia G. Cochrane (Princeton, NJ: Princeton University Press, 1990).
44. Ottavia Niccoli has identified over forty prophetic texts that circulated in Italy between 1480 and 1520—see her "Profezie in piazza: Note sul profetismo popolare nell'Italia del primo Cinquecento," *Quaderni storici* 14 (1979): 500–539. On the circulation of such texts in France, see Denis Crouzet's listing of his sources, many of which were of a prophetic nature, in his *Les guerriers de Dieu: La violence au temps des troubles de religion, vers 1525–vers 1610* (Seyssel: Champ Vallon, 1990), 2:633–98. But the most extensive work on prophetic materials is in Germany. For an orientation to this literature, see Ernst Zinner, *Geschichte und Bibliographie der astronomischen Literatur in Deutschland zur Zeit der Renaissance*, 2nd ed. (Stuttgart: Hiersemann, 1964); and, more recently, the appendix to Green, *Printing and Prophecy*, 156–203. Green's listing, which includes 176 individual works, covers the period down to 1550. Volker Leppin's "Corpus lutherischer apokalyptischer Flugschriften" and "Sonstige Quellen," in his *Antichrist und Jüngster Tag: Das Profil apokalyptischer Flugschriftenpublizistik im deutschen Luthertum, 1548–1618* (Gütersloh: Gütersloher Verlagshaus, 1999), 311–65, also offer an extensive listing of works, while Jürgen Beyer, "List of Prophets (1517–1800)," in *Lay Prophets in Lutheran Europe, c. 1550–1700* (Leiden: Brill, 2017), 238–329, provides a listing of works down to 1648 and Leigh T. I. Penman offers a similar listing in his *Hope and Heresy: The Problem of Chiliasm in Lutheran Confessional Culture, 1570–1630* (Dordrecht: Springer, 2019).

Nor are these the only texts published in early modern German with an apocalyptic edge—for other works of interest on this matter, see the catalog of titles in Adolf Dresler, ed. *Newe Zeitungen: Relationen, Flugschriften, Flugblätter, Einblattdrucke von 1470 bis 1820* (Munich: J. Halle, 1929), as well as the remarkable collection assembled in late sixteenth-century Zurich by Johann Jakob Wick: Wolfgang Harms and Michael Schilling, *Die Sammlung der Zentralbibliothek Zürich: Kommentierte Ausgabe, Die Wickiana*, 2 vols. (Tübingen: Niemeyer, 1997). Of course, given its precocity in the printing trade as well as the central role the press played in the Reformation, Germany may have witnessed more apocalyptic and astrological texts than any other region in Europe. Yet Crouzet's work makes it clear that such works were extremely popular in France, while Keith Thomas has demonstrated that astrological works had a broad appeal in early modern England, noting that more than six hundred almanacs were published before the end of the sixteenth century and another two thousand appeared in the seventeenth, adding that "the figure of some three to four million" copies "which is sometimes suggested as the total production of almanacs in the seventeenth century, is a distinct underestimate." See Thomas, *Religion and the Decline of Magic: Studies in Popular Beliefs in Sixteenth- and Seventeenth-Century England* (New York: Scribner, 1971), 294. Moreover, a search of catalog entries for almanacs in Italy in the sixteenth century makes it clear that such works were popular there as well. In short, while such writings often drew on old traditions and certainly circulated in manuscript in the Middle Ages, the early modern period was marked by an immense intensification in the circulation of texts that helped focus the attention of readers and those who gathered around them on the future and also on the End of History.

45. Ruderman, "Hope against Hope," 393–94.
46. Roberto Rusconi, introduction to *The Book of Prophecies*, ed. Roberto Rusconi, trans. Blair Sullivan (Eugene, OR: Wipf and Stock, 1997), 35.
47. Elisheva Carlebach, *Palaces of Time: Jewish Calendar and Culture in Early Modern Europe* (Cambridge, MA: Harvard University Press, 2011), 203.
48. Girolamo Savonarola, *Il trattato contro li astrologi*, in *Contro gli astrologi*, ed. Claudio Gigante (Rome: Salerno, 2000).
49. Robin B. Barnes, *Astrology and Reformation* (New York: Oxford University Press, 2016), 128.
50. Johannes Lichtenberger, *Pronosticatio in Latino* (Ulm: [Johann Zainer?], 1488).
51. "Qui erit princeps et monarcha tocius Europe, reformabit ecclesias et clerum et post illum nullus amplius imperabit," cited in Marjorie Reeves, *The Influence of Prophecy in the Later Middle Ages: A Study in Joachimism* (Notre Dame, IN: University of Notre Dame Press, 1993), 350. On Lichtenberger, see Dietrich Kurze, *Johannes Lichtenberger: Eine Studie zur Geschichte der Prophetie und Astrologie* (Lübeck: Matthiesen, 1960); and Green, *Printing and Prophecy*, 64–82.
52. Green, *Printing and Prophecy*, esp. 119–24.
53. Baruchson-Arbib, *La culture livresque des juifs*, 159–60.
54. *Alcoranus Arabice* (Venice: Paganino and Alessandro Paganini, 1537–38). On the history of this work, see Angela Nuovo, "A Lost Arabic Koran Rediscovered," *The Library*, 6th ser., 12 (1990): 273–92; and "La scoperta del Corano arabo, ventisei anni

dopo: Un riesame," *Nuovi annali della Scuola Speciale per archivisti e bibliotecari* 27 (2013): 9–23.

55. Angelika Neuwirth, *The Qur'an and Late Antiquity: A Shared Heritage,* trans. Samuel Wilder (Oxford: Oxford University Press, 2019).

56. Geoffrey Roper, "The History of the Book in the Muslim World," in *The Oxford Companion to the Book,* ed. Michael F. Suarez, S. J., and H. R. Woudhuysen (Oxford: Oxford University Press, 2010), 1:330. Already in the sixteenth century European observers expressed bafflement over what they saw as Ottoman resistance to printing. Among modern historians, some have attributed this resistance to political repression; see Metin M. Cosgel, Thomas J. Miceli, and Jared Rubin, "The Political Economy of Mass Printing: Legitimacy and Technological Change in the Ottoman Empire," *Journal of Comparative Economics* 40 (2012): 357–71, who argue that Ottoman rulers "banned" printing because they feared it would undermine their legitimacy. Others have stressed economic factors, above all the self-interests of calligraphers, copyists, and illuminators, whose guilds were especially powerful, to protect their incomes; see Kathryn A. Schwartz, "Did the Ottoman Sultans Ban Print?" *Book History* 20 (2017): 1–39. What is needed is an anthropology of the role of writing and textual transmission in premodern Islamic societies.

57. Jean Bodin, *Colloquium of the Seven about the Secrets of the Sublime,* ed. and trans. Marion Leathers Kuntz (Princeton, NJ: Princeton University Press, 1975), 294. On the authorship of this work, see Noel Malcolm, "Jean Bodin and the Authorship of the 'Colloquium Heptaplomeres,'" *Journal of the Warburg and Courtauld Institutes* 69 (2006): 95–150.

58. On pocket-sized Qur'ans, see François Déroche, "Written Transmission," in *The Blackwell Companion to the Qur'ān,* ed. Andrew Rippin and Jawid Mojaddedi, 2nd ed. (Hoboken, NJ: Wiley, 2017), 181; on the productivity of calligraphers, see Jonathan Bloom, *Paper before Print: The History and Impact of Paper in the Islamic World* (New Haven: Yale University Press, 2001).

59. Tijana Krstić, *Contested Conversions to Islam: Narratives of Religious Change in the Early Modern Ottoman Empire* (Stanford, CA: Stanford University Press, 2011), 35–37.

60. David Cook, "Mainstream Popular Ottoman-Era Muslim Apocalypses," paper delivered to the Providential Modernity Seminar, Duke University, January 9, 2020. I thank Professor Cook for permission to use his paper.

61. Massumeh Farhad with Serpil Bağcı, ed., *Falanama: The Book of Omens* (Washington, DC: Smithsonian Institution, 2009).

62. Gerard A. Wiegers, "Jean de Roquetaillade's Prophecies among the Muslim Minorities of Medieval and Early-Modern Christian Spain: An Islamic Version of the *Vademecum in Tribulatione,*" in *The Transmission and Dynamics of the Textual Sources of Islam: Essays in Honour of Harald Motzki,* ed. Nicolet Boekhoff-van der Voort, Kees Versteegh, and Joas Wagemakers (Leiden: Brill, 2011), 229–47.

63. S. Aslıhan Gürbüzel, "Teachers of the Public, Advisors to the Sultan: Preachers and the Rise of a Political Sphere in Early Modern Istanbul" (PhD diss., Harvard University, 2018), 5.

64. Jonathan Berkey, *Popular Preaching and Religious Authority in the Medieval Islamic Near East* (Seattle: University of Washington Press, 2001).
65. Colin Imber, "A Note on 'Christian' Preachers in the Ottoman Empire," *Osmanlı Araştırmaları / Journal of Ottoman Studies* 10 (1990): 75.
66. Krstić, *Contested Conversions*, 36; Berkey, *Popular Preaching and Religious Authority*.
67. Stefano Dall'Aglio, " 'Faithful to the Spoken Word': Sermons from Orality to Writing in Early Modern Italy," *The Italianist* 34 (2014): 463–77; and his "Voices under Trial: Inquisition, Abjuration, and Preachers' Orality in Sixteenth-Century Italy," *Renaissance Studies* 31 (2015): 25–42. See also Giorgio Caravale, *Preaching and Inquisition in Renaissance Italy: Words on Trial* (Leiden: Brill, 2017).
68. Reeves, *Influence of Prophecy in the Later Middle Ages*.
69. Richard Wunderli, *Peasant Fires: The Drummer of Niklashausen* (Bloomington: Indiana University Press, 1992).
70. Reeves, *Influence of Prophecy in the Later Middle Ages*, 263.
71. Rodney L. Petersen, *Preaching in the Last Days: The Theme of "Two Witnesses" in the Sixteenth and Seventeenth Centuries* (Oxford: Oxford University Press, 1993).
72. On Ibn Shem Tob, see Marc Saperstein, ed., *Jewish Preaching, 1200–1800: An Anthology* (New Haven: Yale University Press, 1989), 181–98; on Molkho, see Moshe Idel, *Messianic Mystics* (New Haven: Yale University Press, 1998), 145–50, 242–52, and 255–69 for examples of sermons from these figures.
73. On Solomon ben Isaac Levi and Judah Moscato, see Saperstein, ed., *Jewish Preaching*, 242–52, 255–69. On the growing importance of Jewish sermons throughout Europe and the Mediterranean in this period, see David B. Ruderman, ed., *Preachers of the Italian Ghetto* (Berkeley: University of California Press, 1992), 3.

CHAPTER THREE. CHRISTOPHER COLUMBUS

1. Bartolomé de las Casas, *Historia de las Indias,* ed. André Saint-Lu (Caracas: Biblioteca Ayacucho, 1956), 1:346.
2. Ibid.
3. Fernando Colombus, *Historie del S. D. Fernando Colombo: nelle quali s'hà particolare, & vera relatione della vita, & de fatti dell'Ammiraglio D. Christoforo Colombo, suo padre, et dello scoprimento, ch'egli fece dell'Indie Occidentali, dette* mondo nuovo, *hora possedute dal Serenissimo Rè Catolico,* trans. Alfonso de Ulloa (Venice: Francesco de' Franceschi, 1571), 84v. The original Spanish version was never published and the manuscript is lost.
4. Las Casas, *Historia de las Indias*, 1:347.
5. Ibid.
6. Ibid., 1:348.
7. Ibid.
8. Christopher Columbus, *The Diario of Christopher Columbus's First Voyage to America: 1492–1493,* ed. and trans. Oliver Dunn and James E. Kelley, Jr. (Norman: University of Oklahoma Press, 1989), 16–17.
9. Ibid., 18–19.

10. Columbus, "Prologue to the King and Queen," in *Diario of Christopher Columbus,* 18–19.
11. Marjorie Reeves, *The Influence of Prophecy in the Later Middle Ages: A Study in Joachimism* (Notre Dame, IN: University of Notre Dame Press, 1993).
12. On Columbus's early years and social milieu see Jacques Heers, *Christophe Colomb* (Paris: Hachette, 1981), 29–73. On his early education, Las Casas, *Historia de las Indias,* 1:29.
13. Columbus, *Historie . . . della vita, & de fatti dell'Ammiraglio,* 11v–12r.
14. Las Casas, *Historia de las Indias,* 1:30.
15. Columbus, "Carta del almyrante al rey e a la rreyna," in Columbus, *Libro de las profecías,* in *The Book of Prophecies,* ed. Roberto Rusconi, trans. Blair Sullivan (Eugene, OR: Wipf and Stock, 1997), 66–67.
16. Pliny the Elder, *Historia naturalis,* trans. Cristoforo Landino (Venice: Bartolomeo de Zani de Portesio, 1489); Plutarch, *Las vidas de los ilustres Varones* [here under the title *Fenecen en dos volumines las vidas de Plutarco*], trans. Alfonso de Palencia (Seville: Paulo de Colonia et Iohannes de Nuremberg et Magno et Thomas Alpanes, 1491). I have consulted the digital and microform versions of these texts, as they were annotated by Columbus, in the Biblioteca Capitular y Colombina in Seville.
17. Marco Polo, *Liber de consuetudinibus et condicionibus orientalium regionum,* trans. Francesco Pipino (Gouda: Gerardus Leeu, 1483–84); for this text, as it was annotated by Columbus, I consulted the facsimile, published by Testimonio in Madrid, held in the Biblioteca Capitular y Colombina in Seville, as well as the companion volume *El libro de Marco Polo: Ejemplar anotado por Cristóbal Colón y que se conserva en la Biblioteca Capitular y Colombina de Sevilla,* ed. Juan Gil (Madrid: Alianza, 1986). Columbus received the Marco Polo as a gift in 1497 from the English merchant John Day. As Gil demonstrates, the volume includes annotations not only by Columbus but also by his friend Gaspar Gorrico and his son Fernando.
18. Aeneas Sylvius Piccolomini (Pius II), *Historia rerum ubique gestarum: cum locorum descriptione non finita* (Venice: Johannes de Colonia et Johannes Manthen, 1477). For this text, as it was annotated by Columbus, I used the facsimile, published by Testimonio in Madrid, held in the Biblioteca Capitular y Colombina.
19. Columbus almost certainly owned a copy of Ptolemy, the 1478 edition published by Arnoldus Buckinck in Rome.
20. Columbus had copied out in his own hand two letters from Toscanelli and these are bound in the back of his copy of Piccolomini's *Historia rerum ubique gestarum.*
21. Pierre d'Ailly, *Tractatus de imagine mundi Petri de Aliaco, et varia ejusdem auctoris et Ioannis Gersonis opuscula* (Louvain: Joannes de Westphalia, 1480–83). I have used the rare facsimile edition (Boston: Massachusetts Historical Society, 1927). Columbus's personal copy is in the Biblioteca Colombina in Seville. There is also a useful French translation, with commentary, of four of the cosmological treatises of the volume: Edmond Buron, *Ymago mundi de Pierre d'Ailly: Texte latin et traduction française des quatre traités cosmographiques de d'Ailly et des notes marginales de Christophe Colomb; Étude sur les sources de l'auteur,* 3 vols. (Paris: Maisonneuve Frères, 1930). And with its more than eight hundred annotations, Columbus's personal copy of

this volume provides a rare glimpse of a reader reacting to, pondering, and mastering a set of texts.
22. Las Casas, *Historia de las Indias,* 1:61.
23. "Nota quod hoc anno de .88. in mense decembri . . . Bartholomeus didacus . . . renunciavit ipso Serenissimo regi prout navigaviter ultra Yan navigatum . . . usque uno promuntorio per ipsum nominatum 'cabo de boa esperança' . . . quem viagium pictavit & scripsit de leucha in leucha in una carta navigacionis ut occuli visui ostenderet ipso serenissimo regi in quibus omnibus interfui," Columbus's annotations to d'Ailly "Ymago mundi," in his *Tractatus de imagine mundi,* fol. 13r ("Postille ai trattati di P. d'Ailly," in vol. 2.1 of *Raccolta di documenti e studi pubblicati dalla R. Commissione Colombiana,* ed. Cesare de Lollis [Rome: Ministero della Pubblica Istruzione, 1894], no. 23). I thank Juan Gil for his help in the reading of "yan" as an unusual spelling of *iam* in this passage.
24. "Inter finem ispanie et principium indie est mare paruum et navigabile in paucis diebus." *Tractatus de imagine mundi Petri de Aliaco,* fol. 10 (de Lollis, *Raccolta di documenti,* no. 23). Pauline Moffitt Watts argues on the basis of internal evidence that Columbus read d'Ailly prior to his first voyage; see her "Prophecy and Discovery: On the Spiritual Origins of Christopher Columbus's 'Enterprise of the Indies,'" *American Historical Review* 90 (1985): 85–86. By contrast Juan Gil maintains that Columbus did not read d'Ailly until or after 1497; see his *Colón y su tiempo,* vol. 1 of *Mitos y utopías del descubrimiento* (Madrid: Alianza, 1989), chap. 4. The argument in favor of the earlier dating for Columbus's reading of the cosmological treatises seems persuasive, but Columbus may have read and annotated the astrological texts at a later date. There is no consensus on the dates when Columbus read and annotated the texts.
25. William M. Reddy, "The Eurasian Origins of Empty Time and Space: Modernity as Temporality Reconsidered," *History and Theory* 55 (2016): 325–56.
26. Columbus's annotations to d'Ailly "Tractatus de legibus et sectis" in his *Tractatus de imagine mundi,* 50r and 51r (de Lollis, *Raccolta di documenti,* nos. 555 and 561).
27. Ernst Cassirer, *The Individual and the Cosmos in the Renaissance,* trans. Mario Domandi (Chicago: University of Chicago Press, 1963).
28. Columbus, "La cuenta de la creación del mondo segondo los Judios," bound in the back of his copy of Piccolomini, *Historia rerum ubique gestarum.*
29. Laura Ackerman Smoller, *History, Prophecy, and the Stars: The Christian Astrology of Pierre d'Ailly, 1350–1420* (Princeton, NJ: Princeton University Press, 1994), offers an insightful analysis of d'Ailly's astrology and apocalypticism.
30. Columbus's annotations to d'Ailly, *Tractatus de legibus et sectis,* 45v (de Lollis, *Raccolta di documenti,* no. 524).
31. Watts was the first to underscore the centrality of Pseudo-Methodius to Columbus's eschatology; see her "Prophecy and Discovery," 73–102. But Columbus's providentialism and apocalypticism have received considerable attention. See, for example, John Leddy Phelan, *The Millennial Kingdom of the Franciscans in the New World: A Study of the Writings of Gerónimo de Mendieta, 1525–1604* (Berkeley: University of California Press, 1970), esp. chap. 2; Alain Milhou, *Colón y su mentalidad mesiánica en el ambiente franciscanista español* (Valladolid: Casa-Museo de Colón, 1983); Gil,

Mitos y utopias del descubrimiento; Delno C. West and August Kling, *The Libro de las profecías of Christopher Columbus* (Gainesville: University of Florida Press, 1991), esp. the three introductory chapters; Carol Lowery Delaney, *Columbus and the Quest for Jerusalem* (New York: Free Press, 2011); and Denis Crouzet, *Christophe Colomb: Héraut de l'Apocalypse* (Paris: Presses Universitaires de France, 2018).

32. D'Ailly, "Tractatus de concordia astronomice veritatis et narrationis hystorice," 120v.
33. Christopher Columbus, *Relación del tercer viaje,* in *Cristóbal Colón: Textos y documentos completos,* ed. Consuelo Varela and Juan Gil (Madrid: Alianza, 2003), 376–77.
34. Consuelo Varela, *La caída de Cristóbal Colón: El juicio de Bobadilla,* ed. Isabel Aguirre (Madrid: Marcial Pons Historia, 2006).
35. Columbus, *Libro de las profecías,* 70–71.
36. Ibid., 166–67.
37. Ibid., 294–95.
38. Ibid., 70–71.
39. "Liber s[ive manipulus de au]ctoritatibus, dictis ac sententiis et p[rophetiis circa] materiam recuperande sancta civitatis et montis Dei Syon ac inventionis et conversionis insularum Indye et omnium gentium atque nationum, ad Ferdinandum et Helysabeth reges nostros hyspanos." Ibid., 58–59.
40. Ibid., 211.
41. Ibid., 238–49.
42. Ibid., 209–10.
43. St. Augustine, *City of God,* 20.7.
44. Columbus, *Libro de las profecías,* 70–71.
45. Columbus, "Carta del almyrante al rey e a la reyna," 66–69.
46. Columbus, *Libro de las profecías,* 104–5.
47. Ibid., 136–37. Not St. Augustine as Columbus believed but rather the *Soliloquium animae ad Deum,* attributed to Augustine.
48. Ibid., 316–17; also 76–77.
49. Christopher Columbus, *Relación del cuarto viaje,* in Varela and Gil, eds., *Cristóbal Colón: Textos y documentos completos,* 491–92.
50. Columbus, *Libro de las profecías,* 290–91, from Seneca's *Medea.* Columbus owned a copy printed in Venice in 1492.
51. Cited in Watts, "Prophecy and Discovery," 94.
52. On Taíno resistance to Columbus and other early settlers, see Irving Rouse, *The Taínos: Rise and Decline of the People Who Greeted Columbus* (New Haven: Yale University Press, 1993), 150–61.
53. Christopher Columbus, "Relación del viaje a Cuba y Jamaica," in Varela and Gil, eds., *Cristóbal Colón: Textos y documentos completos,* 305–6.
54. Crouzet also suggests that Columbus invented the speech to convince Ferdinand and Isabella that it would not be difficult to convert the Indians—see his *Christophe Colomb: Héraut de l'Apocalypse,* 299.
55. Columbus, *Historie . . . della vita, & de fatti dell'Ammiraglio,* 116v; Las Casas, *Historia de las Indias,* 1:409.

56. Pietro Martire d'Anghiera, *De orbe novo Petri Martyris Angelerii Mediolanensis protonotarij c[a]esaris senatoris decades* (Alcalá de Henares: Miguel de Eguía, 1530), fol. 10r; *De Orbe Novo: The Eight Decades,* trans. Francis Augustus MacNutt (New York: Putnam, 1912), 1:270–72.
57. "Este fue la primera injusticia . . . y el comienzo del derramamiento de sangre, que despues tan copioso fue en esta isla." Las Casas, *Historia de las Indias,* 1:397.
58. Michele da Cuneo, *De novitatibus insularum occeani Hesperii repertarum a don Christoforo Columbo Genuensi,* in *Nuovo mondo: Gli Italiani, 1492–1565,* ed. Paolo Collo and Pier Luigi Crovetto (Turin: Einaudi, 1991), 118; *News of the Islands of the Hesperian Ocean Discovered by Don Christopher Columbus of Genoa,* in *Italian Reports on America (1493–1522): Accounts by Contemporary Observers,* ed. Geoffrey Symcox and Luciano Formisano, trans. Theodore J. Cachey Jr. and John C. McLucas (Turnhout: Brepols, 2002), 188.
59. Las Casas, *Historia de las Indias,* 1:398.

CHAPTER FOUR. CONQUEST AND UTOPIA

1. Christopher Columbus, *The Diario of Christopher Columbus's First Voyage to America: 1492–1493,* ed. and trans. Oliver Dunn and James E. Kelley, Jr. (Norman: University of Oklahoma Press, 1989), 64–65.
2. *Capitulacion de Santa Fe,* in *The Book of Privileges Issued to Christopher Columbus by King Fernando and Queen Isabel, 1492–1502,* ed. and trans. Helen Nader (Berkeley: University of California Press, 1996), 267; *Capitulation of Santa Fe* in the same volume, 63. On the function of gold in Columbus's enterprise, see Elvira Vilches, *New World Gold: Cultural Anxiety and Monetary Disorder in Early Modern Spain* (Chicago: University of Chicago Press, 2010), 53–94.
3. Christopher Columbus, "Carta a Luis de Santángel," in *Cristóbal Colón: Textos y documentos completos,* ed. Consuelo Varela and Juan Gil (Madrid: Alianza, 2003), 221–24; "Letter of Columbus to Various Persons Describing the Results of his First Voyage and Written on the Return Journey," in *The Four Voyages of Christopher Columbus,* ed. and trans. J. M. Cohen (Harmondsworth: Penguin, 1969), 117–21.
4. First published in Castilian in Barcelona in 1493, the work was quickly translated into Latin, Italian, and German, with a total of at least fifteen editions in the 1490s alone and was best known under its Latin title, *Epistola de insulis nuper inventis.* See David Henige, "Finding Columbus: Implications of a Newly Discovered Text," in *The European Outthrust and Encounter: The First Phase, c. 1400–c. 1700,* ed. Cecil H. Clough and Paul E. H. Hair (Liverpool: Liverpool University Press, 1994), 141–65.
5. Frank E. Manuel and Fritzie P. Manuel, *Utopian Thought in the Western World* (Cambridge, MA: Harvard University Press, 1979), 64–92.
6. Adriano Prosperi, "America e apocalisse: Note sulla 'conquista spirituale' del Nuovo Mondo," in *America e Apocalisse e altri saggi* (Pisa: Istituti Editoriali e Poligrafici Internazionali, 1999), 15–63. For the humanist framing of European overseas expansion, see, among others, Andrew Fitzmaurice, *Humanism and America: An Intellectual History of English Colonisation, 1500–1625* (Cambridge: Cambridge University Press, 2003).

7. Pietro Martire d'Anghiera, *De orbe novo Petri Martyris Anglerii Mediolanensis protonotarij c[a]esaris senatoris decades* (Alcalá de Henares: Miguel de Eguía, 1530), fol. 6v; *De orbe novo: The Eight Decades,* trans. Francis Augustus MacNutt (New York: Putnam, 1912), 1:103–4. While the first complete edition of Peter Martyr's history appeared in 1530, the *First Decade* was published in 1511 followed by an edition of the *First Three Decades* in 1516, though a version of the first three decades had also appeared in Italian under the title *Libretto da tutta la navigatione de re Spagna de le isole et terreni novamente trovati* in Venice in 1504. Then a version of Martyr's work was included in the *Paesi novamente retrovati et Novo Mondo di Alberico Vesputio* in Vicenza in 1507, with a Latin version published the following year.
8. D'Anghiera, *De orbe novo,* fol. 10r; MacNutt, *De orbe novo,* 94.
9. Marjorie Reeves, *The Influence of Prophecy in the Later Middle Ages: A Study in Joachimism* (Notre Dame, IN: University of Notre Dame Press, 1993), 505–9; Yosef Hayim Yerushalmi, *Zakhor: Jewish History and Jewish Memory* (Seattle: University of Washington Press, 1982), 5–26.
10. Amerigo Vespucci, *Mundus novus* (Augsburg: Johannes Otmar, 1504). Two earlier editions of this text were published: one in either 1502 or 1503; the other in 1503; for the English translation, "Letter V: Mundus Novus," in *Letters from a New World: Amerigo Vespucci's Discovery of America,* ed. Luciano Formisano, trans. David Jacobson (New York: Marsilio, 1992), 45–56.
11. "Superioribus diebus satis ample scripsi de reditu meo ab novis illis regionibus, quas . . . perquisivimus et invenimus quasque novum mundum appellare licet." Vespucci, *Mundus novus,* fol. 1r (unpaginated); "Letter V: Mundus Novus," 45.
12. Vespucci, *Mundus novus,* fol. 1v; "Letter V: Mundus Novus," 49.
13. Vespucci, *Mundus novus,* fol. 2r; "Letter V: Mundus Novus," 50.
14. Vespucci, *Mundus novus,* fol. 2r–2v; "Letter V: Mundus Novus," 49–50.
15. Vespucci, *Mundus novus,* fol. 2r; "Letter V: Mundus Novus," 49. Vespucci himself had not included this passage, but it was present in almost all the versions of the text that circulated and did much to frame the discoveries, as Columbus had done, in an apocalyptic key. See Rosario Romeo, *Le scoperte americane nella coscienza italiana del Cinquecento* (Milan: Ricciardi, 1954), 15–16.
16. Martin Waldseemüller, *Cosmographiae introductio cum quibusdam geometriae ac astronomiae principiis ad eam rem necessariis* (Saint-Dié: Gualterius and Nikolaus Lud, 1507). I have used *The Cosmographiae introductio of Martin Waldseemüller in facsimile,* ed. Charles George Herbermann, trans. Edward Burke and Mario Emilio Cosenza (New York: United States Catholic Historical Society, 1907), n.p.; "From Martin Waldseemüller's *Cosmographiae Introductio,*" in Formisano and Jacobson, *Letters from a New World,* 116.
17. Quentin Skinner, "Sir Thomas More's *Utopia* and the Language of Renaissance Humanism," in *The Languages of Political Theory in Early Modern Europe,* ed. Anthony Pagden (Cambridge: Cambridge University Press, 1987), 123–58.
18. Manuel and Manuel, *Utopian Thought in the Western World,* esp. chap. 1.
19. Felipe Fernández-Armesto, *Amerigo: The Man Who Gave His Name to America* (London: Weidenfeld and Nicolson, 2006).

20. "Lettera di Amerigo Vespucci delle isole nuovamente trovate in quattro suoi viaggi," in *Nuovo mondo: Gli Italiani, 1492–1565,* ed. Paolo Collo and Pier Luigi Crovetto (Turin: Einaudi, 1991), 267; "Letter of Amerigo Vespucci Concerning the Islands Newly Discovered on his Four Voyages," in Formisano and Jacobson, *Letters from a New World,* 96.
21. Thomas More, *De optimo reipublicae statu deque nova insula Utopia* [*The Best State of a Commonwealth and the New Island of Utopia*] (Louvain: Martens, 1516). This first edition was immediately followed by a Paris edition of 1517 and the two definitive Basel editions, both published by Froben, of 1518.
22. I have used the critical Latin edition of *Utopia,* with the English translation on the facing page, in vol. 4 of *The Complete Works of St. Thomas More,* ed. Edward Surtz, S.J., and J. H. Hexter (New Haven: Yale University Press, 1965). For this citation, see *Utopia,* 53.
23. Perhaps, with an appreciation for its ambiguity, the word *Utopia* hovered between two possibilities in Greek: not only *ou-topos* ("no place") but also *eu-topos* ("good place"). Above all, the coinage was part of the playfulness with which More, among other things a scholar of classical Greek, gave names to various features of this imagined island with the goal, in part, of underscoring that he was imagining a place that did not exist. Writing in Latin for a learned, humanist audience, More called the main river of the island *Anhyder* ("waterless"), its rulers *Ademos* ("people-less"), and its capital *Amaurot* ("phantom"). And he called his interlocutor in the dialogue *Hythloday* ("peddler of nonsense").
24. More, *Utopia,* 102–5.
25. Ibid., 104–5.
26. Ibid., 106–7.
27. Aristotle, *Politics,* 2.5 (1263a8–21).
28. On the relation between the two books, see J. H. Hexter's classic study *More's Utopia: The Biography of an Idea* (Princeton, NJ: Princeton University Press, 1952).
29. More, *Utopia,* 150.
30. Ibid., 152
31. Ibid., 156.
32. Richard Marius, *Thomas More: A Biography* (Cambridge, MA: Harvard University Press, 1984).
33. Cited in David Wootton, "Friendship Portrayed: A New Account of *Utopia,*" *History Workshop Journal* 45 (1998): 32.
34. Ibid.; see also David Harris Sacks, "*Utopia* as a Gift: More and Erasmus on the Horns of a Dilemma," *Moreana* 54 (2017): 157–71, for a trenchant analysis of the debate in *Utopia* over private property.
35. Thomas More, *A Treatyce (unfynyshed) vppon these wordes of holye Scrypture,* Memorare nouvissima, & in eternum non peccabis, *Remember the last thynges,* in *The Complete Works of Thomas More,* ed. Anthony S. G. Edwards, Katherine Gardiner Rodgers, and Clarence H. Miller (New Haven: Yale University Press, 1997), 1:127–82.
36. "Hanc Reipublicae formam, quam omnibus libenter optarim, Vtopiensibus saltem contigisse gaudeo, qui ea vitae sunt instituta sequuti, quibus Reipublicae fundamenta

iecerunt non modo felicissime, verumetiam quantum humana praesagiri coniectura contigit, aeternum duratura." More, *Utopia,* 244–45.

37. Manuel and Manuel, *Utopian Thought in the Western World,* again chap. 1 especially.
38. Cited in John Leddy Phelan, *The Millennial Kingdom of the Franciscans in the New World: A Study of the Writings of Gerónimo de Mendieta, 1525–1604* (Berkeley: University of California Press, 1970), 23.
39. Ibid., 48–49.
40. José A. Maravall, "La utopía politíca-religiosa de los franciscanos en Nueva España," *Estudios Americanos* 2 (1949): 199–227.
41. "Y como floreció el principio la iglesia en oriente, que es el principio del mundo, bien así ahora en el fin de los siglos tiene de florecer en occidente, que es fin del mundo." Toribio de Motolinía, *Historia de los indios de la Nueva España,* ed. Edmundo O'Gorman (Mexico City: Editorial Porrua, 1973), 25; *History of the Indians of New Spain,* trans. Elizabeth Andros Foster (Berkeley: Cortés Society, 1950), 222.
42. Phelan, *Millennial Kingdom;* and Beatriz Pastor Bodmer, *El jardín y el peregrino: Ensayos sobre el pensamiento utópico latinoamericano, 1492–1695* (Amsterdam: Rodopi, 1996).
43. Cited in Henry Kamen, *Empire: How Spain Became a World Power, 1492–1763* (New York: HarperCollins, 2003), 149.
44. Particularly useful on the *encomienda* system is Luis Arranz Márquez, ed., *Repartimientos y encomiendas en la Isla Española: El Repartimiento de Alburquerque de 1514* (Madrid: Fundación García Arévalo, 1991); see also Inga Clendinnen, *Ambivalent Conquests: Maya and Spaniard in Yucatan, 1517–1570,* 2nd ed. (Cambridge: Cambridge University Press, 2005), 38–40, 60–62, 102.
45. On Las Casas's early life, see Helen Rand Parish and Harold E. Weidman, *Las Casas en México: Historia y obras desconocidas* (Mexico City: Fondo de Cultura Económica, 1992).
46. On this event, see Las Casas, *Historia de las Indias,* 1:282–83.
47. Las Casas, *Brevissima relacion de la destruyción de las Indias* (Seville: Sebastian Trugillo, 1552), 2r (unpaginated); *A Short Account of the Destruction of the Indies,* trans. Nigel Griffin (New York: Penguin, 1992), 11.
48. Las Casas, *Brecissima relacion,* 11r; *Short Account,* 27–28.
49. Bartolomé de Las Casas, *Memorial de Remedios para las Indias,* in *Cartas y memoriales,* vol. 13 of *Obras completas,* ed. Paulino Castañeda Delgado (Madrid: Alianza, 1995). Here and below I have benefited from the English translation of this memorial in Victor N. Baptiste, *Bartolomé de las Casas and Thomas More's* Utopia*: Connections and Similarities* (Culver City, CA: Labyrinthos, 1990), 14–59.
50. Ibid., 34.
51. Ibid., 41–42.
52. Ibid., 39–40.
53. Ibid., 40–41.
54. Ibid., 59.
55. Marcel Bataillon, "La Vera Paz: Roman et histoire," *Études sur Bartolomé de Las Casas* (Paris: Centre de Recherches de l'Institut d'Études Hispaniques, 1965), 137–202.

CHAPTER FIVE. THE LAST WORLD EMPEROR

1. On Ottoman policy in the Indian Ocean, see Giancarlo Casale, *The Ottoman Age of Exploration* (Oxford: Oxford University Press, 2010), 55–67; on the Battle of Diu, see Diogo do Couto, *Décadas,* ed. António Baião (Lisbon: Sá Da Costa, 1947); Lopo de Souza Coutinho, *O primeiro cerco de Diu,* ed. Luis de Albuquerque and Maria da Graça Pericão (Lisbon: Publicações Alfa, 1989); Damião da Góis, *Opúsculos Históricos,* trans. Dias de Carvalho (Porto: Livrario Civilização, 1945); and Fernão Mendes Pinto, *The Travels of Mendes Pinto,* ed. and trans. Rebecca D. Catz. (Chicago: University of Chicago Press, 1989).
2. Marshall G. S. Hodgson, *The Gunpowder Empires and Modern Times,* vol. 3 of *The Venture of Islam: Conscience and History in a World Civilization* (Chicago: University of Chicago Press, 1974), 1–133.
3. Leopold von Ranke, *The Ottoman and the Spanish Empires, in the Sixteenth and Seventeenth Centuries,* trans. Walter Keating Kelly (London: Whittaker, 1843), offered an early history of these two empires, but for more recent comparative analyses see Jane Burbank and Frederick Cooper, *Empires in World History: Power and the Politics of Difference* (Princeton, NJ: Princeton University Press, 2010), 117–148; and Noel Malcolm, *Useful Enemies: Islam and the Ottoman Empire in Western Political Thought, 1450–1750* (Oxford: Oxford University Press, 2019), 57–75.
4. Mona Hassan, *Longing for the Lost Caliphate: A Transregional History* (Princeton, NJ: Princeton University Press, 2016), 145; Alan Mikhail, *God's Shadow: Sultan Selim, His Ottoman Empire, and the Making of the Modern World* (New York: Norton, 2020), 309.
5. Gunter Düriegl, ed., *Wien 1529: Die erste Türkenbelagerung: Sonderausstellung des Historischen Museums der Stadt Wien* (Vienna: Böhlau, 1979), 1:7–33.
6. Daniel Goffman, *The Ottoman Empire and Early Modern Europe* (Cambridge: Cambridge University Press, 2002).
7. On the demographic and economic shifts that underlay Ottoman power, see Fernand Braudel, *The Mediterranean and the Mediterranean World in the Age of Philip II,* trans. Siân Reynolds (New York: Harper and Row, 1973), 2:660–72; and Halil Inalcık with Donald Quataert, *An Economic and Social History of the Ottoman Empire, 1300–1600* (Cambridge: Cambridge University Press, 1994).
8. Gülru Necipoğlu, *The Age of Sinan: Architectural Culture in the Ottoman Empire* (Princeton, NJ: Princeton University Press, 2005).
9. Colin Imber, *The Ottoman Empire, 1300–1600: The Structure of Power* (Houndsmill, Basingstoke: Palgrave Macmillan, 2002), 267–75.
10. Malcolm, *Useful Enemies,* 63.
11. Gilles Veinstein, "On the Ottoman Janissaries (Fourteenth-Nineteenth Centuries)," in *Fighting for a Living: A Comparative Study of Military Labour, 1500–2000,* ed. Erik-Jan Zürcher (Amsterdam: Amsterdam University Press, 2013), 115–34.
12. Rhoads Murphey, "Ottoman Imperial Identity in the Post-Foundation Era," *Archivum Ottomanicum* 26 (2009–2010): 86; Suraiya Faroqhi, *Ottoman Empire and the World Around It* (London: I. B. Tauris, 2004), 75–97.

13. Timur Kuran, ed., *Social and Economic Life in Seventeenth-Century Istanbul: Glimpses from Court Records*, 10 vols. (Istanbul: İş Bankası Kültür Yayınları, 2010–13), while emphasizing cases involving economic matters, provides English summaries and opens a window onto the Islamic courts in early modern Istanbul.
14. Benjamin Braude, "Foundation Myths of the Millet System," in *Christians and Jews in the Ottoman Empire: The Functioning of a Plural Society*, ed. Benjamin Braude and Bernard Lewis (New York: Holmes and Meier, 1982), 1:69–88.
15. Leslie P. Peirce, *The Imperial Harem: Women and Sovereignty in the Ottoman Empire* (Oxford: Oxford University Press, 1993).
16. Leslie P. Peirce, *Empress of the East: How a European Slave Girl Became Queen of the Ottoman Empire* (New York: Basic Books, 2017).
17. Beatrice Forbes Manz, *The Rise and Rule of Tamerlane* (Cambridge: Cambridge University Press, 1989).
18. Ibn Khaldūn, *The Muqaddimah: An Introduction to History*, trans. Franz Rosenthal, 3 vols. (Princeton, NJ: Pantheon, 1958), 2:213.
19. On the theorization of sovereignty in India and Iran after Tamerlane, see A. Azfar Moin, *The Millennial Sovereign: Sacred Kingship and Sainthood in Islam* (New York: Columbia University Press, 2012); the citation concerning the tent peoples is from p. 27. For a sense of the wider context, see Matthew Melvin-Koushki, "Astrology, Lettrism, Geomancy: The Occult-Scientific Methods of Post-Mongol Islamicate Imperialism," *Medieval History Journal* 19 (2016): 142–50.
20. Cornell H. Fleischer, "A Mediterranean Apocalypse: Prophecies of Empire in the Fifteenth and Sixteenth Centuries," *Journal of the Economic and Social History of the Orient* 61 (2018): 55. (The passage from Mevlānā 'Isā in the chapter epigraph is taken from p. 65 of this article.)
21. Ibid., 71. In his article, Fleischer offers a "summary" translation of the introductory portion of Haydar's *New Year Prognostication of 1535:* 69–72. On Haydar, see also Fleischer, "Shadow of Shadows: Prophecy in Politics in 1530s Istanbul," *International Journal of Turkish Studies* 13 (2007): 51–62.
22. On Mevlānā 'Isa, see Cornell H. Fleischer, "The Lawgiver as Messiah: The Making of the Imperial Image in the Reign of Süleymân," in *Soliman le Magnifique et son temps*, ed. Gilles Veinstein (Paris: Documentation Française, 1992), 165–66; and Fleischer, "Mediterranean Apocalypse," 63–69; Barbara Flemming, "*Ṣāḥib-kirān* und *Mahdī*: Türkische Endzeiterwartungen im ersten Jahrzehnt der Regierung Süleymāns," in *Between the Danube and the Caucasus: A Collection of Papers Concerning the Oriental Sources on the History of the Peoples of Central and South-Eastern Europe*, ed. György Kara (Budapest: Akadémiai Kiadó, 1987), 43–62; and Flemming, "Der Ğâmi ülmeknūnāt: eine Quelle 'Alîs aus der Zeit Sultan Süleymâns," in *Studien zur Geschichte und Kultur des vorderen Orients: Festschrift für Bertold Spuler zum Siebzigsten Geburtstag*, ed. Hans Robert Roemer and Albrecht Noth (Leiden: Brill, 1981): 79–92.
23. Fleischer, "Mediterranean Apocalypse," 23.
24. Hanna Sohrweide, "Der Sieg der Safaviden in Persien und seine Rückwirkungen auf die Schiiten Anatoliens im 16. Jahrhundert," *Der Islam: Zeitschrift für Geschichte und Kultur des Islamischen Orients* 41 (1965): esp. 164–86.

25. Molla Kâbiz in 1527, quoted in Tijana Krstić, *Contested Conversions to Islam: Narratives of Religious Change in the Early Modern Ottoman Empire* (Stanford, CA: Stanford University Press, 2011), 93.
26. Moin, *Millennial Sovereign.*
27. Sanjay Subrahmanyam, "Turning the Stones Over: Sixteenth-Century Millenarianism from the Tagus to the Ganges," *Indian Economic and Social History Review* 40 (2003): 129–61.
28. Mark Greengrass, *Christendom Destroyed: Europe, 1517–1648* (New York: Viking, 2014), 12.
29. Blaise de Monluc, *Commentaires, 1521–1576,* ed. Paul Courteault (Paris: Gallimard, 1964), 34–35.
30. J. H. Elliott, "A Europe of Composite Monarchies," *Past and Present* 137 (1992): 53.
31. At the time of Maximilian's death the electors included three archbishops, one king, and three princes.
32. Peter H. Wilson, *Heart of Europe: A History of the Holy Roman Empire* (Cambridge, MA: Harvard University Press, 2016), 407.
33. Niccolò Machiavelli, *Discourses on Livy,* trans. Harvey C. Mansfield and Nathan Tarcov (Chicago: University of Chicago Press, 1996), 173.
34. Benjamin Arbel, "Venice's Maritime Empire in the Early Modern Period," in *A Companion to Venetian History, 1400–1797,* ed. Eric R. Dursteler (Leiden: Brill, 2013), 125–254.
35. Henry Kamen, *Empire: How Spain Became a World Power, 1492–1763* (New York: HarperCollins, 2003), 89, 352–53.
36. Pseudo-Methodius, *Apocalypse of Pseudo-Methodius: An Alexandrian World Chronicle,* trans. Benjamin Garstad (Cambridge, MA: Harvard University Press, 2012).
37. Marjorie Reeves, *The Influence of Prophecy in the Later Middle Ages: A Study in Joachimism* (Notre Dame, IN: University of Notre Dame Press, 1993), 347–74.
38. Ibid., 354.
39. Mercurino Arborio di Gattinara, *The Autobiography,* trans. Rebecca Ard Boone, in her *Mercurino di Gattinara and the Creation of the Spanish Empire* (London: Pickering and Chatto, 2014), 75–136.
40. Cited in Boone, *Mercurino di Gattinara,* 27.
41. Cited in John M. Headley, "The Habsburg World Empire and the Revival of Ghibellinism," in *Theories of Empire, 1450–1800,* ed. David Armitage (Aldershot: Ashgate, 1998), 98.
42. Cited in Reeves, *Influence of Prophecy in the Later Middle Ages,* 363.
43. Ibid., 365. For the broader context of this millenarian theology, see Robert J. Wilkinson, *Orientalism, Aramaic, and Kabbalah in the Catholic Reformation: The First Printing of the Syriac New Testament* (Leiden: Brill, 2007); Marjorie Reeves, "Cardinal Egidio da Viterbo: A Prophetic Interpretation of History," in *Prophetic Rome in the High Renaissance Period: Essays* (Oxford: Oxford University Press, 1992), 91–177; and Brian P. Copenhaver and Daniel Stein Kokin, "Egidio da Viterbo's Book on Hebrew Letters: Christian Kabbalah in Papal Rome," *Renaissance Quarterly* 67 (2014): 1–42.
44. Ariosto, *Orlando furioso,* 15.25–26.

45. Vicente de Cadenas y Vicent, ed., *Doble coronación de Carlos V en Bolonia* (Madrid: Hidalguía, 1985). See also Konrad Eisenbichler, "Charles V in Bologna: The Self-Fashioning of a Man and a City," *Renaissance Studies* 13 (1999): 430–39; Thomas J. Dandelet, *The Renaissance of Empire in Early Modern Europe* (Cambridge: Cambridge University Press, 2014).
46. Cited in James D. Tracy, *Emperor Charles V, Impresario of War: Campaign Strategy, International Finance, and Domestic Politics* (Cambridge: Cambridge University Press, 2002), 27, n. 26.
47. Fleischer, "Mediterranean Apocalypse," 64.
48. Cited in Roger Crowley, *Empires of the Sea: The Siege of Malta, the Battle of Lepanto, and the Contest for the Center of the World* (New York: Random House, 2008), 45.
49. Crowley, *Empires of the Sea*, 26–33.
50. Tracy, *Emperor Charles V, Impresario of War*, 139.
51. Andrew C. Hess, *The Forgotten Frontier: A History of the Sixteenth-Century Ibero-African Frontier* (Chicago: University of Chicago Press, 1978), 73.
52. Noel Malcolm, *Agents of Empire: Knights, Corsairs, Jesuits, and Spies in the Sixteenth-Century Mediterranean World* (Oxford: Oxford University Press, 2015), 35–37.
53. On messianism in the Portuguese court—though in a slightly earlier period—see Sanjay Subrahmanyam, *The Career and Legend of Vasco da Gama* (Cambridge: Cambridge University Press, 1997), 54–57.
54. Rebekka Voß, "Charles V as Last World Emperor and Jewish Hero," *Jewish History* 30 (2016): 81–106.
55. Harris Lenowitz, *The Jewish Messiahs: From the Galilee to Crown Heights* (Oxford: Oxford University Press, 1998), 93–123, offers an overview of these figures, with important excerpts from contemporary sources. For an updated, scholarly treatment of Molkho, see Moshe Idel, *Messianic Mystics* (New Haven: Yale University Press, 1998), 140–52. The key source for ha-Reuveni is his diary, which survives only in a nineteenth-century facsimile following the disappearance of the original manuscript from the Bodleian Library at Oxford; for an English translation see, "*Diary* [of David Reubeni]," in *Jewish Travelers in the Middle Ages,* ed. Elkan Nathan Adler (New York: Dover, 2016), 251–328. Two recent studies, which I draw on here, have done much to help contextualize ha-Reuveni's mission to Christian Europe: Martin Jacobs, "David ha-Re'uveni: Ein 'zionistisches Experiment' im Kontext der europäischen Expansion des 16. Jahrhunderts?" in *An der Schwelle zur Moderne: Juden in der Renaissance,* ed. Giuseppe Veltri and Annette Winkelmann (Leiden: Brill, 2003), 191–206; and Moti Benmelech, "History, Politics, and Messianism: David Ha-Reuveni's Origin and Mission," *Association of Jewish Studies Review* 35 (2011): 35–60.
56. Cited in Reeves, *Influence of Prophecy in the Later Middle Ages,* 366. The *Scechina* is the Shekhinah or feminine presence of God as described in kabbalistic writings.
57. Cited in Hugh Thomas, *Rivers of Gold: The Rise of the Spanish Empire, from Columbus to Magellan* (New York: Random House, 2004), 512.
58. Cited in Jean-Benoît Nadeau and Julie Barlow, *The Story of Spanish* (New York: St. Martin's Press, 2013), 172.

59. Hugh Thomas, *The Golden Age: The Spanish Empire of Charles V* (London: Allen Lane, 2010), 316, 428.
60. J. H. Elliott, *Empires of the Atlantic World: Britain and Spain in America, 1492–1830* (New Haven: Yale University Press, 2006), 117–52.
61. Earl Rosenthal, "*Plus Ultra, Non plus Ultra,* and the Columnar Device of Emperor Charles V," *Journal of the Warburg and Courtauld Institutes* 34 (1971): 204–28; Peter Burke, "Presenting and Re-Presenting Charles V," in *Charles V, 1500–1558, and His Time,* ed. Hugo Soly and Willem Pieter Blockmans (Antwerp: Mercatorfonds, 1999), 393–476. On the use of the device in 1517, see Roger Bigelow Merriman, *The Rise of the Spanish Empire in the Old World and the New* (New York: Macmillan, 1918–34), 3:27.

CHAPTER SIX. ANTICHRIST AND REFORMATION

1. Martin Luther and Philip Melanchthon, *Passional Christi und Antichristi* (Wittenberg: J. Grunenberg, 1521). I have used the facsimile edition: *Passional Christi und Antichristi von Dr. Martin Luther mit Bildern von Lukas Kranach dem Aelteren,* ed. C. F. W. Walther (St. Louis: Deutsche Evangelische Synode von Missouri, 1878). On this work, see especially Robert W. Scribner, *For the Sake of Simple Folk: Popular Propaganda for the German Reformation* (Cambridge: Cambridge University Press, 1981), 148–89. The translator of the chapter epigraph from Luther is August Crull in his *Hymn Book for the Use of the Lutheran Schools and Congregations* (Decorah, IA: Lutheran Publishing House, 1884).
2. David C. Steinmetz, *Luther and Staupitz: An Essay in the Intellectual Origins of the Protestant Reformation* (Durham, NC: Duke University Press, 1980).
3. Erich Vogelsang, *Die Anfänge von Luthers Christologie nach der ersten Psalmenvorlesung* (Berlin: De Gruyter, 1929).
4. Martin Luther, *Luthers Werke: Kritische Gesamtausgabe; Tischreden* (Weimar: Herman Böhlau, 1912–21), 5:5247.
5. Luther, *Autobiographical Fragment,* appended to the 1545 edition of his complete works, in *D. Martin Luthers Werke: Kritische Gesamtausgabe; Schriften* (Weimar: Herman Böhlau, 1883–2009), 54:185, 17–20; *Luther's Works,* ed. Jaroslav Pelikan, Helmut T. Lehmann, Christopher Boyd Brown, and Hilton C. Oswald (Philadelphia: Muhlenberg Press, 1955–2016), 34:336–37.
6. Luther, *Autobiographical Fragment,* 186:5–6; *Luther's Works,* 34:336–37.
7. Luther, *Tischreden,* 5:5247.
8. Luther, *An den Christlichen Adel deutscher Nation,* in *Luthers Werke, Schriften,* 6:409; *To the Christian Nobility of the German Nation,* trans. Charles M. Jacobs, in *Luther's Works,* 44:144.
9. Luther, *De captivitate Bablyonica ecclesiae praeludium,* in *D. Martin Luthers Werke,* 6:484–573; *Babylonian Captivity of the Church* in *Luther's Works,* 36:3–126.
10. Martin Luther, *Freiheit eines Christenmenschen,* in *Luthers Werke, Schriften,* vol. 7; Luther, *On the Freedom of a Christian,* trans. W. A. Lambert, in *Luther's Works,* 31:367. On the publishing history of this work as well as of Luther's two other major treatises

of this year, see Andrew Pettegree, *Brand Luther: How an Unheralded Monk Turned His Small Town into a Center of Publishing, Made Himself the Most Famous Man in Europe—and Started the Protestant Reformation* (New York: Penguin, 2015), 124–31.
11. Luther, *Freiheit eines Christenmenschen; On the Freedom of a Christian*, 371.
12. Cited in Heinz Schilling, *Martin Luther: Rebel in an Age of Upheaval*, trans. Rona Johnston (Oxford: Oxford University Press, 2017), 455.
13. "Gute gerechte Werke machen niemals einen guten gerechten Menschen, sondern ein guter gerechte Mensch tut gute gerechte Werke"; Luther, *Freiheit eines Christenmenschen*, § 23; *On the Freedom of a Christian*, 361.
14. Much of the most exciting research on the Reformation in the last two generations has focused not on the history of theology but rather on social history, as historians have sought to explain why so many members of Europe's urban elites—patricians, professionals, merchants, and wealthy artisans—were attracted to the new teachings of Luther and other major reformers in this era. For an orientation, see Max Steinmetz, ed., *Die frühbürgerliche Revolution in Deutschland: Referat und Diskussion zum Thema Probleme der frühbürgerlichen Revolution in Deutschland, 1476–1535* (Berlin: Akademie-Verlag, 1961), 8–48; Bernd Moeller, *Imperial Cities and the Reformation: Three Essays*, trans. H. C. Erik Midelfort and Mark U. Edwards, Jr. (Philadelphia: Fortress Press, 1972); Rainer Wohlfeil, *Reformation oder frühbürgerliche Revolution* (Munich: Nymphenburger Verlag, 1972); and Steven E. Ozment, *The Reformation in the Cities: The Appeal of Protestantism to Sixteenth-Century Germany and Switzerland* (New Haven: Yale University Press, 1975).
15. Richard Wunderli, *Peasant Fires: The Drummer of Niklashausen* (Bloomington: Indiana University Press, 1992).
16. Norman Cohn, *The Pursuit of the Millennium: Revolutionary Millenarians and Mystical Anarchists of the Middle Ages*, rev. ed. (Oxford: Oxford University Press, 1970), 228.
17. Lyndal Roper, *Martin Luther: Renegade and Prophet* (London: Penguin, 2016), 42–43.
18. Gerald Strauss, ed. and trans., *Book of One Hundred Chapters and Forty Statutes*, in *Manifestations of Discontent on the Eve of the Reformation in Germany* (Bloomington: Indiana University Press, 1971), 235.
19. Luther, *Wider die Bulle des Endchrists*, in *Luthers Werke, Schriften*, 6:629.
20. Winfried Vogel, "The Eschatological Theology of Martin Luther, Part I: Luther's Basic Concepts," *Andrews University Seminary Studies* 24 (1986): 251, 264; and "Part II: Luther's Exposition of Daniel and Revelation," *Andrews University Seminary* 25 (1987): 186.
21. Robin B. Barnes, *Prophecy and Gnosis: Apocalypticism in the Wake of the Lutheran Reformation* (Stanford, CA: Stanford University Press, 1988), 54, 188–201.
22. Peter Blickle, *The Revolution of 1525: The German Peasants' War from a New Perspective*, trans. Thomas A. Brady, Jr. and H. C. Erik Midelfort (Baltimore: Johns Hopkins University Press, 1981). For an important critique of Blickle, see Govind P. Sreenivasan, "The Social Origins of the Peasants' War of 1525 in Upper Swabia," *Past and Present* 171 (2001): 30–65.

23. Cohn, *Pursuit of the Millennium*, 236.
24. Thomas Müntzer, "Außlegung des andern unterschydes Danielis, des propheten, gepredigt auffm Schlos zu Alstet vor den tetigen, thewern herzogen und vorstehern zu Sachsen, 1424," in *Thomas Müntzer, Schriften und Briefe: Kritische Gesamtausgabe*, ed. Günter Franz and Paul Kirn (Gütersloh: Gerd Mohn, 1968), 247; "Interpretation of the Second Chapter of Daniel," in *The Collected Works of Thomas Müntzer*, trans. Peter Matheson (Edinburgh: T. and T. Clark, 1988), 235–36. On Müntzer, see, in addition to the fundamental study of Ernst Bloch, *Thomas Münzer als Theologe der Revolution* (Frankfurt: Suhrkamp Verlag, 1969; the first edition of this work appeared in 1921); Tom Scott, *Thomas Müntzer: Theology and Revolution in the German Reformation* (London: Macmillan, 1989); Hans-Jürgen Goertz, *Thomas Müntzer: Revolutionär am Ende der Zeiten; Eine Biographie* (Munich: C. H. Beck, 2015); and Matthias Riedl, "Apocalyptic Violence and Revolutionary Action: Thomas Müntzer's Sermon to the Princes," in *A Companion to Premodern Apocalypse*, ed. Michael A. Ryan (Leiden: Brill, 2016), 260–96.
25. Müntzer, "Außlegung des andern unterschydes Danielis," 255; "Interpretation of the Second Chapter of Daniel," 255.
26. Müntzer, "An den Schösser Zeiß, 22 July 1524," in Franz and Kirn, *Schriften und Briefe*, 420; "To the Intendant, Zeiss, Allstedt," in *Collected Works*, trans. Matheson, 100.
27. Müntzer, "An die Allsteder, Mühlhausen, ca. April 26 or 27," in Franz and Kirn, *Schriften und Briefe*, 455; "To the People of Allstedt, 1525," in *Collected Works*, trans. Matheson, 142.
28. Müntzer, "An die Eisenacher, 9 May 1525," in *Schriften und Briefe*, 463–64; "To the People of Eisenach," in *Collected Works*, trans. Matheson, 150–51.
29. Müntzer, "Bekenntnis, 16 May 1525," in Franz and Kirn, *Schriften und Briefe*, 548; "Interrogation and 'Recantation,'" in *Collected Works*, trans. Matheson, 436–37.
30. Müntzer, "An die Mühlhäuser, 17 May 1525," in Franz and Kirn, *Schriften und Briefe*, 474; "Müntzer to the people of Mülhausen, 17 May 1525," in *Collected Works*, trans. Matheson, 161.
31. James M. Stayer, *The German Peasants' War and Anabaptist Community of Goods* (Montreal: McGill-Queen's University Press, 1991); and Walter Klaassen, *Living at the End of Ages: Apocalyptic Expectation in the Radical Reformation* (Lanham, MD: University Press of America, 1992). In their analysis of Anabaptism, scholars have traditionally emphasized their opposition to infant baptism, undoubtedly a defining trait, but many Anabaptists were also, as Carlos Eire has shown, "apocalyptic activists"—see Carlos M. N. Eire, *Reformations: The Early Modern World, 1450–1650* (New Haven: Yale University Press, 2016), chart, 255.
32. "Die zwölf Artickel der oberschwäbischen Bauren," in *Quellen zur Geschichte des Bauernkrieges*, ed. Günther Franz (Munich: R. Oldenbourg, 1963), 174–79; *The Twelve Articles of the Upper Swabian Peasants*, in *The Radical Reformation*, ed. and trans. Michael G. Baylor (Cambridge: Cambridge University Press, 1991), 231–38.
33. Michael Gaismair, *Territorial Constitution for Tyrol*, in *Radical Reformation*, ed. and trans. Baylor, 254–60; Adolf Laube and Hans Werner Seiffert, ed., *Flugschriften der*

Bauernkriegszeit, 2nd ed. (Berlin: Akademie-Verlag, 1978), 139–43; Walter Klaassen, *Michael Gaismair: Revolutionary and Reformer* (Leiden: Brill, 1978), 5, 47, and 58.

34. Hans Hergot, *Von der newen Wandlung eynes christlichen Lebens*, in *Hans Hergot und die Flugschrift* (Leipzig: VEB Fachbuchverlag, 1977), 107; *On the New Transformation of the Christian Life*, in *Radical Reformation*, ed. and trans. Baylor, 210–12.

35. Hergot, *Von der newen Wandlung*, 108; *On the New Transformation of the Christian Life*, 210–11.

36. Hergot, *Von der newen Wandlung*, 107; *On the New Transformation of the Christian Life*, 210.

37. Gottfried Seebass, *Müntzers Erbe: Werke, Leben und Theologie des Hans Hut* (Göttingen: Gütersloher Verlagshaus, 2002).

38. Anselm Schubert, *Täufertum und Kabbalah: Augustin Bader und die Grenzen der Radikalen Reformation* (Gütersloh: Gütersloher Verlagshaus, 2008); Rebekka Voß, *Umstrittene Erlöser: Politik, Ideologie und jüdisch-christlicher Messianismus in Deutschland, 1500–1600* (Göttingen: Vandenhoeck and Ruprecht, 2011), 138–52; and Robert Bast, "The Messianic Kingship of Augustin Bader as Anti-Habsburg Polemic: Prophecy and Politics in Reformation Germany," *Journal of the Economic and Social History of the Orient* 61 (2018): 147–71.

39. Lorna Jane Abray, *The People's Reformation: Magistrates, Clergy, and Commons in Strasbourg, 1500–1598* (Ithaca, NY: Cornell University Press, 1985).

40. Jonathan Green, "The Lost Book of the Strasbourg Prophets: Orality, Literacy, and Enactment in Lienhard Jost's Visions," *Sixteenth-Century Journal* 46 (2015): 317–18.

41. Klaus Deppermann, *Melchior Hoffman: Social Unrest and Apocalyptic Visions in the Age of Reformation*, ed. Benjamin Drewery, trans. Malcolm Wren (Edinburgh: T. and T. Clark, 1987).

42. Hermann von Kerssenbroch, *Narrative of the Anabaptist Madness: The Overthrow of Münster, the Famous Metropolis of Westphalia*, trans. Christopher S. Mackay, 2 vols. (Leiden: Brill, 2007); Henry Gresbeck, *False Prophets and Preachers: Henry Gresbeck's Account of the Anabaptist Kingdom of Münster*, trans. Christopher S. Mackay (Kirksville, MO: Truman State University Press, 2016), 89–91.

43. "Report of the Execution of Jan of Leiden and the Münster Anabaptist Leaders," in John A. Wagner, ed., *Documents of the Reformation* (Santa Barbara: ABC-CLIO, 2018), 127–29; Hans J. Hillerbrand, *The Reformation: A Narrative History Related by Contemporary Observers and Participants* (New York: Harper and Row, 1964), 265–66.

44. Cited in Schilling, *Martin Luther*, 172.

45. Luther developed his *Zweireicheslehre* in his treatise *Von weltlicher Oberkeit, wie weit man ihr Gehorsam schuldig sei*, in *Luthers Werke, Schriften*, 11:229–81; "Temporal Authority: To What Extent it Should be Obeyed," trans. J. J. Schindel, in *Luther's Works*, 44:75–129. My analysis here draws primarily on Quentin Skinner, "The Principles of Lutheranism," in *The Age of Reformation*, vol. 2 of *The Foundations of Modern Political Thought* (Cambridge: Cambridge University Press, 1978), 3–19.

46. Luther, *Wider die räuberischen und mörderischen Rotten der Bauern*, in *Luthers Werke, Schriften*, 18:311–61; *Against the Robbing and Murdering Hordes of Peasants*, in *Luther's Works*, 46:45–55; and *Eine schreckliche Geschichte und ein Gericht Gottes über Thomas*

Müntzer, in *Luthers Werke, Schriften*, 18:362–74; *A Shocking History and God's Judgment on Thomas Müntzer*, in *Luther's Works*, 46:63–85.
47. Luther, *Vorrede auf die offenbarung S. Joannis*, in *Luthers Werke: Kritische Gesamtausgabe; Die Deutsche Bibel* (Weimar: Herman Böhlau, 1931), 7:404; *Preface to the Revelation of St. John I (1522)*, in *Luther's Works*, 6:399.
48. Luther, *Vorrede auf die offenbarung S. Joannis*, in *Luthers Werke: Deutsche Bibel*, 7:406–21; *Preface to the Revelation of St. John II 1546 (1530)*, in *Luther's Works*, 6:399–411. On Luther's reassessment of this book, see Barnes, *Prophecy and Gnosis*, 41. See also John M. Headley, *Luther's View of Church History* (New Haven: Yale University Press, 1963), 240–50.
49. Luther, *Vorrede uber den Propheten Daniel*, in *Luthers Werke: Deutsche Bibel*, 11:2:129; *Preface to the Book of Daniel*, in *Luther's Works*, 35:316.
50. Scribner, *For the Sake of Simple Folk*, 148–49.
51. Luther, *Vorrede auf die offenbarung S. Joannis*, in *Luthers Werke: Deutsche Bibel*, 7:414; *Preface to the Revelation of St. John II 1546 (1530)*, in *Luther's Works*, 35:406.
52. Cited in Roper, *Martin Luther*, 382.
53. Noel Malcolm, *Useful Enemies: Islam and the Ottoman Empire in Western Political Thought, 1450–1750* (Oxford: Oxford University Press, 2019), 77–92.
54. Cited in Roper, *Martin Luther*, 388.
55. Luther, *Daß Jesus Christus ein geborner Jude sei*, in *Luthers Werke, Schriften*, 11:314–36; *That Jesus Christ Was Born a Jew*, in *Luther's Works*, 45: 99–229; *Von den Juden und ihren Lügen*, in *Luthers Werke, Schriften*, 53:412–552; *On the Jews and Their Lies*, in *Luther's Works*, 47:121–305. In addition to these two well-known works, Luther published numerous other anti-Jewish writings over the course of his career. On Luther and the Jews, see most recently, Kenneth Austin, *The Jews and the Reformation* (New Haven: Yale University Press, 2020), 64–68.
56. Cited in Pettegree, *Brand Luther*, 294.
57. Cited in Schilling, *Martin Luther*, 498–99.
58. Robert Kolb, *Martin Luther as Prophet, Teacher, Hero: Images of the Reformer, 1520–1620* (Grand Rapids, MI: Baker Academic, 1999), 34–38.
59. Scribner, *For the Sake of Simple Folk*, 19–21.

CHAPTER SEVEN. "NO ONE KNOWS THE HOUR"

1. Servetus, *De trinitatis erroribus libri septem* (Hagenau: Johann Setzer, 1531); Servetus, *On the Errors of the Trinity, Seven Books*, in *The Two Treatises of Servetus on the Trinity*, trans. Earl Morse Wilbur (Cambridge, MA: Harvard University Press, 1932), 1–184. On the debate that never happened, see Roland Bainton, *Hunted Heretic: The Life and Death of Michael Servetus, 1511–1553* (Boston: Beacon Press, 1953), 81.
2. "Aspectus, facies, effigies, signum, character, sigillum, insignis nota, insculptura"; Servetus, *De trinitatis erroribus*, 106v; *On the Errors of the Trinity*, 165.
3. Servetus, *De trinitatis erroribus*, 109v; *On the Errors of the Trinity*, 169.
4. Servetus, *De trinitatis erroribus*, 102r; *On the Errors of the Trinity*, 157. No brief summary can do justice to Servetus's complex and often contradictory views of the Trinity.

His debt to Kabbalah, which I see as implicit in the *De trinitatis erroribus*, is explicit in his *Christianismi restitutio: Totius ecclesiae apostolicae est ad sua limina vocatio, in integram restituta cognitione Dei, fidei Christi, iustificationis nostrae, regenerationis baptismi, et caenae domini manducationis. Restituto denique nobis regno coelesti, Babylonis impia captivitate soluta, et Antichristo cum suis penitus destructo* ([Vienna]: [Balthazar Arnoullet], 1553; facsimile: Frankfurt-am-Main: Minerva, 1966)—see 74, where Servetus writes, "The rabbis call divinity *schechina* from the verb *sachan*, which means 'to inhabit.' The divinity of Christ, therefore, is the inhabitation of God." "Divinitatem rabbini vocant *sechina* a verbo *sachan*, quod inhabitare significat. Ergo divinitas Christi est inhabitatio Dei." For an English translation of the first half of this volume, see *The Restoration of Christianity*, trans. Christopher A. Hoffman and Marian Hillar (Lewiston, NY: Edwin Mellen Press, 2007), and here, 195.

5. Servetus, *De trinitatis erroribus*, 42v–43r; *On the Errors of the Trinity*, 66–67.
6. Jerome Friedman, *Michael Servetus: A Case Study in Total Heresy* (Geneva: Librairie Droz, 1978), offers an important analysis of Servetus's use of Jewish sources. On Servetus's knowledge of Islam, see Peter Hughes, "Servetus and the Quran," *Journal of Unitarian Universalist History* 30 (2005): 55–70; and Jaume De Marcos Andreu, "Servet y el Islam," *Bandue* 5 (2011): 119–30.
7. Servetus, *De trinitatis erroribus*, 43r; *On the Errors of the Trinity*, 67.
8. For the religious environment in Spain in this period, Stefania Pastore, *Un'eresia spagnola: Spiritualità conversa, alumbradismo e inquisizione, 1449–1559* (Florence: Leo S. Olschki, 2004), and on Servetus, 258–62; see also, for an important perspective not only on the interaction but even on the occasional hybridity of Islam and Christianity, Mercedes García-Arenal, "A Catholic-Muslim Prophet: Agustín de Ribera, 'the Boy Who Saw Angels,' " *Common Knowledge* 18 (2012): 207–91.
9. In their *Miguel Servet en España, 1506–1527*, expanded ed. (Tudela: Imprenta Castilla, 2017), Miguel González Ancin and Otis Towns provide considerable evidence, though not definitive, that Servetus was actually the son of *converso* parents—the de Villanueva—and was later adopted by the Servetus family when his birth parents were tried by the Inquisition in 1515. The book also places Servetus's birth in 1506, three to five years earlier than in the standard accounts.
10. Richard H. Popkin, "Savonarola and Cardinal Ximenes: Millenarian Thinkers and Actors at the Eve of the Reformation," in *Millenarianism: From Savonarola to the Abbé Gregoire*, vol. 2 of *Millenarianism and Messianism in Early Modern European Culture*, ed. Karl A. Kottman (Dordrecht: Kluwer, 2001), 15–26.
11. Michael Servetus, *A Translation of His Geographical, Medical, and Astrological Writings, with Introductions and Notes*, trans. Charles Donald O'Malley (Philadelphia: American Philosophical Society, 1953); on the Pagnini Polyglot, see Bainton, *Hunted Heretic*, 96–100.
12. Servetus, *Christianismi restitutio*, 409–10.
13. Ibid.
14. Ibid., 196; *Restoration of Christianity*, 281.
15. Servetus included the letters "Epistolae triginta ad Ioannem Calvinum" in his *Christianismi restitutio*, 577–664. For an English translation, see Servetus, *Thirty*

Letters to Calvin, Preacher to the Genevans; And Sixty Signs of the Kingdom of the Antichrist and His Revelation Which Is Now at Hand, in *Restoration of Christianity.*
16. Bainton, *Hunted Heretic,* 211–12.
17. François Wendel, *Calvin: The Origins and Development of His Religious Thought,* trans. Philip Mairet (Durham, NC: Labyrinth Press, 1987).
18. John Calvin, *Institutio christianae religionis* (1559), in vol. 2 of *Ioannis Calvini Opera quae supersunt omnia,* ed. Johann-Wilhelm Baum, Edouard Cunitz, and Eduard Reuss, 59 vols. (Braunschweig: C. A. Schwetschke, 1892)—hereafter *Institutio; The Institutes of the Christian Religion,* ed. John T. McNeill, trans. Ford Lewis Battles, 2 vols. (Philadelphia: Westminster Press, 1960). On the history of this volume and its multiple editions and revisions, see Bruce Gordon, *John Calvin's "Institutes of the Christian Religion": A Biography* (Princeton, NJ: Princeton University Press, 2016).
19. Ulrich Zwingli, "Reproduction from Memory of a Sermon on the Providence of God," in *On Providence and Other Essays,* ed. William John Hinke (Durham, NC: Labyrinth Press, 1983), 128–234.
20. "Deus alios in spem vitae adoptat, alios adiudicat aeternae morti. . . . Praedestionem vocamus aeternum Dei decretum, quo apud se constitutum habuit, quid de unoque homine fieri vellet. Nam enim pari conditione creantur omnes, sed aliis vita aeterna, aliis damnatio aeterna praeordinatur," Calvin, *Institutio,* 3.21.5, col. 682–83; Calvin, *Institutes* 3.21.5, 2:926.
21. Calvin, *Institutio,* 1.26.6, col. 703; Calvin, *Institutes,* 3.23.6, 2:954.
22. Calvin, *Institutio,* 3.22.4, col. 690; *Institutes,* 3.22.4, 2:936–37. On earlier Christian variations of this theme, see Gerson D. Cohen, "Esau as Symbol in Early Medieval Thought," in *Jewish Medieval and Renaissance Studies,* ed. Alexander Altmann (Cambridge, MA: Harvard University Press, 1967), 19–48. On the prominence of this theme in Jewish thought, see Malachi Haim Hacohen, *Jacob and Esau: Jewish European History between Nation and Empire* (Cambridge: Cambridge University Press, 2018).
23. "Est enim praedestinatio Dei vere labyrinthus unde hominis ingenium nullo modo se explicare queat," Calvin, *Commentarius in Epistolam Pauli ad Romanos, Opera,* 49, col. 180; *The Epistles of Paul the Apostle to the Romans and to the Thessalonians,* vol. 8 of *Commentaries,* ed. David W. Torrance and Thomas F. Torrance, trans. Ross Mackenzie (Grand Rapids, MI: Eerdmans, 1960), 202.
24. Calvin, *Institutio,* 3.21.1, col. 680; *Institutes,* 3.21.1, 2:922–23. On the image of the labyrinth in Calvin, see William J. Bouwsma, *John Calvin: A Sixteenth-Century Portrait* (New York: Oxford University Press, 1988), 45–48.
25. Calvin, *Commentarius in Epistolam Pauli ad Romanos, Opera,* 49, col. 177; *Epistles of Paul the Apostle to the Romans and to the Thessalonians,* 199.
26. Calvin, *Commentarius in Epistolam Pauli ad Romanos, Opera,* 49, col. 191; *Epistles of Paul the Apostle to the Romans and to the Thessalonians,* 214–15.p.
27. Cited in Jaroslav Pelikan, "Some Uses of Apocalypse in the Magisterial Reformers," in *The Apocalypse in English Renaissance Thought and Literature: Patterns, Antecedents, and Repercussions,* ed. Constantinos A. Patrides and Joseph Wittreich (Ithaca, NY: Cornell University Press, 1984), 87.

28. Calvin, *Institutio*, 3.25.10, col. 741; *Institutes*, 3.25.10, 2:1004.
29. Calvin, *Institutio*, 2.16.17, col. 383–84; Calvin, *Institutes*, 2.16.17, 1:525.
30. Calvin, "Epistola ad Thessalonicenses I," in *Opera*, 52, col. 168; *Epistles of Paul the Apostle to the Romans and to the Thessalonians*, 367. On Calvin's eschatology, see Heinrich Quistorp, *Calvin's Doctrine of the Last Things*, trans. Harold Knight (London: Lutterworth Press, 1955).
31. Calvin, *Institutio*, 3.25.5, col. 734; *Institutes*, 3.25.5, 2:995.
32. Denis Crouzet, *Les guerriers de Dieu: La violence au temps des troubles de religion, vers 1525–vers 1610* (Seyssel: Champ Vallon, 1990), 1:219.
33. Calvin, *Institutio*, 1.17.11, col. 163; *Institutes*, 1.17.11, 1:224.
34. Nelson H. Minnich, "Prophecy and the Fifth Lateran Council (1512–1517)," in *Prophetic Rome in the High Renaissance Period: Essays*, ed. Marjorie Reeves (Oxford: Oxford University Press, 1992), 63–87.
35. Christopher F. Black, *The Italian Inquisition* (New Haven: Yale University Press, 2009); and, for the broader context, Gustav Henningsen, John Tedeschi, and Charles Amiel, eds., *The Inquisition in Early Modern Europe: Studies on Sources and Methods* (Dekalb: Northern Illinois University Press, 1986).
36. Marion Leathers Kuntz, *The Anointment of Dionisio: Prophecy and Politics in Renaissance Italy* (University Park: Pennsylvania State University Press, 2001); and my "Popular Heresies and Dreams of Political Transformation in Sixteenth-Century Venice," in *Popular Politics in an Aristocratic Republic: Political Conflict and Social Contestation in Late Medieval and Early Modern Venice*, ed. Maartje van Gelder and Claire Judde de LaRivière (Abingdon: Routledge, 2020), 67–87.
37. John W. O'Malley, *Trent: What Happened at the Council* (Cambridge, MA: Harvard University Press, 2013).
38. Reinhart Koselleck, *Futures Past: On the Semantics of Historical Time*, trans. Keith Tribe (Cambridge, MA: MIT Press, 1985), 12–13, developing Carl Schmitt's concept of the *katechon* as central to political power in the European Middle Ages—see Schmitt, *Nomos of the Earth in the International Law of the Jus Publicum Europaeum*, trans. G. L. Ulmen (New York: Telos, 2003).
39. H. Outram Evennett, *The Spirit of the Counter-Reformation*, ed. John Bossy (Cambridge: Cambridge University Press, 1968); R. Po-Chia Hsia, *The World of Catholic Renewal, 1540–1770*, 2nd ed. (Cambridge: Cambridge University Press, 2005), 26–42.
40. John W. O'Malley, *The First Jesuits* (Cambridge, MA: Harvard University Press, 1993), 79; Philip Caraman, *Ignatius Loyola: A Biography of the Founder of the Jesuits* (San Francisco: Harper and Row, 1990); W. W. Meissner, *Ignatius of Loyola: The Psychology of a Saint* (New Haven: Yale University Press, 1992).
41. Cited in B. J. Kidd, *The Counter Reformation, 1550–1600* (London: Society for the Promotion of Christian Knowledge, 1933), 28.
42. Stefania Pastore, "Mozas Criollas and New Government: Francis Borgia, Prophetism, and the *Spiritual Exercises* in Spain and Peru," in *Visions, Prophecies, and Divinations: Early Modern Messianism and Millenarianism in Iberian America, Spain, and Portugal*, ed. Luís Filípe Silvério Lima and Ana Paula Torres Megiani (Leiden: Brill,

2016), 59–73; and Cándido de Dalmases, S.J., *Francis Borgia: Grandee of Spain, Jesuit, Saint,* trans. Cornélius-Michaël Buckley, S.J. (St. Louis: Institute of Jesuit Sources, 1991).

43. Ignatius of Loyola, "Judicium de quibusdam opinionibus quae falso revelationis credebantur," in *Epistolae et instructiones,* vol. 12 of *Monumenta Ignatiana, ex autographis vel ex antiquioribus exemplis collecta* (Madrid: Gabriel Lopez del Horno, 1911), appendix, 635–38; "Ignatius to Francis Borgia, S.J. Duke of Gandia, 1549," in Ignatius Loyola, *Personal Writings: Reminiscences, Spiritual Diary, Select Letters including the Text of the Spiritual Exercises,* trans. Joseph A. Munitiz and Philip Endean (New York: Penguin, 1996), 213–15.
44. Ignatius of Loyola, "Judicium de quibusdam opinionibus," 648; "Ignatius to Francis Borgia," 223.
45. Ignatius of Loyola, *Ejercicios espirituales,* in *Exercitia spiritualia . . . et eorum directoria,* in *Monumenta Ignatiana, ex autographis vel ex antiquioribus exemplis collecta: Series secunda* (Madrid: Ribadeneira, 1919), 348; *Spiritual Exercises,* in *Personal Writings,* 310–11.
46. Ignatius of Loyola, *Ejercicios espirituales,* 566; *Rules to Understand Movements of the Soul,* in *Personal Writings,* 358.
47. St. Augustine, *The Trinity,* trans. Stephen McKenna (Washington, DC: Catholic University of America Press, 1963).
48. See my *Venice's Hidden Enemies: Italian Heretics in a Renaissance City,* rev. ed. (Baltimore: Johns Hopkins University Press, 2004), 107–11.
49. Ignatius of Loyola, *Acta P. Ignatii ut primum scripsit P. Ludovicus Gonzales excipiens ex ore ipsius Patris,* in *Scripta de sancto Ignatio de Loyola Monumenta Ignatiana,* vol. 1 of *Series quarta* (Madrid: Gabriel Lopez de Horno, 1904), 53; "Reminiscences (Autobiography)," in *Personal Writings,* 25–26.

CHAPTER EIGHT. BATTLES FOR GOD

1. Frederic J. Baumgartner, *Henry II: King of France, 1547–1559* (Durham, NC: Duke University Press, 1988), 248–52; see also Philip Benedict, Lawrence M. Bryant, and Kristen B. Neuschel, "Graphic History: What Readers Knew and Were Taught in the *Quarante Tableaux* of Perrissin and Tortorel," *French Historical Studies* 28 (2005): 208–20.
2. Stuart Carroll, *Noble Power during the French Wars of Religion: The Guise Affinity and the Catholic Cause in Normandy* (Cambridge: Cambridge University Press, 1998), esp. 8, 14–52; and his *Martyrs and Murderers: The Guise Family and the Making of Europe* (Oxford: Oxford University Press, 2009).
3. Lucien Febvre, "Une question mal posée: Les origines de la Réforme française et le problème général des causes de la Réforme," *Revue historique* 161 (1929): 1–73.
4. *Articles veritables sur les horribles / grandz et importables abuz de la Messe papalle: Inventée directement contre la saincte Cene de Jesus Christ,* in Gabrielle Berthoud, *Antoine Marcourt, réformateur et pamphlétaire: Du 'Livre des marchans' aux placards de 1534* (Geneva: Librairie Droz, 1973), 287–89.

5. "Toute congnoissance de Jesus Christ est efface /predications de l'evangile rejectée et empechée / le temps occupé en sonneries / urlemens / chanteries / ceremonies / luminaires / encensmens / desguisements et telle manière de singeries"; *Articles veritables*, 289.
6. Ibid.
7. Lucien Febvre, "L'origine des placards de 1534," *Bibliothèque d'humanisme et Renaissance* 7 (1945): 62–75; Donald R. Kelley, *The Beginning of Ideology: Consciousness and Society in the French Reformation* (Cambridge: Cambridge University Press, 1981).
8. Robert M. Kingdon, *Geneva and the Coming of the Wars of Religion in France, 1555–1563* (Geneva: Librairie Droz, 1956); Francis M. Higman, *Lire et découvrir: La circulation des ideés au temps de la Réforme* (Geneva: Librairie Droz, 1998).
9. For the social composition of the Huguenots, see the classic study by Henri Hauser, "La Réforme et les classes populaires en France au XVIe siècle," *Revue d'histoire moderne et contemporaine* 1 (1899–1900): 24–34. For more recent analyses of the social composition of this movement, see Emmanuel Le Roy Ladurie, *The Peasants of Languedoc*, trans. John Day (Urbana: University of Illinois Press, 1974); Natalie Zemon Davis, "Strikes and Salvation at Lyon," in *Society and Culture in Early Modern France: Eight Essays* (Stanford, CA: Stanford University Press, 1975); and Philip Benedict, *Christ's Churches Purely Reformed: A Social History of Calvinism* (New Haven: Yale University Press, 2002). Benedict estimates that approximately 10 percent of the French population supported the Reform (137).
10. Katherine Crawford, "Catherine de Médicis and the Performance of Political Motherhood," *Sixteenth-Century Journal* 31 (2000): 643–73, offers an important gendered analysis of Catherine's rule.
11. My analysis of the role of apocalypticism in the French Wars of Religion is deeply indebted to Denis Crouzet's *Les guerriers de Dieu: La violence au temps des trouble de religion, vers 1525–vers 1610,* 2 vols. (Paris: Champs Vallon, 1990). Crouzet, in my view, overstates what he describes as the "eschatological panic" among Roman Catholics in this period, and he underestimates Catholic hopes for a Beautiful Ending. For a thoughtful response to Crouzet, see Larissa Juliet Taylor, *Heresy and Orthodoxy in Sixteenth-Century Paris: François Le Picart and the Beginnings of the Catholic Reformation* (Leiden: Brill, 1999), esp. 210–12.
12. Crouzet, *Les guerriers de Dieu*, 1:524–26.
13. Denis Crouzet and Jonathan Good, "Circa 1533: Anxieties, Desires, and Dreams," *Journal of Early Modern History* 5 (2001): 24–61.
14. Artus Desiré, *Les disputes de Guillot le porcher, et de la Bergère de S. Denis en France, contre Jehan Calvin, prédicant de Genesve, sur la verité de nostre saincte Foy Catholicque* (Paris: Pierre Gaultier, 1559).
15. Frank S. Giese, *Artus Désiré: Priest and Pamphleteer of the Sixteenth Century* (Chapel Hill, NC: UNC Department of Romance Languages, 1973).
16. Artus Desiré, *Les articles du traicté de la paix entre Dieu et les hommes* (Paris: Pierre Gaultier, 1558), 9, unpaginated.
17. Cited in Crouzet, *Les guerriers de Dieu*, 1:209–10.

18. Like the Guise, the Bourbon were a ducal family—they had been dukes of Vendôme since 1515—and they too proved enormously adept at making connections. In 1548 Louis' brother Antoine de Bourbon had married Jeanne D'Albret, the queen of Navarre, a kingdom that sat just above the Pyrenees along the Bay of Biscay. But Jeanne was also closely connected to the French royal family. Her uncle was King Francis I; and she was first cousin to King Henri II and was thus also closely related, in 1559, to the new king Francis II. Yet, unlike the Guise, this family was often supportive of Calvinists. Jeanne's mother, Marguerite, while herself an evangelical who hoped to see reform from within the Church, opened up some space for the Reform movement in France, and her daughter Jeanne was in fact a Calvinist. And Jeanne and Antoine's son Henri, who would become king of Navarre, was raised a Calvinist.
19. For a powerful, near contemporary report of the massacre, see Giovanni Michiel's ambassadorial report, reprinted in James C. Davis, *Pursuit of Power: Venetian Ambassadors' Reports on Spain, Turkey, and France in the Age of Philip II, 1560–1600* (New York: Harper and Row, 1970), 72–76, 78–79. Classic studies of the massacre include Jannine Garrisson-Estebe, *Tocsin pour un massacre: La saison des Saint-Barthélémy* (Paris: Le Centurion, 1975); Barbara B. Diefendorf, *Beneath the Cross: Catholics and Huguenots in Sixteenth-Century Paris* (Oxford: Oxford University Press, 1991); and Denis Crouzet, *La nuit de Saint-Barthélémy: Un rêve perdu de la Renaissance* (Paris: Fayard, 1994).
20. Cited in Crouzet, *Les guerriers de Dieu*, 2:93.
21. Natalie Zemon Davis offers an important analysis of the violence of this period in her "Rites of Violence: Religious Riot in Sixteenth-Century France," *Past and Present* 59 (1973): 51–91.
22. Crouzet, *Les guerriers de Dieu*, 2:361–423.
23. Helmut G. Koenigsberger, "Rule from Madrid: The Regency of Margaret of Parma (1559–1567)," in *Monarchies, States Generals, and Parliaments: The Netherlands in the Fifteenth and Sixteenth Centuries* (Cambridge: Cambridge University Press, 2001), 193–219.
24. Geoffrey Parker, *The Dutch Revolt* (Ithaca, NY: Cornell University Press, 1977), 54–67.
25. Ibid.; Alastair Duke, *Reformation and Revolt in the Low Countries* (London: Hambledon, 1990); on Philip's own longings for the End of History, albeit in a different context, see also Geoffrey Parker, "The Place of Tudor England in the Messianic Vision of Philip II of Spain: The Prothero Lecture," *Transactions of the Royal Historical Society* 12 (2002): 167–221.
26. Cited in Mark Greengrass, *Christendom Destroyed: Europe, 1517–1648* (New York: Viking, 2014), 411.
27. E. H. Kossmann and A. F. Mellink, eds., *Edict of the States-General of the United Netherlands, 26 July 1581* [*Act of Abjuration*] in *Texts Concerning the Revolt of the Netherlands* (Cambridge: Cambridge University Press, 1974), 216–28.
28. Tijana Krstić, *Contested Conversions to Islam: Narratives of Religious Change in the Early Modern Ottoman Empire* (Stanford, CA: Stanford University Press, 2011), 79–80.

29. Krstić, *Contested Conversions to Islam*; see also Cornell H. Fleischer, *Bureaucrat and Intellectual in the Ottoman Empire: The Historian Mustafa Âli (1541–1600)* (Princeton, NJ: Princeton University Press, 1986).
30. Andrew C. Hess, *The Forgotten Frontier: A History of the Sixteenth-Century Ibero-African Frontier* (Chicago: University of Chicago Press, 1978), 82–84.
31. David Abulafia, *The Great Sea: A Human History of the Mediterranean* (Oxford: Oxford University Press, 2013), 448–51; Noel Malcolm, *Agents of Empire: Knights, Corsairs, Jesuits, and Spies in the Sixteenth-Century Mediterranean World* (Oxford: Oxford University Press, 2015), 162–74; and Roger Crowley, *Empires of the Sea: The Siege of Malta, the Battle of Lepanto, and the Contest for the Center of the World* (New York: Random House, 2008), 252–90.
32. Benjamin Paul, " 'And the Moon Has Started to Bleed': Apocalypticism and Religious Reform in Venetian Art at the Time of the Battle of Lepanto," in *The Turk and Islam in the Western Eye, 1450–1750: Visual Imagery before Orientalism,* ed. James G. Harper (Farnham: Ashgate, 2011), 67–94.
33. Reem F. Iversen, "Prophecy and Politics: Moriscos and Christians in Sixteenth- and Seventeenth-Century Spain" (PhD diss., Princeton University, 2002); and Parker, "Place of Tudor England in the Messianic Vision of Philip II of Spain."
34. Gerard A. Wiegers, "Jean de Roquetaillade's Prophecies among the Muslim Minorities of Medieval and Early-Modern Christian Spain: An Islamic Version of the *Vademecum in Tribulatione*," in *The Transmission and Dynamics of the Textual Sources of Islam: Essays in Honour of Harald Motzki,* ed. Nicolet Boekhoff-van der Voort, Kees Versteegh, and Joas Wagemakers (Leiden: Brill, 2011), 229–47.
35. Mayte T. Green-Mercado, "The Mahdī in Valencia: Messianism, Apocalypticism, and Morisco Rebellions in Late Sixteenth-Century Spain," *Medieval Encounters* 19 (2013): 193–220.
36. Green-Mercado, "The Mahdī in Valencia," and for the broader context, her *Visions of Deliverance: Moriscos and the Politics of Prophecy in the Early Modern Mediterranean* (Ithaca, NY: Cornell University Press, 2019).
37. Ruth MacKay, "The Tragedy of Alcazarquivir: The Collapse of Kingship, Empire, and Narrative," *Bulletin of the Society for Spanish and Portuguese Historical Studies* 40 (2015): 1–20; and Mercedes García-Arenal, *Ahmad al-Mansur: The Beginnings of Modern Morocco* (Oxford: OneWorld, 2009).
38. "Conteneva solo di Profecie cioè che haveva da essere la rotta del Turcho, la finitione dell'Imperio in Casa d'Austria, et che s'haveva a pigliare Constantinopoli, et che il Turcho haveva a venir alla fede christiana et che haveva ad essere uno solo Ovile et un sol Pastore." Archivio Patriarcale di Venezia, *Criminalia Sanctae Inquisitionis,* b. 2, dossier "Contra Joannem Baptistam Ravajoli Friulanum," testimony of October 6, 1573.
39. Archivio Patriarcale di Venezia, *Criminalia Sanctae Inquisitionis,* b. 2, dossier "Contra Joannem Baptistam Ravajoli Friulanum," letter of Lunardo to Zuanbattista, April 10, 1573. The biblical references in this passage are loosely from the Song of Songs 1 and 2. On the problematic question of the identity of the "nobleman" who either was or pretended to be a member of the Priuli family, see my *Venice's Hidden Enemies: Italian*

Heretics in a Renaissance City (Baltimore: Johns Hopkins University Press, 2004), 208–9, esp. n. 20.

40. Archivio di Stato di Venezia, *Sant'Uffizio,* b. 35, dossier "Contra Dominico di Lorenzo callegaro et al.," testimony of September 7, 1573.

CHAPTER NINE. THE SPIRITUAL GLOBE

1. Christopher Columbus, "Relácion del tercer viaje," in *Cristóbal Colón: Textos y documentos completos,* ed. Consuelo Varela and Juan Gil (Madrid: Alianza, 2003), 380, 382; "Narrative of the Third Voyage," in *The Four Voyages of Christopher Columbus,* ed. and trans. J. M. Cohen (Harmondsworth: Penguin, 1969), 222, 224.
2. Pierre d'Ailly, *Tractatus de imagine mundi Petri de Aliaco, et varia ejusdem auctoris et Ioannis Gersonis opuscula* (Louvain: Joannes de Westphalia, 1480–83; facsimile, Boston: Massachusetts Historical Society, 1927), 36r; *Ymago mundi de Pierre d'Ailly,* ed. and trans. Edmond Buron (Paris: Maisonneuve Frères, 1930), 2:470–71.
3. Juan Gil, *Colón y su tiempo,* vol. 1 of *Mitos y utopías del descubrimiento* (Madrid: Alianza, 1989).
4. Scott D. Westrem, *The Hereford Map: A Transcription and Translation of the Legends with Commentary* (Turnhout: Brepols, 2001); Evelyn Edson, *Mapping Time and Space: How Medieval Mapmakers Viewed Their World* (London: British Library, 1997).
5. Chet van Duzer and Ilya Dines, *Apocalyptic Cartography: Thematic Maps and the End of the World in a Fifteenth-Century Manuscript* (Leiden: Brill, 2015), 173–76.
6. Giancarlo Casale, "Did Alexander the Great Discover America? Debating Space and Time in Renaissance Istanbul," *Renaissance Quarterly* 72 (2019): 874; Thomas D. Goodrich, *The Ottoman Turks and the New World: A Study of the "Tarih-i Hind-i Garbi" and Sixteenth-Century Ottoman Americana* (Wiesbaden: Otto Harrassowitz, 1990), 8–11; and M. Pinar Emiralioğlu, *Geographical Knowledge and Imperial Culture in the Early Modern Ottoman Empire* (Farnham: Ashgate, 2014), 24–27. Finally, on apocalyptic currents in the Ottoman Empire in the early sixteenth century, see Cornell H. Fleischer, "A Mediterranean Apocalypse: Prophecies of Empire in the Fifteenth and Sixteenth Centuries," *Journal of the Economic and Social History of the Orient* 61 (2008): 18–80.
7. The *Iggeret orhot 'olam* was first published in Venice in 1586. I have consulted Thomas Hyde's Latin translation: *Iggeret orhot 'olam, id est, Itinera mundi sic dicta nempe Cosmographia* (Oxford: Sheldonian Theatre, 1691). On Farissol, the best modern study is David B. Ruderman, *The World of a Renaissance Jew: The Life and Thought of Abraham ben Mordecai Farissol* (Cincinnati: Hebrew Union College Press, 1981); but see also Noah J. Efron, "Knowledge of Newly Discovered Lands among Jewish Communities of Europe (from 1492 to the Thirty Years' War)," in *The Jews and the Expansion of Europe to the West, 1450–1800,* ed. Paolo Bernardini and Norman Fiering (New York: Berghahn Books, 2001), 54.
8. Claudius Ptolemy, *Geographia,* was first translated into Latin by Jacopo d'Angeli in 1406; the first printed edition appeared in 1475; and the first printed edition to include maps appeared in Bologna in 1477; see Patrick Gautier Dalché, *La* Géographie

de Ptolémée en Occident, IVe-XVIe siècle (Turnhout: Brepols, 2009); see also Gautier Dalché, "The Reception of Ptolemy's *Geography* (End of the Fourteenth to the Beginning of the Sixteenth Century)," in *Cartography in the European Renaissance*, vol. 3.1 of *The History of Cartography*, ed. J. B. Harley and David Woodward (Chicago: University of Chicago Press, 2007), 285–364, where he lists forty-five editions printed from 1475 to 1624.

9. Gerhard Bott, ed., *Focus Behaim-Globus*, 2 vols. (Nuremberg: Germanisches Nationalmuseums Verlag, 1992); Martin Waldseemüller, *Cosmographiae introductio cum quibusdam geometriae ac astronomiae principiis ad eam rem necessariis* (Saint-Dié: Gualterius and Nikolaus Lud, 1507; *The Cosmographiae introductio of Martin Waldseemüller in facsimile*, edited by Charles George Herbermann, trans. Edward Burke and Mario Emilio Cosenza (New York: United States Catholic Historical Society, 1907); The conflation of America with Asia was widespread; see Elizabeth Horodowich and Alexander Nagel, "Amerasia: European Reflections of an Emergent World, 1492–ca. 1700," *Journal of Early Modern History* 23 (2019): 257–95.

10. Michel de Montaigne, "Apologie de Raimond Sebond," in *Les essais*, ed. Pierre Villey, Verdun L. Saulnier, and Marcel Conche (Paris: Quadrige / Presses Universitaires de France, 2004), xii, 572; "Apology for Raymond Sebond," in *The Complete Essays of Montaigne*, trans. Donald M. Frame (Stanford, CA: Stanford University Press, 1958), 430.

11. Luís de Camões, *Os Lusíadas*, ed. Emanuel Paulo Ramos (Porto: Porto Editora, 1972), 10:77, 79; *The Lusiads*, trans. William C. Atkinson (Harmondsworth: Penguin, 1973), 233.

12. Abraham Ortelius, *Theatrum orbis terrarum* (Antwerp: Aegidius Coppens van Diest, 1570).

13. Cited in Denis E. Cosgrove, *Apollo's Eye: A Cartographic Genealogy of the Earth in the Western Imagination* (Baltimore: Johns Hopkins University Press, 2001), 133.

14. Cited in Ortelius, *Theatrum*, fol. 1.

15. Cosgrove, *Apollo's Eye*, 133.

16. See especially Alastair Hamilton, *The Family of Love* (Cambridge: James Clarke, 1981).

17. Guillaume Postel, "Letter to Abraham Ortelius, 1579," in *Abrahami Ortelii et virorum eruditorum ad eundem et ad Jacobum Colium Ortelianum Epistulae*, ed. Jan Hendrik Hessels (Osnabrück: Otto Zeller, 1969), 187–92.

18. Giorgio Mangani, *Il "mondo" di Abramo Ortelio: Misticismo, geografia, e collezionismo nel Rinascimento dei Paesi Bassi* (Modena: Franco Cosimo Panini, 1998), 228–29.

19. Nicolas Crane, *Mercator: The Man Who Mapped the Planet* (New York: Henry Holt, 2003), 258.

20. Gerard Mercator, *Atlas sive cosmographicae meditationes de fabrica mundi et fabricati figura*, 3 vols. (Duisburg, 1595–1602).

21. Mercator, *Atlas*, 1, 3r.

22. Ignatius of Loyola, *Ejercicios espirituales*, in *Exercitia spiritualia . . . et eorum directoria*, in *Monumenta Ignatiana, ex autographis vel ex antiquioribus exemplis collecta: Series secunda* (Madrid: Ribadeneira, 1919), 222–24; *The Spiritual Exercises of St. Ignatius*, trans. Louis J. Puhl, S.J. (Chicago: Loyola Press, 1951), 49–51.

23. Sumathi Ramaswamy, *Terrestrial Lessons: The Conquest of the World as Globe* (Chicago: University of Chicago Press, 2017), 20.
24. R. Po-chia Hsia, *A Jesuit in the Forbidden City: Matteo Ricci, 1552–1610* (Oxford: Oxford University Press, 2010), 87. Nicolo Longobardi, Ricci's successor as superior general of the Chinese mission, and his fellow Jesuit Manuel Dias continued to introduce cartographical and cosmological studies from Europe in their mission work; see Helen M. Wallis and E. D. Grinstead, "A Chinese Terrestrial Globe, A.D. 1623," *British Museum Quarterly* 25 (1962): 83–91.
25. Joannes Fredericus Lumnius, *De extremo Dei iudicio, et Indorum vocatione libri duo* (Venice: Dominico Farri, 1569), 86.
26. José de Acosta, *De temporibus novissimis: Libri quatuor* (Rome: Iacobus Tornerius, 1590), 453–54.
27. This text is also known as the *Tarih-i Hind-i Garbi* (A History of India in the West), but this is a title given to it in the eighteenth century, when it was first published. For an English translation of this work, see Goodrich, *Ottoman Turks and the New World*, 71–423. For a recent discussion of this text, see Casale, "Did Alexander the Great Discover America?" 892–901.
28. Ptolemy had been well known to Muslim geographers throughout the Middle Ages. But even while there was no Ptolemaic Renaissance in the Ottoman world, his geography remained important for the incorporation of the discoveries in the period into Ottoman cosmologies.
29. *Tarih-i Hind-i Garbi*, trans. Goodrich in *Ottoman Turks and the New World*, 86.
30. Ibid., 173.
31. Ibid., 253.
32. Ibid., 339. This is in the Beyazit MS (1583/1584); see Goodrich, *Ottoman Turks and the New World*, 19. Casale's study is based on a copy at the Newberry Library in Chicago: MS Ayers 612.
33. Al-Hajari, *Kitāb nāṣir al-din ʿalāʾl-qawm al-kāfirīn* (The Supporter of Religion against the Infidels), ed. and trans. Pieter Sjoerd van Koningsveld, Qāsim al-Samarrāʾī, and Gerard A. Wiegers (Madrid: Consejo Superior de Investigaciones Científicas / Agencia Española de Cooperación Internacional, 1997), 175–76. On Ahmad Zarruq, see Scott Alan Kugle, *Rebel between Spirit and Law: Ahmad Zarruq, Sainthood, and Authority in Islam* (Bloomington: Indiana University Press, 2006).
34. Al-Hajari, *Kitāb nāṣir al-din*, 179.
35. Joseph ha-Kohen, *Divre ha-yamin le-malke Tsarefet u-malke bet Otoman ha-Togar* (Sabbionetta: Adelkind, 1554). I have used the English translation: *The Chronicles of Rabbi Joseph ben Joshua ben Meir, the Sphardi*, trans. Friedrich Bialloblotzky, 2 vols. (London: Published for the Oriental Translation Fund of Great Britain and Ireland by Richard Bentley, New Burlington Street, 1835–36); for the discussion of Europe's expansion into the East Indies and the New World, see 2:1–12.
36. *Masiv gevulot ʿammim*, unpublished manuscript, with copies in Berlin, Paris, New York, and Moscow. I have drawn on Martin Jacobs's analysis of this text in his *Islamische Geschichte in jüdischen Chroniken: Hebräische Historiographie des 16. und 17. Jahrhunderts* (Tübingen: Mohr Siebeck, 2004), 106–8. I also thank Flora Cassen for her insights into ha-Kohen.

37. On messianic themes in the *Divre ha-yamin*, see Yosef Hayim Yerushalmi, "Messianic Impulses in Joseph ha-Kohen," in *Jewish Thought in the Sixteenth Century*, ed. Bernard Dov Cooperman (Cambridge, MA: Harvard University Press, 1983), 460–87.
38. The first Dutch edition was published in 1571; the French and German translations were printed in 1572; the Spanish in 1588; and the Italian in 1608—from Mangani, *Il "mondo" di Abramo Ortelio*, 27.
39. Luke Clossey, *Salvation and Globalization in the Early Jesuit Missions* (New York: Cambridge University Press, 2008), 75–76.
40. Cited in Anthony Pagden, *Spanish Imperialism and the Political Imagination: Studies in European and Spanish-American Social and Political Theory, 1513–1830* (New Haven: Yale University Press, 1990), 40.
41. Luigi Amabile, *La sua congiura, i suoi processi, e la sua pazzia*, vol. 2 of *Fra Tommaso Campanella* (Naples: Antonio Morano, 1882), 195–96; and Frank E. Manuel and Fritzie P. Manuel, *Utopian Thought in the Western World* (Cambridge, MA: Harvard University Press, 1979), 266.
42. Noel Malcolm, "The Crescent and the City of the Sun: Islam and the Renaissance Utopia of Tommaso Campanella," *Proceedings of the British Academy* 125 (2005): 41–67.
43. Cited in Pagden, *Spanish Imperialism*, 50.
44. Tommaso Campanella, *Discorsi universali del governo ecclesiastico per far una gregge e un pastore*, in *Scritti scelti di Giordano Bruno e di Tommaso Campanella*, ed. Luigi Firpo, 2nd ed. (Turin: Unione Tipografico-Editrice, 1968), 469–70.
45. Tommaso Campanella, *La monarchia del Messia*, ed. Vittorio Frajese (Rome: Edizioni di Storia e di letteratura, 1995), 61.
46. Ibid., 62.
47. Ibid., 63.
48. Campanella first composed *The City of the Sun* in Italian in 1602 but it was the Latin version he produced in 1612–13 that was published in his lifetime: it appeared as an appendix to part 3 of a larger philosophical work on politics in 1623: *Realis philosophiae epilogisticae partes quatuor: hoc est: de rerum natura, hominum moribus, politica (cui Civitas solis iuncta est) & oeconomica* (Frankfurt: Godefridus Tampachius, 1623), 415–64; the first Italian edition of the *Città del Sole* did not appear until 1941: I have used the Italian edition in Firpo, *Scritti scelti di Giordano Bruno e di Tommaso Campanella*, 405–63.
49. Campanella, *La città del Sole*, in Firpo, *Scritti scelti di Giordano Bruno e di Tommaso Campanella*, 411; *La città del Sole: Dialogo poetico / The City of the Sun: A Poetical Dialogue*, trans. Daniel J. Donno (Berkeley: University of California Press, 1981), 31.
50. Campanella, *La città del Sole*, 414; *City of the Sun*, 37.
51. Campanella, *La città del Sole*, 415; *City of the Sun*, 38–39.
52. Campanella, *La città del Sole*, 423; *City of the Sun*, 54–55.
53. Campanella, *La città del Sole*, 425; *City of the Sun*, 56–59.
54. Campanella, *La città del Sole*, 423–24; *City of the Sun*, 54–57. Campanella's source for the practice of gazing upon statues prior to intercourse was, in all likelihood, Ocellus Lucanus, *De universi natura libellus*, printed in Venice in 1559; see John M. Headley,

Tommaso Campanella and the Transformation of the World (Princeton, NJ: Princeton University Press, 1997), 304–5.
55. Campanella, *La città del Sole*, 426; *City of the Sun*, 58–59.
56. Campanella, *La città del Sole*, 448; *City of the Sun*, 102–3.
57. Noel Malcolm identifies Postel's *De orbis terrae concordia* as the source for Campanella's notion of the Solarian Trinity and the names of the three officers: Potentia, Scientia, and Amor. See Malcolm, "Crescent and the City of the Sun," 61.
58. Campanella, *La città del Sole*, 451; *City of the Sun*, 108–9.
59. Campanella, *La città del Sole*, 458–59; *City of the Sun*, 120–21.
60. Campanella, *La città del Sole*, 459; *City of the Sun*, 108–9.
61. Campanella, *La città del Sole*, 463; *City of the Sun*, 126–27.
62. Sanjay Subrahmanyam, "A Tale of Three Empires: Mughals, Ottomans, and Habsburgs in a Comparative Context," *Common Knowledge* 12 (2006): 66–92.
63. Malcolm, "Crescent and the City of the Sun," 55.

CHAPTER TEN. CANNIBALS

1. Michel de Montaigne, "Des Coches," in *Les Essais*, ed. Pierre Villey, Verdun L. Saulnier, and Marcel Conche (Paris: Quadrige / Presses Universitaires de France, 1999), 3.6:913–14; "On Coaches," in *The Complete Essays of Montaigne*, trans. Donald M. Frame (Stanford, CA: Stanford University Press, 1958), 698.
2. Montaigne read Francisco López de Gómara's *Historia* in a French translation: *Histoire génerale de Indes occidentales*, trans. Martin Fumée, Seigneur de Genillé (Paris: Chez Michel Sonnius, 1568); the myth of the suns, which Montaigne follows closely, appears on 157–59. Many European observers recorded the myth—see Wayne Elzey, "The Nahua Myth of the Suns: History and Cosmology in Pre-Hispanic Mexican Religions," *Numen* 23 (1976): 114–35.
3. Alain Legros, "Montaigne between Fortune and Providence," in *Chance, Literature, and Culture in Early Modern France*, ed. John D. Lyons and Kathleen Wine (Farnham: Ashgate, 2009), 18–30.
4. Montaigne, "Des prognostications," in *Les Essais*, 1.11:41; "Of Prognostications," in *Essays*, 27.
5. Montaigne, "Des prognostications," in *Les Essais*, 1.11:43; "Of Prognostications," in *Essays*, 29.
6. Montaigne, "Des prognostications," in *Les Essais*, 1.11:44; "Of Prognostications," in *Essays*, 29.
7. Montaigne, "Des Coches," in *Les Essais*, 3.6:908; "Of Coaches," in *Essays*, 693.
8. Columbus's role in shaping the European perception of cannibals is well studied. See William Arens, *The Man-Eating Myth: Anthropology and Anthropophagy* (New York: Oxford University Press, 1979); but, above all, Peter Hulme, "Columbus and the Cannibals: A Study of the Reports of Anthropophagy in the Journal of Christopher Columbus," *Ibero-amerikanisches Archiv* 4 (1978): 115–39.
9. Christopher Columbus, *The Diario of Christopher Columbus's First Voyage to America: 1492–1493*, ed. and trans. Oliver Dunn and James E. Kelly, Jr. (Norman: University of Oklahoma Press, 1989), 132–33.

10. Ibid., 166–67.
11. Ibid., 216–17.
12. Ibid., 236–37.
13. Ibid., 330–31; 334–35.
14. Columbus, "Carta a Luis de Santángel," in *Cristóbal Colón: Textos y documentos completos*, ed. Consuelo Varela and Juan Gil (Madrid: Alianza, 2003), 221.
15. Columbus, "Memorial . . . para los Reyes Católicos . . . sobre el suceso de su segundo viage a las Indias," in *Cristóbal Colón: Textos y documentos completos*, 259–60.
16. Amerigo Vespucci, "Lettera di Amerigo Vespucci delle isole nuovamente trovate in quattro suoi viaggi," in *Nuovo Mondo: Gli italiani, 1492–1565*, ed. Paolo Collo and Pier Luigi Crovetto (Turin: Einaudi, 1991), 242; "Letter of Amerigo Vespucci concerning the Islands Newly Discovered on His Four Voyages," in *Letters from a New World: Amerigo Vespucci's Discovery of America*, ed. Luciano Formisano, trans. David Jacobson (New York: Marsilio, 1992), 66–67.
17. Amerigo Vespucci, *Mundus novus* (Augsburg: Johannes Otmar, 1504), fol. 2v (unpaginated); "*Letter V: Mundus Novus*," in Formisano, *Letters from a New World*, 50.
18. Pietro Martire d'Anghiera, *De orbe novo Petri Martyris Angelerii Mediolanensis protonotarij c[a]esaris senatoris decades* (Alcalá de Henares: Miguel de Eguía, 1530), fols 3v–4r, 6r, 17v; *De orbe novo: The Eight Decades*, ed. and trans. Francis Augustus MacNutt (New York: Putnam, 1912), 1:63, 76–77, 155–56.
19. Cited in Neil L. Whitehead, "Carib Cannibalism: The Historical Evidence," *Journal de la Société des américanistes* 70 (1984): 70.
20. Montaigne, "Des cannibales," in *Les Essais*, 1.31:209; "On Cannibals," in *Essays*, 155.
21. Montaigne, "Des cannibales," in *Les Essais*, 1.31:206; "On Cannibals," in *Essays*, 152–53.
22. Montaigne, "Des cannibales," in *Les Essais*, 1.31:206–7; "On Cannibals," in *Essays*, 153.
23. Montaigne, "Des cannibales," in *Les Essais*, 1.31:205; "On Cannibals," in *Essays*, 152–53.
24. Montaigne, "Des cannibales," in *Les Essais*, 1.31:210; "On Cannibals," in *Essays*, 156. See P. Kenneth Himmelman, "The Medicinal Body: An Analysis of Medicinal Cannibalism in Europe, 1300–1700," *Dialectical Anthropology* 22 (1997): 183–203.
25. In an interesting essay on Montaigne, Carlo Ginzburg has argued that "ethnography emerged when the curiosity and methods of the antiquarians was transferred from the study of people who had lived long before, such as the Greeks and Romans, to those who lived far away, such as the peoples of the New World." Ginzburg, "Montaigne, i cannibali, e le grotte," in *Il filo e le tracce: Vero falso finto* (Milan: Feltrinelli, 2006), 76; see also Ginzburg, *Rapporti di forza: Storia, retorica, prova* (Milan: Feltrinelli, 2000), 101–3.
26. André Thevet, *Les singularitez de la France Antarctique* (Paris: Maurice de la Porte and Heirs, 1557).
27. Jean de Léry, *Histoire d'un voyage fait en la terre de Brésil* (La Rochelle: Antoine Chuppin, 1578), 256–58; Léry, *History of a Voyage to the Land of Brazil*, trans. Janet Whatley (Berkeley: University of California Press, 1990), 131–33; though note that,

while Montaigne used the 1578 edition of the *Histoire,* Whatley's translation is of the 1580 edition of the *Histoire,* which Léry had extensively revised.

28. Léry, *Histoire d'un voyage,* 218–20; *History of a Voyage,* 112.
29. The reading of cannibalism as a central element in a feud was not original to either Thevet or Léry. Other sixteenth-century authors had offered a similar account. This was a strong current among such Portuguese observers as Manuel da Nóbrega, José de Anchieta, and Gabriel Soares de Sousa—see Neil L. Whitehead, introduction to *Hans Staden's True History: An Account of Cannibal Captivity in Brazil,* trans. Michael Harbsmeier (Durham, NC: Duke University Press, 2008).
30. Léry, *Histoire d'un voyage,* 407 ff.; *History of a Voyage,* 211 ff.
31. Claudio Povolo, *L'intrigo dell'onore: Poteri e istituzioni nella Repubblica di Venezia tra Cinque e Seicento* (Verona: Cierre, 1997); Edward Muir, *Mad Blood Stirring: Vendetta and Factions in Friuli during the Renaissance* (Baltimore: Johns Hopkins University Press, 1993). Intriguingly, Muir's work, while focused primarily on the dynamics of the vendetta in the celebrated revolt of Giovedì Grasso, 1511, also offers an account of how feuding parties in the Friuli inscribed this revolt within an apocalyptic framework (122–26).
32. Stuart Carroll, "The Peace in the Feud in Sixteenth- and Seventeenth-Century France," *Past and Present* 178 (2003): 84.
33. Innocent Gentillet, *Discours, sur les moyens de bien gouverner . . . contre Nicolas Machiavel Florentin* (Geneva, 1576).
34. David Quint, *Montaigne and the Quality of Mercy: Ethical and Political Themes in the Essais* (Princeton, NJ: Princeton University Press, 1998).
35. Hans Staden, *Warhafftige Historia und Beschreibung einer Landtschafft der Wilden, Nacketen, Grimmigen Menschfresser Leuthen in der Newen Welt America gelegen* (Frankfurt: Weigand Han, 1557); Whitehead and Harbsmeier, *Hans Staden's True History.* On Staden, see Eve M. Duffy and Alida C. Metcalf, *The Return of Hans Staden: A Go-Between in the Atlantic World* (Baltimore: Johns Hopkins University Press, 2011).
36. Theodore de Bry, ed., *Americae tertia pars: Memorabilem provinciae Brasiliae historiam continens* (Frankfurt-am-Main: Johannes Wechelus, 1592).
37. See Henry Kamen, *Empire: How Spain Became a World Power, 1492–1763* (New York: HarperCollins, 2003), 368 ff.
38. Cited in Stefania Pastore, "Mozas Criollas and New Government: Francis Borgia, Prophetism, and the *Spiritual Exercises* in Spain and Peru," in *Visions, Prophecies, and Divinations: Early Modern Messianism and Millenarianism in Iberian America, Spain, and Portugal,* ed. Luís Filípe Silvério Lima and Anna Paula Torres Megiani (Leiden: Brill, 2016), 70.
39. Adriano Prosperi, "America e apocalisse: Note sulla 'conquista spirituale' del Nuovo Mondo," in *America e Apocalisse e altri saggi* (Pisa: Istituti Editoriali e Poligrafici Internazionali 1999), 18, 60.
40. José de Acosta, *De temporibus novissimis: Libri quatuor* (Rome: Jacobus Tornerius, 1590), 453–54.
41. José de Acosta, *Historia natural y moral de las Indias* (Seville: Casa de Juan de Leon, 1590), 305; *Natural and Moral History of the Indies,* trans. Frances López-Morillas

(Durham, NC: Duke University Press, 2002), 329. On this work, see especially Edmundo O'Gorman, "Prólogo" to José de Acosta, *Historia natural y moral de las Indias,* ed. Edmundo O'Gorman (Mexico City: Fondo de Cultura Económica, 1962), xvii–lxvii; and Anthony Pagden, *The Fall of Natural Man: The American Indian and the Origins of Comparative Ethnology* (Cambridge: Cambridge University Press, 1982).

42. Acosta, *Historia natural y moral,* 510; *Natural and Moral History,* 429.
43. Alan Durston, *Pastoral Quechua: The History of Christian Translation in Colonial Peru, 1550–1654* (Notre Dame, IN: University of Notre Dame Press, 2007), 33–34, 174. This makes the pertinent point that it is not accurate that the Council of Trent had banned vernacular Bible translations, rather expressing a preference for the Vulgate as most "authentic."
44. Acosta, *Historia natural y moral,* 410–11; *Natural and Moral History,* 343.
45. Elizabeth Hill Boone and Walter D. Mignolo, eds., *Writing without Words: Alternative Literacies in Mesoamerica and the Andes* (Durham, NC: Duke University Press, 1994).
46. Acosta, *Historia natural y moral,* 411; *Natural and Moral History,* 343.
47. Acosta, *Historia natural y moral,* 412; *Natural and Moral History,* 344.
48. Acosta, *Historia natural y moral,* 408; *Natural and Moral History,* 340.
49. Ibid.
50. J. H. Elliott, *The Old World and the New, 1492–1650* (Cambridge: Cambridge University Press, 1970), 35.
51. Acosta, *Historia natural y moral,* 408; *Natural and Moral History,* 341.
52. Acosta, *Historia natural y moral,* 404; *Natural and Moral History,* 337. On Acosta's encounter with Sanchez, see Serge Gruzinski, *The Eagle and the Dragon: Globalization and European Dreams of Conquest in China and America in the Sixteenth Century,* trans. Jean Birrell (Cambridge: Polity, 2014), 236–37.
53. "Muy barbaros y silvestres"; Acosta, *Historia natural y moral,* 453–54; *Natural and Moral History,* 380–81. See also Anthony Pagden, *European Encounters with the New World: From Renaissance to Romanticism* (New Haven: Yale University Press, 1993), 171.
54. Walter D. Mignolo, *The Darker Side of the Renaissance: Literacy, Territoriality, and Colonization* (Ann Arbor: University of Michigan Press, 1995), 125 ff.
55. Philippe Desan, *Montaigne: A Life,* trans. Steven Rendall and Lisa Neal (Princeton, NJ: Princeton University Press, 2017), 170.
56. Montaigne, "Des cannibales," in *Les Essais,* 1.31:213; "On Cannibals," in *Essays,* 159.
57. Montaigne, "Des cannibales," in *Les Essais,* 1.31:213–14; "On Cannibals," in *Essays,* 159.
58. Montaigne, "Des Coches," in *Les Essais,* 3.7:909; "On Coaches," in *Essays,* 693.
59. François Rigolot, "Curiosity, Contingency, and Cultural Diversity: Montaigne's Readings at the Vatican Library," *Renaissance Quarterly* 64 (2011): 856–57.
60. Montaigne, "De l'experience," in *Les Essais,* 3.13:1071; "Of Experience," in *Essays,* 820.
61. Montaigne, "Des Coches," in *Les Essais,* 3.7:908; "On Coaches," in *Essays,* 693.
62. Montaigne, "Des Cannibales," in *Les Essais,* 1.31:214; "On Cannibals," in *Essays,* 159.

CHAPTER ELEVEN. THE RESTITUTION OF ALL THINGS

1. Moshe Idel, *Messianic Mystics* (New Haven: Yale University Press, 1998), 162–82.
2. In the mid-twentieth century, Gershom Scholem emphasized the messianic dimensions of Luria's teachings; see especially his *Major Trends in Jewish Mysticism*, 3rd ed. (New York: Schocken, 1971; the first Hebrew edition appeared in 1941), 244–86; and his *Sabbatai Ṣevi: The Mystical Messiah, 1626–1676*, trans. R. J. Zwi Werblowsky (Princeton, NJ: Princeton University Press, 1973; originally published in Hebrew in 1957), 22–77. In his *Messianic Mystics,* Moshe Idel has offered a major revaluation of Luria, making it clear that messianism was a less pronounced feature of Luria's teachings than Scholem had assumed. On Luria's life and teachings, see also Lawrence Fine, *Physician of the Soul, Healer of the Cosmos: Isaac Luria and His Kabbalistic Fellowship* (Stanford, CA: Stanford University Press, 2003).
3. *The Zohar,* trans. Daniel C. Matt, 12 vols. (Stanford, CA: Stanford University Press, 2004–17), 5:33–35.
4. Christoph Schulte, *Zimzum: Gott und Weltursprung* (Berlin: Jüdischer Verlag, 2014).
5. Scholem, *Sabbati Ṣevi,* 45.
6. Ibid., 46; see also Scholem, "The Messianic Idea in Kabbalism," in *The Messianic Idea in Judaism and Other Essays on Jewish Spirituality* (New York: Schocken, 1971), 47.
7. Scholem, *Sabbati Ṣevi,* 37–42.
8. Chayyim Vital, *Window of the Soul: The Kabbalah of Rabbi Isaac Luria (1534–1572): Selections from Chayyim Vital,* ed. James David Dunn, trans. Nathan Snyder (San Francisco: Weiser, 2008); *The Palace of Adam Kadmon,* vol. 1 of *The Tree of Life: Chayyim Vital's Introduction to the Kabbalah of Isaac Luria,* trans. Donald Wilder Menzi and Zwe Padeh (New York: Arizal, 2008).
9. Pico della Mirandola, "Conclusiones Cabalisticae numero 71 secundum opinionem proprium, ex ipsis Hebracorum sapientum fundamentis Christianam religionem maxime confirmantes," in his *Conclusiones nongentae in omni genere scientiarum* ([Nuremberg]: [Johannes Petrejus], 1532), 151–52. On Pico's Kabbalah, see most recently, Brian P. Copenhaver, *Magic and the Dignity of Man: Pico della Mirandola and His Oration in Modern Memory* (Cambridge, MA: Harvard University Press, 2019). On Reuchlin, see Franz Posset, *Johann Reuchlin (1455–1522): A Theological Biography* (Boston: De Gruyter, 2015). On Giles of Viterbo, the essay by Marjorie Reeves, "Cardinal Egidio of Viterbo: A Prophetic Interpretation of History," in *Prophetic Rome in the High Renaissance Period: Essays* (Oxford: Oxford University Press, 1992), 91–110, is indispensable. See also Moshe Idel, *Kabbalah in Italy, 1280–1510: A Survey* (New Haven: Yale University Press, 2011).
10. William Bouwsma, *Concordia Mundi: The Career and Thought of Guillaume Postel (1510–1581)* (Cambridge, MA: Harvard University Press, 1957); and Marion Leathers Kuntz, *Guillaume Postel: Prophet of the Restitution of All Things, His Life and Thought* (The Hague: Martinus Nijhoff, 1981).
11. "Con l'operatione di millari d'Angeliche virtutis & operationi," Guillaume Postel, *Le prime nove del altro mondo* (Venice: published by the author, 1555), 24v (not paginated).
12. Bouwsma, *Concordia Mundi,* 161.

13. Kuntz, *Guillaume Postel*, 74, n. 237.
14. "Cosi sendendola raggionare, io restava come morto & fuori di me stesso, & conderando come io che mi parèva per haver letto molti & valentissimi Theologi & con il beneficio della contemplattone havea (merce della bontà infinita) gustato qualche cosa della divine gratie, non era mai arrivato à tanto raggionmenti ne altri concetti." Postel, *Le prime nove del altro mondo*, 7v (not paginated).
15. "E quant à parler de sçavoir feminine, si tres-grand & eminent estoit en elle. . . . icelle que n'apprint oncques Latin, ne Grec, n'Hebreu, ne autre langue ou lecture, me sçavoit tellement ouvrir & declarer quand je tournoys le Zohar, livre tres-difficile & contenant l'ancienne Doctrine Evangelique en Latin," Guillaume Postel, *Les trèsmerveilleuses victoires des femmes du nouveau-monde* (Paris: Jehan Ruelle, 1553; facsimile: Geneva: Slatkine Reprints, 1970), 19.
16. On Hebrew and the unity of languages, see Guillaume Postel, *De originibus seu de Hebraicae linguae et gentis antiquitate, déque variarum linguarum affinitate, liber* (Paris, 1538); on his cosmology, Postel, *De orbis terrae concordia: Libri quatuor* (Basel: J. Oporinus, 1544).
17. Guillaume Postel, *Histoire et considérations de l'origine, loy et costumes des Tartares, Persiens, Arabes, Turcs, et tous autres Ismaelites ou Muhamediques, dicts par nous Mahometains ou Sarrazins* (Poitiers: Enguilbert de Marnef, 1560), 52–53. This was one of three treatises Postel published on the Islamic world in 1560, though it is possible that he had composed the texts as early as 1547—see Noel Malcolm, *Useful Enemies: Islam and the Ottoman Empire in Western Political Thought, 1450–1750* (Oxford: Oxford University Press, 2019), 132.
18. On Postel's eschatology, see above all Bouwsma, *Concordia Mundi*, 276–98; and Kuntz, *Guillaume Postel*.
19. Bouwsma, *Concordia Mundi*, 278–81.
20. Allison P. Coudert stresses not only the parallels but also the interactions between early modern kabbalists and alchemists; see her "Kabbalistic Messianism versus Kabbalistic Enlightenment," in *Jewish Messianism in the Early Modern World*, vol. 1 of *Millenarianism and Messianism in Early Modern European Culture*, ed. Matt Goldish and Richard H. Popkin (Dordrecht: Kluwer, 2001), 107–24; see also Gershom Scholem, *Alchemy and Kabbalah*, trans. Klaus Ottmann (Putnam, CT: Spring, 2006).
21. Paracelsus, *Das Buch Paragranum . . . Inn welchem die Vier Columnae, als nemlich Philosophia, Astronomia, Alchimia, vnd Virtus darauff er seine Medicin fundiret beschieben werben*, in Paracelsus, *Essential Theoretical Writings*, ed. and trans. Andrew Weeks (Leiden: Brill, 2008), 84–85.
22. Paracelsus, *Volumen medicinae paramirum*, in *Essential Readings*, ed. Nicholas Goodrick-Clarke (Berkeley, CA: North Atlantic, 1999), 45–46.
23. Paracelsus, *Das Buch Paragranum*, 133.
24. Ibid., 106–209, for Paracelsus's discussion of "Philosophia" and "Astronomia."
25. On the centrality of Galen in the Renaissance, see Nancy G. Siraisi, *Medieval and Early Renaissance Medicine: An Introduction to Knowledge and Practice* (Chicago: University of Chicago Press, 1990).

26. Andrew Cunningham and Ole Peter Grell, *The Four Horsemen of the Apocalypse: Religion, War, Famine, and Death in Reformation Europe* (Cambridge: Cambridge University Press, 2000), 308–9.
27. Walter Pagel, *Paracelsus: An Introduction to Philosophical Medicine in the Era of the Renaissance,* 2nd ed. (Basel: Karger, 1982), 126–222.
28. "Tanta & talis futura est rerum omnium renovatio & mutatio, ut plane aurea saecula rediise videātur, ubi plane puerilis candor, simplicitas ac integritas regnabit explosis omnibus versutiis, astutiis, & insidiis hominum." Paracelsus, *Prognosticatio ad vigesimum quartum usque annum duratura,* trans. Marcus Tatius Alpinus (Augsburg: Heinrich Steyner, 1536), not paginated. Paracelsus originally published this text in German: *Prognostication auff xxiiii jar zükunfftig* (1536).
29. The best overview of Paracelsus's eschatology remains Kurt Goldammer, "Paracelsische Eschatologie: Zum Verständnis der Anthropologie und Kosmologie Hohenheims," *Nova Acta Paracelsica* 5 (1948): 45–95, and "Paracelsische Eschatologie II," *Nova Acta Paracelsica* 6 (1952): 68–102; but see also Leigh T. I. Penman, *Hope and Heresy: The Problem of Chiliasm in Lutheran Confessional Culture, 1570–1630* (Dordrecht: Springer, 2019). For the major texts in which Paracelsus addressed the End of Time, see his *Buch der natürlichen Dingen* in vol. 2.1 of *Sämtliche Werke,* ed. Karl Sudhoff (Munich: Oldenbourg, 1930); his *De imaginibus;* his *Philosophia sagax,* and his writings published in the volume *Theologische Werke 1: Vita Beata—Vom seligen Leben,* ed. Urs Leo Gantenbein, Michael Baumann, and Detlef Roth (Berlin: De Gruyter, 2008).
30. Cited in Andrew Weeks, *Paracelsus: Speculative Theory and the Crisis of the Early Reformation* (Albany: State University of New York Press, 1997), 74–75.
31. Walter Pagel, "The Paracelsian *Elias Artista* and the Alchemical Tradition," *Medizinhistorisches Journal* 16 (1981): 6–19.
32. Paracelsus, *Buch der natürlichen Dingen,* in vol. 2.1 of *Sämtliche Werke,* ed. Karl Sudhoff (Munich: Oldenbourg, 1930), 163.
33. Weeks, *Paracelsus: Speculative Theory and the Crisis of the Early Reformation,* 14 ff.
34. Cited in Pamela H. Smith, *The Body of the Artisan: Art and Experience in the Scientific Revolution* (Chicago: University of Chicago Press, 2004), 85.
35. Hugh R. Trevor-Roper, "The Paracelsian Movement," in *Renaissance Essays* (Chicago: University of Chicago Press, 1985), 149–99.
36. As Andrew Weeks notes, "theory is an apt characterization for a writing that can alternate in a single fragment between the projects of explaining nature, healing human life, interpreting the meanings of things, and imagining their cosmic, metaphysical and divine contexts." Weeks, *Paracelsus: Speculative Theory and the Crisis of the Early Reformation,* ix.
37. Robin B. Barnes, *Prophecy and Gnosis: Apocalypticism in the Wake of the Lutheran Reformation* (Stanford, CA: Stanford University Press, 1988).
38. Melchior Ambach, *Vom Ende der Welt: Und Zukunfft des Endtchrists* (Frankfurt: Herman Gülfferich, 1545).
39. Tara E. Nummedal, *Anna Zieglerin and the Lion's Blood: Alchemy and End Times in Reformation Germany* (Philadelphia: University of Pennsylvania Press, 2019), 131.
40. Ibid., 2.

41. Cited in Marjorie Reeves, *The Influence of Prophecy in the Later Middle Ages: A Study in Joachimism* (Notre Dame, IN: University of Notre Dame Press, 1993), 499. See also Jürgen Moltmann, "Jacob Brocard als Vorläufer der Reich-Gottes-Theologie und der symbolisch-prophetischen Schriftauslegung des Johann Coccejus," *Zeitschrift für Kirchengeschichte* 71 (1960): 110–29; and Antonio Rotondò, "Iacopo Brocardo," in *Dizionario biografico degli Italiani* (Rome: Istituto della Enciclopedia Italiana, 1972), 14:384–89.
42. For modern editions, see Richard van Dülmen, ed., *Fama Fraternitatis (1614), Confessio Fraternitatis (1615), Chymische Hochzeit: Christiani Rosencreutz Anno 1459 (1616)* (Stuttgart: Calwer Verlag, 1973).
43. *Confessio Fraternitatis*, 46.
44. Carlos Gilly, *Cimelia Rhodostaurotica: Die Rosenkreuzer im Spiegel der zwischen 1610 und 1660 entstandenen Handschriften und Drucke* (Amsterdam: In de Pelikaan, 1995), 43–44.
45. Frances A. Yates, *The Rosicrucian Enlightenment* (London: Routledge and Kegan Paul, 1972); Walter Sparn, " 'Chiliasmus crassus' und 'Chiliasmus subtilis' im Jahrhundert Comenius," in *Johannes Amos Comenius und die Genese des modernen Europa*, ed. Norbert Kotowski and Jan B. Lášek (Fürth: Flacius, 1992): 122–29; Hartmut T. Lehmann, *Das Zeitalter des Absolutismus: Gottesgnadentum und Kriegsnot* (Stuttgart: Kohlhammer, 1980), 105–13.
46. Barnes, *Prophecy and Gnosis*, 171–81, 199–201.
47. Leigh T. I. Penman, *Hope and Heresy: The Problem of Chiliasm in Lutheran Confessional Culture, 1570–1630* (Dordrecht: Springer, 2019), xxii, with a list of the titles provided in the appendix, "Printed Works Concerning Optimistic Apocalyptic Expectations, 1600–1630," 197–210.
48. Ibid., 38–46.
49. Hans Schneider, *Der fremde Arndt: Studien zu Leben, Werk und Wirkung Johann Arndts, 1555–1621* (Göttingen: Vandenhoeck and Ruprecht, 2006).
50. Cited in Penman, *Heresy and Hope*, 50.
51. Frank E. Manuel and Fritzie P. Manuel, *Utopian Thought in the Western World* (Cambridge, MA: Harvard University Press, 1979), 298–99.
52. Johann Valentin Andreae, *Christianopolis*, ed. and trans. Edward H. Thompson (Dordrecht: Kluwer, 1999).
53. Andreae, *Christianopolis*, 145.
54. Ibid., 147.
55. Ibid., 190.

CHAPTER TWELVE. CROSSING THE PILLARS OF HERCULES

1. Cited in Robert Zaller, *The Parliament of 1621: A Study in Constitutional Conflict* (Berkeley: University of California Press, 1971), 84. On the ambassador Amerigo Salvetti, see Stefano Villani, "Per la progettata edizione della corrispondenza dei rappresentanti toscani a Londra: Amerigo Salvetti e Giovanni Salvetti Antelminelli durante il 'Commonwealth' e il Protettorato, 1649–1660," *Archivio storico italiano* 162 (2004): 109–25.

2. On Bacon's trial, see Zaller, *Parliament of 1621*. On his biography, see James Spedding, *An Account of the Life and Times of Francis Bacon: Extracted from the Edition of his Occasional Writings*, 2 vols. (Boston: Houghton, Osgood, 1878), a work that remains foundational for almost all subsequent biographical studies of Bacon. Among the more recent biographies, Lisa Jardine and Alan Stewart, *Hostage to Fortune: The Troubled Life of Francis Bacon* (New York: Hill and Wang, 1998), is especially useful.
3. *Francisci de Verulamio, Summi Angliae Cancellarii Instauratio magna* (London: Apud Ioannem Billium, 1620). For this text, with a facing English translation, I have used *The* Instauratio magna *Part II:* Novum Organum *and Associated Texts*, ed. and trans. Graham Rees and Maria Wakely (Oxford: Clarendon, 2004); the citation is from the *Novum Organum*, bk. 1, aphorism 92, included in this volume at 150–51. It is important to note that, with the title *Instauratio magna*, Bacon envisioned a multi-part work, never completed.
4. Bacon, *Novum Organum*, bk. 1, aphorism 85, in *Instauratio magna / Great Instauration*, 136–37.
5. Paul F. Grendler, *The Universities of the Italian Renaissance* (Baltimore: Johns Hopkins University Press, 2002), 260; on the continued use of Aristotle's *Organon* at Cambridge, see William T. Costello, S.J., *The Scholastic Curriculum at Early Seventeenth-Century Cambridge* (Cambridge, MA: Harvard University Press, 1958).
6. Bacon, *Novum Organum*, bk. 1, aphorism 9, in *Instauratio magna / Great Instauration*, 66–67.
7. Francis Bacon, *The Advancement of Learning*, ed. Michael Kiernan (Oxford: Clarendon, 2000), 116.
8. Bacon, *Novum Organum*, bk. 1, aphorism 39, in *Instauratio magna / Great Instauration*, 78–79.
9. Bacon, *Novum Organum*, bk. 2, aphorisms 11–13, in *Instauratio magna / Great Instauration*, 216–37.
10. Bacon, *Novum Organum*, bk. 2, aphorisms 11–20, in *Instauratio magna / Great Instauration*, 216–63. On Bacon's theory of "eliminative induction," see especially Stephen Gaukroger, *Francis Bacon and the Transformation of Early-Modern Philosophy* (Cambridge: Cambridge University Press, 2001), 132–65.
11. J. B. Bury, *The Idea of Progress: An Inquiry into Its Origin and Growth* (New York: Macmillan, 1932), chap. 2.
12. Thomas Babington Macaulay, "Lord Bacon," in vol. 2 of *Critical, Historical, and Miscellaneous Essays and Poems* (New York: Armstrong, 1860); Carolyn Merchant, *The Death of Nature: Women, Ecology, and the Scientific Revolution* (New York: HarperCollins, 1980), identifies Bacon as a major source in some of the more destructive aspects of modernity.
13. David Harris Sacks offers a penetrating analysis of the frontispiece in "Richard Hakluyt's Navigations in Time: History, Epic, and Empire," *Modern Language Quarterly* 67 (2006): 31–62; but see also his "Rebuilding Solomon's Temple: Richard Hakluyt's Great Instauration," in *New Worlds Reflected: Travel and Utopia in the Early Modern Period*, ed. Chloë Houston (Farnham: Ashgate, 2010), 48–55, and his "On Mending the Peace of the World: Sir Francis Bacon's Apocalyptic Irenicism," *New Global Studies* 16 (2022), forthcoming.

14. John Sugden, *Sir Francis Drake* (New York: Henry Holt, 1990).
15. Cited in Raleigh Trevelyan, *Sir Walter Raleigh* (New York: Penguin, 2002), 70–71.
16. Alan Gallay, *Walter Ralegh: Architect of Empire* (New York: Basic Books, 2019).
17. Karen Ordahl Kupperman, *The Jamestown Project* (Cambridge, MA: Harvard University Press, 2007).
18. John Winthrop, *A Modell of Christian Charity* (Boston: Massachusetts Historical Society, 1838). Mark Peterson has cautioned against the reading of this "sermon" in a prophetic key, noting that we can't even be certain it was delivered; see his *The City-State of Boston: The Rise and Fall of an Atlantic Power, 1630–1865* (Princeton, NJ: Princeton University Press, 2019), 9–10.
19. Philip J. Stern, *The Company-State: Corporate Sovereignty and the Early Modern Foundations of the British Empire in India* (Oxford: Oxford University Press, 2011).
20. Anthony Grafton, with April Shelford and Nancy G. Siraisi, *New Worlds, Ancient Texts: The Power of Tradition and the Shock of Discovery* (Cambridge, MA: Harvard University Press, 1992); Francisco Bethencourt, "European Expansion and the New Order of Knowledge," in John Jeffries Martin, ed., *The Renaissance World* (London: Routledge, 2007), 118–39; and Joan-Pau Rubiés, *Travel and Ethnology in the Renaissance: South India through European Eyes* (Cambridge: Cambridge University Press, 2000).
21. Bacon, *Novum Organum,* bk. 1, aphorism 84, in *Instauratio magna / Great Instauration,* 132–33.
22. José de Acosta, *Historia natural y moral de las Indias* (Seville: Casa de Juan de Leon, 1590), 33–36; *Natural and Moral History of the Indies,* trans. Frances López-Morillas (Durham, NC: Duke University Press, 2002), 31–33.
23. Bacon, *Advancement of Learning,* 55.
24. I say "derived," because Bacon modified the passage slightly; David Burnett has pointed this out. In the Vulgate one reads, "Plurimi pertransibunt, et multiplex erit scientia"; in Tremellius's Protestant translation, "Percurrent multi, et augebitur cognitio." David Burnett, *The Engraved Title-Page of Bacon's* Instauratio magna*: An Icon and Paradigm of Science and Its Wider Implications* (Durham, NC: Thomas Harriot Seminar, 1998), 4.
25. Francis Bacon, *Valerius terminus,* in *Philosophical Works,* vol. 3 of *The Works of Francis Bacon,* ed. James Spedding, Robert Leslie Ellis, and Douglas Denon Heath (London: Longmans, 1857), 3:220–21.
26. Bacon, *Novum Organum,* bk. 1, aphorism 93, in *Instauratio magna / Great Instauration,* 150–51.
27. Bacon, *Novum Organum,* bk. 2, aphorism 52, in *Instauratio magna / Great Instauration,* 446–47.
28. On the religious inspiration of Bacon's thought, the literature is extensive. See Charles Webster, *The Great Instauration: Science, Medicine and Reform, 1626–1660* (New York: Holmes and Meier, 1975); though for an important critique of Webster, see Mordechai Feingold, " 'And Knowledge Shall be Increased': Millenarianism and the *Advancement of Learning* Revisited," *Seventeenth Century* 28 (2013): 363–93. But see also Achsah Guibbory, "Francis Bacon's View of History: The Cycles of Error and the Progress of

Truth," *Journal of English and Germanic Philology* 74 (1975): 336–50; Stephen A. McKnight, *The Religious Foundations of Francis Bacon's Thought* (Columbia: University of Missouri Press, 2006); and, most recently, Peter Harrison, *The Fall of Man and the Foundations of Science* (Cambridge: Cambridge University Press, 2007), esp. 183: "The seventeenth-century quest to re-establish human dominion over the natural world—often associated with that exploitative stance thought to typify the modern West's attitude towards nature—was thus originally conceived as a restorative project designed to return the world to its prelapsarian perfection."

29. Bacon, *Novum Organum*, bk. 1, aphorism 93, in *Instauratio magna / Great Instauration*, 150–51.
30. Bacon, preface to *Instauratio magna / Great Instauration*, 22–23.
31. Bacon, "Plan of the Work," in *Instauratio magna / Great Instauration*, 44–47.
32. Bacon, *Novum Organum*, bk. 1, aphorism 92, in *Instauratio magna / Great Instauration*, 148–51.
33. Bacon, *Instauratio magna / Great Instauration*, 166–67.
34. *New Atlantis* first appeared in the volume *Sylva sylvarum: or A naturall historie, in ten centuries. Written by the Right Honourable Francis Lo. Verulam Viscount St. Alban. Published after the author's death, by William Rawley Doctor of Divinitie, late his Lordship's chaplaine.* (London: Printed by J[ohn Haviland and Augustine Mathewes] for William Lee at the Turks Head in Fleet-Street, next to the Miter, 1627). For a modern edition, see Francis Bacon, *The Major Works Including the* New Atlantis *and the* Essays, ed. Brian Vickers (Oxford: Oxford University Press, 2002).
35. Howard B. White, *Peace among the Willows: The Political Philosophy of Francis Bacon* (The Hague: Martinus Nijhoff, 1968). Bacon had likely read Campanella: Eleanor Dickinson Blodgett, "Bacon's *New Atlantis* and Campanella's *Civitas Solis:* A Study in Relationships," *PMLA* 46 (1931): 764.
36. Bacon, *New Atlantis*, 480.
37. Ibid., 481.
38. Ibid., 487.
39. Ibid., 467.
40. Ibid., 471.
41. Bacon, *Novum Organum*, bk. 1, aphorism 129, in *Instauratio magna / Great Instauration*, 194–95. Indeed, there is a large literature on each of these inventions and their far-reaching effects. On the printing press, see the classic study by Elizabeth L. Eisenstein, *The Printing Press as an Agent of Change: Communications and Cultural Transformations in Early-Modern Europe* (Cambridge: Cambridge University Press, 1979); on gunpowder and its role in reshaping empires, Marshall G. S. Hodgson, *The Gunpowder Empires and Modern Times*, vol. 3 of *The Venture of Islam: Conscience and History in a World Civilization* (Chicago: University of Chicago Press, 1974); and on the mariner's compass and navigation, Barbara M. Kreutz, "Mediterranean Contributions to the Medieval Mariner's Compass," *Technology and Culture* 14 (1973): 367–83. With the magisterial work of Joseph Needham—his *Science and Civilisation in China*, 7 vols. (Cambridge: Cambridge University Press, 1954–2001), it is clear that these

technologies must also be examined in a global context. On Needham's work, I have benefited from the "Review Symposia: Science in China" by Lynn White, Jr. and Jonathan Spence—which covers the first five volumes—in *Isis* 75 (1984): 171–89.
42. Johannes Stradanus [Jan van de Straet], *Nova reperta* (Antwerp: P. Galle, ca. 1590).
43. On the relation of hope to religious faith, see Ernst Bloch, *The Principle of Hope*, trans. Neville Plaice, Stephen Plaice, and Paul Knight, 3 vols. (Cambridge, MA: MIT Press, 1986–95). Bloch offers a kaleidoscopic sociology of hope, with rich insights into the early modern period.
44. Bacon, *Instauratio magna / Great Instauration*, 18–19.
45. Eisenstein, *The Printing Press as an Agent of Change*; see also the works cited above in chapter 3, "The Apocalyptic Chorus."

EPILOGUE

1. On the continuing centrality of astrology in this period, see Keith Thomas, *Religion and the Decline of Magic: Studies in Popular Beliefs in Sixteenth- and Seventeenth-Century England* (New York: Scribner, 1971), 285–383; on the role of prophecy in the Thirty Years' War, Georg Schmidt, *Die Reiter der Apokalypse: Geschichte des Dreißigjährigen Krieges* (Munich: Beck, 2018), 343–428; and on prophecy in the English Civil War, Christopher Hill, *Antichrist in Seventeenth-Century England* (London: Verso, 1990); finally, on the Jewish Messiah, Gershom Scholem, *Sabbatai Ṣevi: The Mystical Messiah, 1626–1676*, trans. R. J. Zwi Werblowsky (Princeton, NJ: Princeton University Press, 1973), and Matt Goldish, *The Sabbatean Prophets* (Cambridge, MA: Harvard University Press, 2004).
2. Thomas Hobbes, *Leviathan, or the Matter, Forme, & Power of a Common-Wealth Ecclesiasticall and Civill* (London: for Andrew Crookes, at the Green Dragon in St. Paul's Church-yard, 1651). I have used the critical edition of Noel Malcolm, *Leviathan*, 3 vols. (Oxford: Oxford University Press, 2012–14), which includes both Hobbes's original English edition as well as his Latin edition published in Amsterdam in his *Opera philosophica* in 1668.
3. On Hobbes's separation of the state from religion, see Mark Lilla, *The Stillborn God: Religion, Politics, and the Modern West* (New York: Vintage, 2007), 75–91.
4. Hobbes's celebrated expression—a "war of all against all," *bellum omnium contra omnes*, appears in his *De cive* of 1642, though close approximations of this phrase also appear in *Leviathan*. See especially Hobbes, *Leviathan*, 2:194 for this and for his characterization of life as "solitary, poore, nasty, brutish, and short."
5. Hobbes, *Leviathan*, 2:260. Among the vast literature on Hobbes's political theory, see S. A. Lloyd, *Ideals as Interests in Hobbes's* Leviathan*: The Power of Mind over Matter* (Cambridge: Cambridge University Press, 1992); and Quentin Skinner, *Hobbes and Republican Liberty* (Cambridge: Cambridge University Press, 2008). On Hobbes's life and intellectual milieu, see especially Noel Malcolm, "Hobbes and the European Republic of Letters," in his *Aspects of Hobbes* (Oxford: Oxford University Press, 2003), 457–545; and A. P. Martinich, *Hobbes: A Biography* (Cambridge: Cambridge University Press, 1999).

6. Hobbes, *Leviathan,* 2:170–87.
7. On Hobbes's critique of the apocalyptic, see above all John G. A. Pocock, "Time, History and Eschatology in the Thought of Thomas Hobbes," in *The Diversity of History: Essays in Honour of Sir Herbert Butterfield,* ed. J. H. Elliott and Helmut G. Koenigsberger (Ithaca, NY: Cornell University Press, 1970): 149–98; and, more recently, Alison McQueen, *Political Realism in Apocalyptic Times* (Cambridge: Cambridge University Press, 2017), 133 ff.
8. On early modern "science," see especially Stephen Gaukroger, *The Emergence of a Scientific Culture: Science and the Shaping of Modernity, 1210–1685* (Oxford: Oxford University Press, 2006); on the separation of politics from science, again with particular attention to Hobbes, see Bruno Latour, *We Have Never Been Modern,* trans. Catherine Porter (Cambridge, MA: Harvard University Press, 1993), 15–35; and on the emergence of "science" and "religion" as new, separate categories in this period, Peter Harrison, *The Territories of Science and Religion* (Chicago: University of Chicago Press, 2015).
9. Galileo Galilei, *Lettera a Cristina di Lorena,* ed. Ottavio Besomi in collaboration with Daniele Besomi (Padua: Antenore, 2012), 57; "Letter to the Grand Duchess Christina," in *Discoveries and Opinions of Galileo,* trans. Stillman Drake (New York: Doubleday, 1957), 186.
10. Steven Shapin and Simon Schaffer, *Leviathan and the Air-Pump: Hobbes, Boyle, and the Experimental Life* (Princeton, NJ: Princeton University Press, 1985).
11. Peter Gay, *The Rise of Modern Paganism,* vol. 1 of *The Enlightenment: An Interpretation* (New York: Vintage, 1966); Jonathan I. Israel, *Radical Enlightenment: Philosophy and the Making of Modernity, 1650–1750* (Oxford: Oxford University Press, 2001); and Charles Taylor, *A Secular Age* (Cambridge, MA: Harvard University Press, 2007). On the separation of politics and science as the hallmark of the "modern constitution," see Latour, *We Have Never Been Modern,* 13–15.
12. McQueen, *Political Realism in Apocalyptic Times,* 133–44. On the religious foundations of modern political theory more generally, see Carl Schmitt, *Political Theology: Four Chapters on the Concept of Sovereignty,* trans. George Schwab (Chicago: University of Chicago Press, 2005); and, with respect to Jewish thought, Eric Nelson, *The Hebrew Republic: Jewish Sources and the Transformation of European Political Thought* (Cambridge, MA: Harvard University Press, 2010).
13. Peter Harrison, *The Fall of Man and the Foundations of Science* (Cambridge: Cambridge University Press, 2007), 183.
14. Karl Löwith, *Meaning in History: The Theological Implications of the Philosophy of History* (Chicago: University of Chicago Press, 1949), 33–51.
15. Ibid., 159.
16. Yuri Slezkine, *The House of Government: A Saga of the Russian Revolution* (Princeton, NJ: Princeton University Press, 2017).
17. Norman Cohn, *Warrant for Genocide: The Myth of the Jewish World-Conspiracy and the Protocols of the Elders of Zion* (New York: Harper and Row, 1967). Finally, for a broad view on the incorporation of Joachim's ideas into the ideologies of the early twentieth century, see Matthias Riedl, "Longing for the Third Age: Revolutionary

Joachism, Communism, and National Socialism," in *A Companion to Joachim of Fiore* (Leiden: Brill, 2017), 267–318.
18. Hans Morgenthau, "Public Affairs: Death in the Nuclear Age," *Commentary* 32 (September 1961): 233. On Morgenthau and the apocalyptic tradition in political thought, see McQueen, *Political Realism in Apocalyptic Times*, 162–91.
19. Francis Fukuyama, *The End of History and the Last Man* (New York: Free Press, 1992).
20. On the early science of climate change, see Nathaniel Rich, *Losing Earth: A Recent History* (New York: Farrar, Strauss & Giroux, 2019); on the dual significance of 1989, see Latour, *We Have Never Been Modern*, 9–10.
21. Roy Scranton, *Learning to Die in the Anthropocene: Reflections on the End of a Civilization* (San Francisco: City Light Books, 2015).
22. John Mecklin, ed., "Closer than Ever: It is 100 Seconds to Midnight," *2020 Doomsday Clock Statement, Bulletin of the Atomic Scientists,* statement released on January 23, 2020.
23. Ernst Bloch, *The Principle of Hope,* trans. Neville Plaice, Stephen Plaice, and Paul Knight, 3 vols. (Cambridge, MA: MIT Press, 1986–95).
24. Prasenjit Duara, *The Crisis of Global Modernity: Asian Traditions and a Sustainable Future* (Cambridge: Cambridge University Press, 2015).
25. Michael S. Northcott, "Ecological Hope," in *Historical and Multidisciplinary Perspectives on Hope,* ed. Steven C. van den Heuvel (Cham, Switzerland: Springer, 2020), 215–38.

Illustration Credits

Page 3: Frontispiece to Francis Bacon's *Instauratio magna;* engraving by Simon van de Passe. Source: Francis Bacon, *Instauratio magna* (London: Johannes Billius, 1620). Credit: Yale University Library; public domain.
Page 11: Paul Klee, *Angelus Novus,* 1920. Credit: The Israel Museum, Jerusalem; public domain.
Page 25: Joachim's *Liber figurarum.* Source: *Religious Texts of Paul the Deacon and Joachim of Fiore,* Corpus Christi College MS 255A. Credit: The Bodleian Library, Corpus Christi College, Oxford; public domain.
Page 31: The Four Horsemen of the Apocalypse. Source: Albrecht Dürer, *Apocalypsis cum figuris* (Nuremberg: Albrecht Dürer, 1498), Typ Inc 2121A. Credit: The Houghton Library, Harvard University; public domain.
Page 33: The Heavenly Jerusalem. Source: Albrecht Dürer, *Apocalypsis cum figuris* (Nuremberg: Albrecht Dürer, 1498), Typ Inc 2121A. Credit: The Houghton Library, Harvard University; public domain.
Page 36: The Last Judgment. Source: *Fragment vom Weltgericht,* ca. 1552–53. Credit: Courtesy of the Gutenberg Museum, Mainz, Germany.
Page 39: *Impressio librorum, Nova reperta.* Illustration designed by Johannes Stradanus. Engraving by Theodor Galle. Published in Antwerp, ca. 1591. Credit: The Metropolitan Museum of Art, New York, Harris Brisbane Dick Fund, 1934; public domain.
Page 45: The opening page of the Torah. Source: *Torah, Nevi'im, Ketuvim* (Venice: Daniel Bomberg, 1524–25). Credit: Courtesy of the Library at the Herbert D. Katz Center for Advanced Judaic Studies, Kislak Center for Special Collections, Rare Books and Manuscripts, University of Pennsylvania, Philadelphia.

ILLUSTRATION CREDITS

Page 49: A chorus of astrologers and prophets. Source: Johannes Lichtenberger, *Prognosticatio* (Ulm: [Johann Zainer?], 1488), 90968. Credit: Courtesy of the Huntington Library, San Marino, California; public domain.

Page 50: The *Sura al-Fātiḥah,* the opening verses of the Qur'an, from the *Corano in arabo.* Source: *Corano in arabo* (Venice: Paganino Paganini & Alessandro Paganini, 1537–38). Credit: Courtesy of the Biblioteca San Francesco della Vigna, Venice, Italy.

Page 52: Prophecy as art. Source: *Album of Calligraphies Including Poetry and Prophetic Traditions (Hadith),* ca. 1500. Credit: Metropolitan Museum of Art, New York, purchase, Edwin Binney 3rd and Edward Ablat Gifts, 1982.

Page 63: Pierre d'Ailly, *Tractatus de imagine mundi Petri de Aliaco, et varia ejusdem auctoris et Ioannis Gersonis opuscula* (Louvain: Joannes de Westphalia, 1480–83). Credit: Courtesy of the Biblioteca Capitular y Colombina, Seville, Spain. Copyright © Seville Cathedral Chapter.

Page 67: Columbus's cryptic signature. Credit: Alamy.

Page 81: A map of Utopia. Source: Thomas More, *Libellus vere aureus nec minus salutaris quam festivus de optimo reip. statu, deq; nova insula Utopia* (Louvain: Thierry Martin, 1516), EC M1835U 1516. Credit: The Houghton Library, Harvard University.

Page 93: The Ottomans and the Indian Ocean. Credit: Mark Thomas, Duke University, Durham, North Carolina.

Page 101: The Habsburg and the Ottoman Empires in the age of Charles V and Süleyman the Magnificent. Credit: Mark Thomas, Duke University, Durham, North Carolina.

Page 110: Spanish possessions in the New World in the age of Charles V. Credit: Mark Thomas, Duke University, Durham, North Carolina.

Page 113: Christ and Antichrist. Source: *Passional Christi und Antichristi von Dr. Martin Luther: mit Bildern von Lukas Cranach dem Aelteren: Auf's Neue aufgelegt und bevorwortet von C. F. W. Walter* (St. Louis: German Synod of Missouri, 1878). Credit: Image courtesy of Concordia Historical Institute, St. Louis, Missouri.

Page 128: The end of the New Jerusalem. Source: *Kontrafactur der Osnabrücker Bischöfe,* pen and ink drawing by Georg Berger, ca. 1607. Credit: Kulturgeschichtliches Museum, Osnabrück, Germany; public domain.

Page 136: Predicting the Apocalypse. Title page of *Christianismi restitutio.* Source: Michael Servetus, *Christianismi restitutio* (Vienna: Balthazar Arnoullet, Guillaume Guéroult, 1553). Credit: Bibliothèque nationale de France, Paris.

Page 157: A Parisian Apocalypse. Source: François Dubois, *Le massacre de la Saint-Barthélemy,* 1572–84. Credit: The Musée Cantonal des Beaux-Arts, Lausanne, Switzerland; public domain.

Page 161: *Allegory of the Tyranny of the Duke of Alba* (1569). Source: *Allegorie op de tirranie van Alva.* Credit: Courtesy of the Museum Catharijneconvent, Utrecht, the Netherlands.

Page 171: A Christian globe. Source: *Cosmography; astrological medicine:* manuscript (Lübeck, 1486–88), MSS HM 83. Credit: The Huntington Library, San Marino, California; public domain.

ILLUSTRATION CREDITS   309

Page 175: Allegory of the five continents. The frontispiece to Abraham Ortelius's *Theatrum orbis terrarum*. Source: Abraham Ortelius, *Theatrum orbis terrarum* (Antwerp: Aegidius Coppenius the Younger of Diesth, 1570), 70496. Credit: The Huntington Library, San Marino, California; public domain.

Page 197: Constructing the image of the savage. Source: *Americae tertia pars Memorabile[m] provinciae Brasiliae historiam contine[n]s* (Frankfurt: Theodor de Bry, 1597). Credit: Private Collection, Rio de Janeiro; public domain.

Page 209: Title page to the *Portae lucis*. Source: Joseph ben Abraham Gikatilla, *Portae lucis; h[a]ec est porta tetragram[m]aton iusti intrabu[n]t p[er] eam,* trans. Paulus Ricius (Augsburg: Officina Millerana, 1516). Credit: Courtesy of the Library at the Herbert D. Katz Center for Advanced Judaic Studies, Kislak Center for Special Collections, Rare Books and Manuscripts, University of Pennsylvania, Philadelphia.

Page 223: Utopia as the Heavenly Jerusalem. Source: Johann Valentin Andreae, *Reipublicae Christianopolitane descriptio* (Strasbourg: Zetzner, 1619). Credit: Courtesy of the Herzog August Bibliothek, Wolfenbüttel, Germany.

Page 238: Modern inventions. Source: Title plate, *Nova reperta,* published in Antwerp, ca. 1591. Credit: Courtesy of the Metropolitan Museum of Art, New York, Harris Brisbane Dick Fund, 1934; public domain.

Page 246: The Heavenly Jerusalem. Source: From the Book of Revelation in the *Historiae celebriores Novi Testamenti,* a pictorial bible, published in Nuremberg by Christoph Weigel in 1712. Personal collection of the author.

Index

Abd al-Malik (sultan, Sa'dian dynasty), 165–66
Abravanel, Isaac, 17–18, 46
Abulafia, Abraham, 27
Acosta, José de, 177, 198–204, 230
Act of Abjuration, 161
Advancement of Learning (Bacon), 230–31
Africa: cartographic representations of, 170, 172, 174, 180; European colonialism in, 8, 186; evangelization of, 145; interpreted in an apocalyptic key, 6, 68, 109; North Africa, 1, 2, 17, 28, 29, 94, 96, 106–7, 163, 164, 165–66, 178–79, 206, 240; source of slaves for Europeans, 60, 76
Age of the Holy Spirit, 24, 148, 219, 247
Ahmad al-Mansur (sultan, Sa'dian dynasty), 166
air pump, 244–45
Akbar (Mughal emperor), 177
Alba, Fernando Álvarez de Toledo, Duke of, 160–61
Albada, Aggaeus van, 162

Alberti, Leon Battista, 40
Albonesi, Teseo Ambrogio degli, 51
Alcazarquivir, Battle of, 165–66, 185
alchemy, 214–16, 218–19, 220
Alexander VI (pope), 40
Alfatimí (El moro, legendary figure), 165
Ali Müezzinzade Pasha (naval commander), 164
Ambach, Melchior, 218
Ambrose of Milan (saint), 64
Americas: Brazil, 78, 80, 87, 186, 190, 193, 195, 197–98, 204–5; Christian missionaries in, 86–87, 90; as "city upon a hill," 230; Columbus' report on cannibals in, 189–91; Columbus' voyages to, 66–67, 71–72; East and West Indies, 198; English colonies in, 229–30; European view of, 77–78; French colonies in, 193; Kingdom of Mexico, 187–88, 201; Las Casas in, 87–91; Montaigne's writings about, 187–89, 191–93; as Paradise, 169; Peru, 198, 199, 203; portrayal on maps and globes,

Americas *(continued)*
 172–74, 176; Portuguese colonies in, 195; quest for gold in, 76, 87; Spanish colonies in, 73–74, 85–87, 109–10; as utopian paradise, 77, 79–80, 85–86; Vespucci's exploration of, 78–80. *See also* Native Americans
Amsterdam, 162, 178, 242
Anabaptists, 122, 126–27, 129, 217, 240
Andreae, Johann Valentin, 220, 222–24
Angelus Novus (Klee), 10–11
Ankara, Battle of, 98
Annio da Viterbo, Giovanni, 104
Antichrist, 104, 120, 122, 129; and the Apocalypse, 137; battle with Christ, 147; Catholic Church as, 120, 160; depictions of, 112–13; Huguenots as, 153; Jews as, 131; pope(s) as, 120, 131, 181; Turks as, 37, 130
Apocalypse: Campanella's view of, 180–81; Christian vision of, 4–5, 19–21, 23–26, 247; Daniel's vision of, 18, 19–20; early visions of, 4–5; Four Horsemen of, 30–32, 137; Islam's vision of, 19, 22–23, 27–28, 247; Judaism's vision of, 4–5, 17–18, 19, 23–24, 26–27, 247; negative implications of, 8; oral culture of, 53–55; Rosicrucian view of, 220; signs and portents of, 46–47, 54, 135, 137. *See also* Book of Revelation; End of History/Time; Last Days
apocalyptic braid, 5, 13–34, 250
apocalyptic imagination, 4–5, 9; early modern, 238–39; in the holy scriptures, 42–43; hopeful message of, 32, 34; and the modern world, 12
apocalypticism, 7–8, 9; in the sixteenth century, 140–41; in the U.S., 249
Aquinas, Thomas (saint), 198
Ariosto, 105
Aristotle, 79, 226, 230
Arndt, Johann, 221–22

Asia: cartographic representations of, 79, 170, 173, 174, 180; Central, 98; Columbus's original destination, 59–60, 61, 62, 67; East, 249; European colonialism in, 8; evangelization of, 145; India, 94, 98, 197, 249; interpreted in an apocalyptic key, 6, 109, 169; Montaigne's interest in, 192, 204. *See also* China
astrology, 46–49, 99, 103, 215, 240, 242; Columbus' interest in, 64–65
astronomy, 47, 60, 64, 182, 185, 223
Augustine of Hippo (saint), 23, 26, 64, 68, 70, 142, 144, 148, 230
Australia, 174

Bacon, Francis, 2, 3, 6, 8, 225; *The Advancement of Learning*, 230–31; on English expansion, 229–30; *The Great Instauration*, 225–26, 228, 232–33, 235–37, 239; *New Atlantis*, 234–37; *Novum Organum*, 226–28, 231; philosophy of hope, 233–34
Bacon, Roger, 65
Bader, Augustin, 124–25
Bar Yochai, Shim'on, 206, 208
Barbarossa, Khair-ad-Din, 107
Barry, Jean du (Lord of La Renaudie), 152–53
Bayezid I (Ottoman sultan), 98
Bayezid II (Ottoman sultan), 16–17, 46, 94, 96, 99
Beeldenstorm, 160
Behaim, Martin, 172
Behem (Böhm), Hans, 119
Benjamin, Walter, 10–12
Besold, Christoph, 222
Bible: Book of Daniel, 2, 4, 18–20, 23, 41, 46, 87, 104, 121–22, 129–30, 132, 178, 231, 236; Book of Revelation, 13, 21, 26, 30, 41–42, 79, 120, 129–30, 132, 136, 137, 154, 236; *Complutensian*, 135; early printed

versions, 41–42; Gutenberg's printing of, 38–40; Hebrew, 43–45, 207; Luther's translation, 118; Rabbinic, 44–46; Santes Pagnini Polyglot, 135; Vulgate translation, 42, 143
Bloch, Ernst, 12
Boabdil (sultan of Granada), 58
Bodin, Jean, 51
Böhm, Johannes, 179
Böhme, Jakob, 221
Bomberg, Daniel, 43–44, 51, 211
Book of Daniel, 2, 4, 18–20, 23, 41, 46, 87, 104, 121–22, 129–30, 132, 178, 231, 236
Book of Hidden Events, The (John of Rupescissa), 26
Book of Prophecies (Columbus), 68–71, 240
Book of Revelation, 13, 21, 26, 30, 41–42, 79, 120, 129–30, 132, 136, 137, 154, 236
Book of Sea Lore (Piri Reis), 170, 172
Book of Sibyls, The (*Sibyllenbuch*), 36–38
Bordeaux, 156, 178, 196, 204–5, 229
Borja, Francisco, 145–46
Bourbon, Louis de (Prince of Condé), 155
Bourbon, Renée de (abbess of Fontevrault), 144
Boyle, Robert, 244
Bridget of Sweden, 48
Brocardo, Jacopo, 219
Bry, Theodor de, 197–98
Bugenhagen, Johannes, 132
Bullinger, Heinrich, 43
Burckhardt, Jacob, 9

Cairo, 206, 242
calendrics, 99
calligraphy, 52–53
Calvin, John, 146–49; early years, 138; influence in France, 152; on the Kingdom of God, 140–41; on predestination, 139–42, 152; vs. Servetus, 133, 137–38
Calvinism and Calvinists, 230; in the Americas, 193; in France 152–59; in the Low Countries, 159–60, 166
Camões, Luís de, 173
Campanella, Tommaso, 8, 180–86, 240; *City of the Sun*, 182–86, 222, 237; *Monarchy of the Messiah*, 181–82, 186
cannibalism, 189–205; Columbus' description of, 189–90; Léry's description of, 193–95; Montaigne's essay on, 191–93, 196, 204; social explanation for, 194–95; stages of the ritual, 195; survivor, 196; Vespucci's description of, 190–91
capitalism, 8, 83, 247, 248
Carafa, Gian Pietro (Pope Paul IV), 144
cartography, 61, 172–80
Catholic Church: as Antichrist, 120, 160; concerns about printing press, 40; Council of Nicea, 137, 147; Council of Trent, 42, 143; Counter-Reformation, 163; Dominican order, 86, 88, 90, 104, 113, 145, 154, 180, 199; and the Eucharist, 151–52; Fifth Lateran Council, 142; Franciscan order, 25–27, 51, 54–55, 85–87, 89, 119, 145, 153, 165, 193, 199; issues of reform within, 143–44; Jesuit order, 144–47, 177, 180, 198–203, 211; on the Millennium, 142–44; opposition to in France, 151–52; and the Reformation, 54–55; Roman Inquisition, 142–43; sale of indulgences by, 113; Tametsi Decree, 143–44. *See also* Christianity
Catholic League, 158
Chaldiran, Battle of, 94
Charles V (Holy Roman Emperor), 100–102, 103, 106–7, 108–11, 118, 140, 145, 240; abdication of, 111; as Last World Emperor, 104–5, 108

Charles VIII (king of France), 15–16, 54, 104
Charles IX (king of France), 156, 204
China, 59, 94, 177, 180, 189–90, 203, 236, 240; Montaigne's praise of, 205
Christianity, 7, 9–10, 12; in the Americas, 86–89, 90, 198–99; conversion to, 18, 108–9, 135, 191–92, 199, 210, 242; cosmologies of, 170–71; and the Crusades, 5, 11, 16, 28, 54, 59, 65, 113, 164, 170; doctrine of the Trinity, 133–35, 137–38, 147–48; early sects at Nag Hammadi, 250; and the fall of Constantinople, 14–15; in Italy, 13–15; military engagement with Muslims, 163–65; in the Ottoman Empire, 96; relationship with Islam, 163; Servetus' attack on, 133; in Spain, 15; vision of the Apocalypse, 4–5, 19–21, 23–26, 247. *See also* Catholic Church; Protestant Reformation
Cicero, 79
City of God (Augustine), 23
City of the Sun, The (Campanella), 182–86, 222, 237
Clement VII (pope), 105, 107, 108
Colloquium of the Seven about the Secrets of the Sublime, 51–52
Columbus, Christopher, 7, 34, 46–47; apocalyptic vision of, 66–74; *Book of Prophecies*, 68–69, 71, 240; as "Christoferens," 66–67, 232; early voyages of, 59–60; interaction with indigenous peoples, 72–77; library of, 60–65; myth/antimyth of, 56; plan for Christian conquest of Jerusalem, 59; refuge in Carthusian monastery, 67–68; reporting on cannibalism, 189–90; in Spain, 57–58; voyages to the Americas, 66–67, 71, 169–70, 189–90
Columbus, Fernando, 57, 60–61, 71, 73

Commentary on Daniel (ben Gershom), 46
Complutensian Bible, The (Ximénes de Cisneros), 135
Constantinople, Ottoman conquest of, 5, 14–15. *See also* Istanbul
Cop, Nicolas, 151
Cordovero, Moses ben Jacob, 206
Cortés, Hernán, 85, 87, 109
Coster, Laurens Janszoon, 37
Council of the Indies, 109
Council of Troubles, 160
Cranach, Lukas, 120, 147
Crespo, Joan (Abrahim Fatimí), 165
Crouzet, Denis, 141, 157
Crusades, 5, 11, 16, 28, 54, 59, 65, 113, 164, 170
Cuneo, Michele da, 74

d'Ailly, Pierre, 62–65, 69, 98, 169, 172
Damascus, 220; siege of, 98
Daniel (biblical), 2, 4, 18, 18–20, 23, 41, 46, 87, 104, 121–22, 129–30, 132, 178, 231, 236
Dante Alighieri, 2, 25
Dati, Gregorio, 40
De orbe novo (*On the New World*; Peter Martyr of Anghiera), 77, 79, 191
democracy, 248
Desiré, Artus, 154, 157
Dias, Bartolomeu, 62, 172
Diet of Augsburg, 118
Diet of Speyer, 140
Diu, First Battle of, 93
Dominican order, 104, 113, 145, 154, 180, 199; in the Americas, 86, 88, 90. *See also* Savonarola, Girolamo
Don Juan of Austria, 164
Doria, Andrea, 163
Drake, Sir Francis, 229
Dreschel, Thomas, 121
Drummer of Niklashausen, 119

Dürer, Albrecht, 30–33, 34, 120; apocalyptic texts, 42; editions of the Bible, 42
Dyvolé, Pierre, 154

Eck, Johann, 116
Edict of Nantes, 158–59
Egmont, Count of (Lamoral), 160
El Dorado, 229
Elijah (or Elias; biblical), 54, 124, 125, 132, 168, 213, 217, 219, 221
Elizabeth I (queen of England), 162, 229
Elizabeth Stuart (Electress Palatine and Queen of Bohemia), 220
el-Su'udi, Mehmed, 177–78
End of History/Time, 5–7, 26; Bacon's view of, 231; as cause of violence, 158; Columbus' view of, 169; Ibn 'Arabi's vision of, 28; Jesus' predictions of, 20–21, 99; Jewish teachings on, 124; kabbalistic, 210; Muslim vision of, 165, 172; and the Ottoman Empire, 54; Paracelsus's view of, 216–217; in printed text, 36; prophecies of, 6–7, 146–47, 166–67, 231, 242; and the Protestant Reformation, 120–21; Second Charlemagne Prophecy, 104; Servetus' view of, 135, 137; signs and portents of, 5, 18, 153, 166, 231; and the triumph of Islam, 163; visions of, 123–24; *See also* Apocalypse; Last Days
End of the World/End Times. *See* End of History/Time
England: colonies in the Americas, 229–30; expansion of empire, 230–31
Erasmus of Rotterdam, 42, 84, 105–6
Erpenius, Thomas, 51
Estienne, Robert, 46
Exposé concerning This Community Passing the Year 1000 (al-Suyūti), 53
Exposition on the Apocalypse (Joachim of Fiore), 24–25, 42–43

Family of Love, 176, 180
Farel, Guillaume, 138
Farissol, Abraham, 172
Fatimí, Abraham (Joan Crespo), 165
Faulhaber, Johann, 221
Felgenhauer, Paul, 221
Felice da Prato, 43
Ferdinand (archduke of Austria), 106
Ferdinand of Aragón, 57–58, 66, 70, 72–73, 75–76, 77, 100, 103
Ficino, Marsilio, 64, 215
Fieschi, Caterina, 144
Fifth Monarchy (Fifth Empire), 87, 104, 121
Firdevsi, Uzun, 17, 99
First World War, 248
Four Horsemen of the Apocalypse, 137
"Four Horsemen of the Apocalypse" (Dürer), 30–32
Fragment vom Weltgericht (Fragment of the Last Judgment), 35–36
France: colonies in the Americas, 193; violent feuding in, 196. *See also* French Wars of Religion
Francis I (king of France), 101–2, 138, 152
Francis II (king of France), 153
Franciscan order, 25–27, 51, 54–55, 119, 145, 153, 165, 193, 199; in the Americas, 85–87, 89, 145
Frederick I (duke of Württemberg), 220
Frederick IV ("Winter King" of the Palatinate), 220
Frederick the Wise (elector of Saxony), 118
Free Spirit heresy, 28
French Wars of Religion, 155–59, 189, 196, 248

Gabriel (archangel), 22–23
Gaismair, Michael, 123
Galatino, Pietro, 146
Galen of Pergamon, 215, 218
Galileo Galilei, 243–44
Gattinara, Mercurino Arborio di, 104–5

General History of the Indies (López de Gómara), 188
Gentillet, Innocent, 196
Geography (Ptolemy), 61–62, 135, 172, 177
geomancy, 99
German Peasants' War, 120, 122–29
Giles of Viterbo, 105, 109, 211
Giovanni da Capistrano, 54, 119
Giovio, Paolo, 95
global warming, 248–49
globalization, 107
globes: Christian, 170–71, 179; early examples, 172–73; as matter of faith, 177. *See also* planispheres
golden age/Golden Age (myth of), 77, 181, 214, 221; within the Last Days, 99; Paracelsus's view of, 216–17
Gorricio, Gaspar, 68
Granvelle, Antoine Perrenot, Lord of, 159–60
Great Instauration, The (Bacon), 225–26, 228, 232–33, 235–37, 239
Gregory of Nazianzus, 230
Guide for the Tribulation (John of Rupescissa), 26, 43, 53
Guise family, 150, 153, 156, 158
Gujarat, Sultan of, 92
gunpowder, 240
Gutenberg, Johannes, 35–38, 40; Bible printed by, 38–40; *Book of Sibyls* (*Sibyllenbuch*), 36–38; "Fragment of the Last Judgment," 35–36; *Turkish Calendar*, 37, 38

Habsburg Empire, 48, 94, 96, 100–102, 104–7, 119, 159, 162, 186, 240; map, 101; military power of, 101–2, 240; religious expectations in, 103–4
al-Hajarī, Ahmad ibn Qāsim, 178–79
ha-Kohen, Joseph, 179
Halevi, Abraham ben Eliezer, 46, 107
Halley's Comet, 5, 46
Hamdullah ibn Mustafa Dede, 52
ha-Reuveni, David, 107–8, 172

harmonia mundi, 213
Haydar-ī Remmal, 99
"Heavenly Jerusalem" (Dürer), 32–33
"Heavenly Jerusalem" (Luiken), 246
Henri I (duke of Guise), 158
Henri II (king of France), 150
Henri III (king of France), 157–58, 159
Henri of Navarre (King Henri IV), 155, 158
Hergot, Hans, 123
Herodotus, 191
Hess, Tobias, 220
Hibetullah b. Ibrahim, 52
Hidden Imam. *See* Twelfth Imam
Hildegard of Bingen, 218
Hilten, Johann, 119
Hispaniola, 67, 69, 73, 76, 77, 87
History of a Journey in the Land of Brazil (Léry), 193–94
History of the Indies (Las Casas), 73–74
Hiyya Me'ir ben David, 43–44
Hobbes, Thomas, 243–45, 248, 249
Hoffman, Melchior, 125–26
Hojeda, Alonso de, 73
Holy Roman Empire, 100–101; apocalyptic ideas in, 119–20; vs. the Ottoman Empire, 106–7; political configuration of, 117; religion in, 117–18
Huguenots, 152–54; violence against, 154–59
humanism, 42, 43, 64, 79, 83, 103, 105, 148
Humayun (Mughal emperor), 92
humoral theory of disease, 215
hurufism, 99
Hut, Hans, 124–25

Ibn Adoniyahu, Jacob ben Hayyim, 43, 44
Ibn al-Wardi, 170, 178
Ibn 'Arabi, Muhyi al-Din, 27–28, 53, 99
Ibn Ezra, Abraham, 44
Ibn Khaldūn, 'Abd al-Rahman, 29–30, 98

Ibn Shem Tod, Shem Tod, 55
Ibn Tumart, Muhammad, 28–29
Ibrahim Pasha (Grand Vizier), 106
Ignatius of Loyola, 145–49, 176–77
Illyricus, Thomas, 153–54
Imago mundi (d'Ailly), 62–64, 169
Indian Ocean, 62, 91–94, 100, 172, 182, 186, 230
Institutes of the Christian Religion, The (Calvin), 138–41
'Isa, 17, 22, 54. *See also* Jesus of Nazareth
Isabella of Castile, 57–58, 66, 70, 72–73, 75–76, 77, 100, 103
Isaiah (biblical), 68–69, 208
Islam, 5, 7, 9–10, 12; and the End of the World, 163; expansion of, 170; and jihad, 97–98; Luther's interest in, 130–31; relationship with Christians, 163–65; sacred cosmologies of, 170, 172; Shi'a, 22; Shi'ite, 96; in Spain, 15, 59; spread of, 179; Sufi, 54; Sunni, 22, 96; and the Trinity, 134; view of Jesus of Nazareth, 22, 239; vision of the Apocalypse, 19, 22–23, 27–28, 247. *See also* Mahdī; Muhammad (prophet); Qu'ran
Istanbul, 16, 95, 106, 164; apocalyptic currents in, 99–100, 163, 242; Islamic preaching in, 53, 54; Jewish community in, 44, 55; Topkapi Palace, 16. *See also* Constantinople

James I (king of England), 220
janissaries, 95
Jerusalem: Antichrist in, 170; Christian conquest of, 16, 59, 65–66, 69; Heavenly, 12, 21, 32, 33, 71, 220, 223, 246; Muslim occupation of, 16, 59, 65; plans for Christian conquest of, 16, 59, 65–66, 69; return of Jews to, 18, 24; as site for Second Coming, 69, 170, 181. *See also* New Jerusalem
Jesuit order, 144–47; mission to the Mughal court, 177

Jesus of Nazareth, 87, 232; divinity of, 142, 147–48; Fifth Monarchy of, 104, 121; Islam's view of, 22, 239; Jewish rejection of, 134; predictions of the End of Time, 20–21. *See also* Messiah; Second Coming
jihad, 97–98
Joachim of Fiore, 26, 148; influence of, 48, 54, 68, 70, 119, 123–24, 137, 154, 180, 189, 213, 217–18, 219, 247; Catholic Church's condemnation of, 142; *Exposition on the Apocalypse*, 24–25, 42–43
Joanna I (Queen of Castile and Aragón), 100
João III (king of Portugal), 107–8
John of Ávila (saint), 144
John of Leiden, 126–27, 128
John of Rupescissa, 25–26, 43, 53, 165, 218
John the Baptist (saint), 20, 132
Jost, Lienhard, 125
Jost, Ursula, 125
Judaism, 7, 9–10, 12; apocalyptic sermons, 55; ascetics at Qumran, 249; cartographies of, 179; and the End of History, 124; eviction from Spain, 17, 58–59, 145; forced conversions of Jews, 17, 108; Jews as Antichrist, 131; and the Messiah, 5, 6, 17–19, 24, 27, 107–8, 210, 249; messianism in, 107–8; in the Ottoman Empire, 96; printed works, 43–44, 47; return of Jews to Jerusalem, 18, 24; in Safed, 206; and the Second World War, 248; and the Trinity, 134; vision of the Apocalypse, 4–5, 17–18, 19, 23–24, 26–27, 247. *See also* Kabbalah
Julius (Duke of Braunschweig-Lüneberg), 218–19
Julius II (pope), 112–13

Kabbalah, 26–27, 105, 108, 134, 206–7, 208, 210, 212–14, 220, 242; and Christianity, 210–11

Kâbiz, Molla, 54
al-Khidr (legendary figure), 165
Klee, Paul, 10–11
Knipperdolling, Bernhard, 128
Krechting, Bernhard, 128

La Cruz, Francisco de, 199–200
Las Casas, Bartolomé de, 56–57, 199; in the Americas, 87–91; on Columbus, 58, 60, 62, 73, 74, 189–90
Last Days, 6, 22, 36, 99, 120, 135, 140, 142, 164, 168, 218; Daniel's vision of, 2, 4; Luther's view of, 129–30; visions of, 125. *See also* Apocalypse; End of History/Time
Last Judgment, 24, 64, 137, 140–41, 199, 216, 217
Last World Emperor, 6, 100, 104, 108, 240
Leber, Oswald, 124–25
Lemlein, Asher, 18–19, 43
Leo X (pope), 113, 118
Leonardo da Vinci, 64
Lepanto, Battle of, 163–64, 166
Léry, Jean de, 193–95
Levi, Solomon ben Isaac, 55
Levi ben Gershom (Gersonides), 46
Leviathan (Hobbes), 243–45
Lichtenberger, Johannes, 48
London, 162, 225, 229
López de Gómara, Francisco, 179, 188
Lorenzo, Domenego di, 166–68
Los Angeles, Francisco de, 86
Luiken, Jan, 246
Lumnius, Joannes Fredericus, 177
Luria, Isaac, 34, 206–8, 210
Luria, Shlomo (the Maharshal), 44
Luther, Martin, 42, 43, 47, 54, 112, 114–18, 217; *Address to the Christian Nobility of the German Nation*, 116, 120; *Against the Execrable Bull of the Antichrist*, 120; *Against the Robbing and Murdering Hordes of Peasants*, 129; *Against the Roman Papacy an Institution of the Devil*, 130; and the Anabaptists, 127, 129; on the Antichrist, 120, 127, 129–30; and the Apocalypse, 120; *Babylonian Captivity of the Church*, 116; *Battle Sermon against the Turks*, 130; on the Beautiful Ending, 131–32; on the faith of the individual, 138–39; influence in France, 151; on Islam, 130–31; and the Last Days, 129–30; *Ninety-five Theses*, 114, 115; *On the Freedom of a Christian*, 116–17; *On the Jews and Their Lies*, 131; *That Jesus Christ Was Born a Jew*, 131
Lutheranism, 218, 222

Machiavelli, Niccoló, 102–3, 196
Madre Zuana (the Venetian Virgin), 212, 213
Mahdī, 5, 17, 22, 28–30, 53–54, 163, 165, 239; Süleyman as, 98, 99–100
Mahdism, 29–30
Maimonides, 18, 206
Mainz, 35, 39, 40, 41, 113
Mamluk Empire, 16, 93–96
Manuel I (king of Portugal), 108
Manutius, Aldus, 41
Margaret of Parma (governor of the Netherlands), 159–60
Marguerite of Valois (queen of Navarre), 155
Marx, Karl, 247
Mary, Duchess of Burgundy, 100
Matthijs, Jan, 126–27
Maximilian I (Habsburg emperor), 100, 102
Meccan Revelations (Ibn 'Arabi), 28
Mede, Joseph, 43
Medici, Catherine de', 150, 153, 156
Melanchthon, Philip, 132
Mendoza, Antonio de, 109
Mercator, Gerardus, 173–74, 176, 180
Messiah, 181; expectations of, 168; Halley's Comet as sign of, 46; Jesus

as, 134; Jewish expectations of, 5, 6, 17–19, 24, 27, 107–8, 210, 238, 242, 249; Jewish people as, 210; Sabbatai as, 242

messianism, 124; and Judaism, 107–8; kabbalistic, 210

Methodius, the pseudo-, 48, 65–66, 69, 104, 213, 218, 240

Meuccio, Silvestro, 54

Mevlānā 'Isa, 99, 106

Mexico, 85–87, 100, 109, 145, 186, 187–88, 201–3, 205

Michael (archangel), 4, 119, 136

millenarianism, 7–8, 28–29, 122, 199–200; in the Americas, 85–86

Millennium: belief in nearness of, 121–22, 126, 137, 141, 188; Calvin's rejection of, 140, 142; Campanella's view of, 181–82; Catholic teachings on, 142–44; Christian hopes for, 6, 21, 85, 122–23, 125, 165–66, 176, 224; conversion to Christianity as essential for, 86, 192; expectations of religious harmony, 186; Ignatius's view of, 146–47; La Cruz's vision of, 199; political implications of 181; preparation for, 54; prophecies of, 242; and the Protestant Reformation, 121; question of timing, 199–200

Mirim Çelebi, 17, 99

Moctezuma II (Aztec Emperor), 201

modernity: as apocalyptic, 7–10, 241; 249–51; as the disentanglement of religion from politics and science, 243–45, 305n8; as a relative concept, 8; Bacon's view of, 237–41; Crossing the Pillars of Hercules as an early modern metaphor for; 1–3, 111, 228, 230–31; Stradanus's view of, 237–38. *See also* capitalism; providential modernity

Mohács, Battle of, 94

Molkho, Shlomo (Diego Pires), 55, 108

Monachus, Franciscus, 173

Monluc, Blaise de, 102

Montaigne, Michel de, 173, 250; on cannibalism, 191–93, 196, 204; on China, 205; essays on the New World, 187–89, 204–5

Montesinos, Antonio de, 88

More, Thomas, 80–85

Morgenthau, Hans, 248

Moriscos, 164–65, 178

Moses ben Nahman, 27

Moses of León, 26–27

Motolinía (Toribio de Benavente), 86–87

Mughal Empire, 94

Muhammad (prophet), 19, 22–23, 99, 100, 163, 166, 178–79, 182, 250

Mundus novus (Vespucci), 78–79

Münster, Anabaptist Kingdom of, 125–27

Münster, Sebastian, 19

Müntzer, Thomas, 121–22, 129

Muqaddimah (Ibn Khaldūn), 29–30

Murad III (Ottoman sultan), 177–78

Murad b. Abdullah, 163

Muslims. *See* Islam

Mustafa (son of Süleyman), 97

al-Mutawakkil, Muhammad, 165–66

Nagel, Paul, 220–21

Nasrid dynasty, 58

Native Americans: Andean, 202; Aztecs, 187–88; Chichimecas, 203; Incas, 200, 202–3; Mexica/Mexicans, 200, 201, 202–3; Taínos, 72–74, 189–90; Tupi, 204–5; Tupi-Guarani, 195; Tupinambá, 192–95, 196; Tupinikin, 195

natural law theory, 198

natural philosophies, 7–8, 41, 218, 226, 233, 234, 241, 244

Neuheuser, Wilhelm Eo, 221

New Atlantis (Bacon), 234–37

New Jerusalem, 16, 21, 69, 125, 128, 176, 221, 250

Nicholas II (pope), 27
Nicolas of Lyra, 68
9/11 terrorist attacks, 249
Novum Organum (Bacon), 226–28, 231–32
nuclear holocaust, 248–49
Nuremberg, 17, 30, 41, 118, 123, 172, 218

Oettingen, Carl von, 219
On the Errors of the Trinity (Servetus), 133
On the Image of the World (d'Ailly), 62–64
On the Necessity of Reforming the Church (Calvin), 140
On the New Transformation of the Christian Life, 123–24
On the Trinity (Augustine), 148
Oratory of Divine Love, 144
Ortelius, Abraham, 173–74, 176, 177, 179–80, 214
Oruch (pirate), 106
Ottoman Empire, 5, 14–16, 162–63; apocalyptic texts in, 52–53; under Bayezid I, 98; under Bayezid II, 16–17, 46, 94, 96, 99; books of omens, 53; conquest of Constantinople by, 5, 14–15; desire for world empire, 97–98; expansion of, 94–96, 163; vs. Holy Roman Empire, 106–7; in the Indian Ocean, 92–94; and Islamic law (sharia), 96; map, 101; military power of, 95, 240; under Murad III, 177–78; policies regarding religion, 96; preaching in, 53–54; printing in, 46; religious texts in, 51–53; Safed, 206; under Selim, 94; sexual politics of, 97; threat to Europe, 131; warring with Spain, 163
Oviedo, Andrés de, 145, 146

Paganini, Paganino, 48, 51
pantheism, 137
Paracelsians, 220–21, 222

Paracelsus (Philippus Aureolus Theoophrastus Bombastus von Hohenheim), 8, 215–19
Paré, Ambroise, 32
Passion of Christ and the Antichrist, The (Cranach), 112–13, 147
Paul III (pope), 142, 145, 146
Paul IV (pope), 144
Paul of Tarsus (saint), 20–21, 84, 115, 144
Pearls of Wonders and Singularity of Marvels (al-Wardi), 170
Penman, Leigh, 221
Peter Comestor, 64
Peter Martyr of Anghiera, 73, 77–78, 79, 191
Pfefferkorn, Johannes, 43
Philip II (king of Spain), 158, 159, 160, 163–64, 166, 174, 186
Philip of Hesse, 122, 127
Philip the Handsome (Duke of Burgundy), 100
Piccolomini, Aeneas Sylvius. *See* Pius II (pope)
Pico della Mirandola, Giovanni, 64, 211
Pillars of Hercules, 1–3, 111, 228, 230–31
Pindar, 1
Piri Reis, 170
Pius II (pope), 15, 39, 61
Pius V (pope), 163–64
Pizarro, Francisco, 85, 87
planispheres, 79, 172, 173, 174. *See also* globes
Plantin, Christophe, 214
Plato, 1, 36, 79, 84, 227, 230, 236
Plessis-Mornay, Philippe du, 161
Pliny, 174, 191
Polanco, Juan, 146
polygamy, 126–27
Portae lucis, 209
Portugal: colonies in the Americas, 195; expansion into the Indian Ocean, 92–94
Portuguese Inquisition, 143
Postel, Guillaume, 51, 146, 176, 211–14

predestination, 138–42, 152
printing: and the Apocalypse, 119–20, 250; of astrological works, 47–49; by Bomberg, 43–44; in China, 37; by Christian authors, 42–43; in Europe, 241; by Gutenberg, 35–38, 40; illustration of print shop, 39; in Italy, 48, 51; in the Jewish community, 43–46, 47; in the Ottoman Empire, 46; of the Qur'an, 48, 50–53; spread of, 40–41
Prognosticatio (Lichtenberger), 48
Proposal on Remedies for the Indies, A (Las Casas), 90
Protestant Reformation, 54–55, 102, 114–18; Catholic response to, 143–44; in France, 151–52; in the Low Countries, 159–62; theological debates, 118–19
Protestantism, 42
providential modernity, 9–10, 247–50
Ptolemy, Claudius, 61, 135, 172, 173, 177, 180, 226
Puritans, 230
Pythagoras, 84

al-Qaeda, 249
Quattuor navigationes (Vespucci), 79–80
Quint, David, 196
Qur'an, 27, 30, 68, 131, 134; printing of, 48, 50–53; vision of the Apocalypse, 22–23

Rabbinic Bible, 44–46. *See also* Bible
Raleigh, Walter, 229
Rashi (Solomon ben Isaac), 44
Rawley, William, 234
Recalcati, Ambrogio, 146
Reformation. *See* Protestant Reformation
Reipublicae Christianopolitane descriptio (Andreae), 222–24
Reperta nova (Stradanus), 237–38

Restitution of Christianity, The (Servetus), 135–37
Reuchlin, Johannes, 211
Ricci, Matteo, 177
Riccio, Paolo, 209
Roanoke Colony, 229
Roman Inquisition, 142–43, 181
Rosenkreuz, Christian, 219–20
Rosicrucian Brotherhood, 8, 220, 221, 222, 232
Roussel, Gérard, 151
Roxelana (wife of Süleyman), 97
Rule of St. Benedict, 79

Sabbatai Ṣevi, 242
Sa'dian dynasty (Morocco), 178–79
Safavid Empire, 94
Sancerre, siege of, 196
Sánchez, Alonso, 203
Savonarola, Girolamo, 13–16, 43, 47, 54, 104, 142, 146, 219, 240
Scechina (Giles of Viterbo), 211
Schmalkaldic League, 118
Schmalkaldic War, 160
Scholem, Gershom, 11–12, 208
Schombach, Heinrich, 219
Sea Beggars, 160–61
Sebastian (king of Portugal), 165–66
Second Charlemagne Prophecy, 104
Second Coming, 5, 6–7, 59, 104, 108–9, 121, 135, 144, 231, 238; Calvin's view of, 140; Campanella's expectation of, 180; conversion to Christianity as necessary for, 18, 108–9, 135, 191–92, 199, 210, 242
Second World War, 248
secularism, 8–9
Selim I (Ottoman sultan), 94
Seneca, 71, 174
sephirot, 207, 208, 209
Servetus, Michael, 133–38
Sibyllenbuch, 36–38
Sibyls, 36–37, 189
Silva e Menezes, João da, 146

Slavery: Africa as a source for slaves, 60, 76; Columbus's use of captives as trophies, 56–58; in the Ottoman Empire, 95, 97, 106; Spanish seizure of Amerindians, 73–74, 76, 89, 189–91
Smith, John, 229
Sömmering, Philipp, 219
Soul of the World (*anima mundi*), 212
Sozzini, Lelio, 148
Spain: Caribbean captives in, 56–58, 72, 74; colonies in the Americas, 73–74, 85–87, 109–10, 179, 186; expansion of, 185–86; under Ferdinand and Isabella, 57–59, 66, 100
Spanish Armada, 162
Spanish Inquisition, 59, 142–43, 165
Spiritual Exercises (Ignatius of Loyola), 176–77
Spiritualism, 217
Staden, Hans, 197–98
Staupitz, Johann von, 114
St. Bartholomew's Day Massacre, 156–57, 166, 167
Stifel, Michael, 120
Storch, Niklas, 121
Stradanus, Johannes, 237–38
Stübner, Thomas, 121
Süleyman the Magnificent (Ottoman sultan), 34, 94–99, 103, 105, 106, 240; as Mahdi, 98, 99–100

Taínos, 72–74, 189–90
Talmud, 23, 27, 44, 46
Tamerlane (Timur), 98–99
Tametsi Decree, 143–44
Tetzel, Johann, 113–14
Theatine Order, 144
Theatrum orbis terrarum (Ortelius), 174–75, 177, 179–80
Thevet, André, 193–94, 195
Thiene, Gaetano da, 144
Third Reich, 247–48
Thirty Years' War, 221, 222, 242, 248

Thomas Aquinas (saint), 198
Timur (Tamerlane), 98–99
Torah, 44–46, 207
Toscanelli, Paolo, 61
Tovar, Juan de, 203
Treatise concerning Laws and Sects (d'Ailly), 65
Treatise on the Agreement of the Truth of Astronomy with the Story of History (d'Ailly), 65–66
Treaty of Cateau-Cambrésis, 159
Trinity, doctrine of, 133–35, 137–38, 147–48
Trithemius, Johannes, 40
Turkish Calendar, 37–38
Twelfth Imam, 22, 238–39
Twelve Articles, 122–23

Ugalde, Francisco de, 109
Union of Utrecht, 161
Utopia (More), 80–85, 182–83
utopianism, 9, 123–24; apocalyptic, 6. *See also* Utopia/utopia
Utopia/utopia, 6, 34, 240; and the millennial kingdom, 85; Andreae's vision of, 222–24; Bacon's vision of, 234–37; Campanella's vision of, 182–86; Las Casas's vision of, 89–91; More's vision of, 80–85. *See also* utopianism

Venice: apocalyptic currents in, 143, 166–68, 211–12, 219; Jewish community in, 43, 55, 107; publishing industry in, 41, 43, 48, 51; wars against Ottomans, 15, 164
Venetian Empire, 103
Vernazza, Ettore, 144
Vespucci, Amerigo, 78–80, 172, 190–91
Virdung, Johann, 48
Virgil, 36
Vital, Chayyim, 208, 210
Viterbo, Giles of, 105, 109, 211

Wahres Christenthum (Arndt), 221
Waldeck, Franz von, 126
Waldseemüller, Martin, 79, 173
Weber, Max, 9
William of Orange, 160–61
Winstanley, Gerrard, 43
Winthrop, John, 230
Wootton, David, 84
writing systems, 202–3

Ximénes de Cisneros, Francisco, 135

Zacuto, Abraham, 47
Zarathustra, 19
Zarruq, Ahmad, 178
Ziegler, Philip, 221
Zieglerin, Anna, 219
Zohar, 206–7, 211–12
Zoroaster, 19
Zuana, Madre, 212, 213
Zwingli, Ulrich, 42, 139, 151